T0190151

BIRKHÄUSER

Pseudo-Differential Operators
Theory and Applications
Vol. 4

Managing Editor

M.W. Wong (York University, Canada)

Editorial Board

Luigi Rodino (Università di Torino, Italy)
Bert-Wolfgang Schulze (Universität Potsdam, Germany)
Johannes Sjöstrand (École Polytechnique, Palaiseau, France)
Sundaram Thangavelu (Indian Institute of Science at Bangalore, India)
Maciej Zworski (University of California at Berkeley, USA)

Pseudo-Differential Operators: Theory and Applications is a series of moderately priced graduate-level textbooks and monographs appealing to students and experts alike. Pseudo-differential operators are understood in a very broad sense and include such topics as harmonic analysis, PDE, geometry, mathematical physics, microlocal analysis, time-frequency analysis, imaging and computations. Modern trends and novel applications in mathematics, natural sciences, medicine, scientific computing, and engineering are highlighted.

Fabio Nicola | Luigi Rodino

Global Pseudo-Differential Calculus on Euclidean Spaces

Birkhäuser

Authors:

Fabio Nicola
Dipartimento di Matematica
Politecnico di Torino
Corso Duca degli Abruzzi 24
10129 Torino
Italy
e-mail: fabio.nicola@polito.it

Luigi Rodino
Dipartimento di Matematica
Università di Torino
Via Carlo Alberto 10
10123 Torino
Italy
e-mail: luigi.rodino@unito.it

2000 Mathematics Subject Classification: 35, 81Q, 81S

Library of Congress Control Number: 2010926370

Bibliographic information published by Die Deutsche Bibliothek
Die Deutsche Bibliothek lists this publication in the Deutsche Nationalbibliografie;
detailed bibliographic data is available in the Internet at <http://dnb.ddb.de>.

ISBN 978-3-7643-8511-8

This work is subject to copyright. All rights are reserved, whether the whole or part of the
material is concerned, specifically the rights of translation, reprinting, re-use of illus-
trations, broadcasting, reproduction on microfilms or in other ways, and storage in data
banks. For any kind of use whatsoever, permission from the copyright owner must be
obtained.

© 2010 Springer Basel AG
P.O. Box 133, CH-4010 Basel, Switzerland
Part of Springer Science+Business Media
Printed on acid-free paper produced of chlorine-free pulp. TCF∞
Printed in Germany

ISBN 978-3-7643-8511-8

e-ISBN 978-3-7643-8512-5

9 8 7 6 5 4 3 2 1

www.birkhauser.ch

Contents

Contents   vii

Contents

Preface

The modern theory of general linear PDEs was largely addressed to local problems, i.e., to the study of solutions in a suitably small neighbourhood of $x_0 \in \mathbb{R}^d$. The analysis of global solutions, in fixed domains, subsets of $\mathbb{R}^d$, or even manifolds, appears as a second step, reconnecting with classical results. A simple but important global situation, where interest goes back to Quantum Mechanics, Signal Analysis and other applications in Physics and Engineering, is represented by the study of solutions in the whole Euclidean space $\mathbb{R}^d$. To such study we devote the present book, making systematic use of the techniques of pseudo-differential operators.

In fact, the Fourier transform and pseudo-differential operators play an essential role in the modern theory of PDEs, both from the local and global points of view. Actually, the pseudo-differential calculus was initially introduced by Kohn-Nirenberg and Hörmander and then developed by other authors, mainly in a local context, to study local regularity and local solvability of PDEs.

On the other hand, the Fourier transform and pseudo-differential calculus find in $\mathbb{R}^d$ their natural setting; for example, the more recent treatment of Hörmander, the so-called Weyl-Hörmander calculus, is formulated in symplectic vector spaces. However, genuine global settings in $\mathbb{R}^d$, providing regularity of solutions in the Schwartz space $\mathcal{S}(\mathbb{R}^d)$ and compactness in $\mathbb{R}^d$ of resolvents of globally-elliptic operators, require symbols with a precise asymptotic control when the variable x goes to infinity. It is exactly these symbols that we address in the present book.

Let us list some topics that we discuss in the volume, and mention several others which we neglect for the sake of brevity. First, our attention is restricted to globally elliptic equations, whereas the important study of the corresponding evolution case, in particular the hyperbolic case, is omitted. Moreover, concerning classes of pseudo-differential operators, we mainly treat symbols possessing a homogeneous structure, namely the so-called classic Γ and G symbols. Nevertheless in the first part of the book, addressed to non-experts, we propose a relatively general calculus, very easy to handle and involving only elementary computations. With respect to the Weyl-Hörmander calculus, we make here a restrictive assumption, namely our weights are bounded from below by positive constants; such a condition is satisfied indeed by Γ and G classes. Also, we omit the study of the symplectic invariance of our pseudo-differential calculus, limiting ourselves to an emphasis on its behaviour under Fourier conjugation, which is of relevance in some applications. Similarly, extensions to more general types of non-compact manifolds, in particular the so-called manifolds with exits, are omitted, as well as general index formulas.

The arguments which are treated in the book reflect the research interests of the authors and their collaborators, and collect results obtained by them in the last 10 years. Our first main line of discourse is devoted to Spectral Theory. We pay particular attention to complex powers and asymptotics for the counting function (without sharp remainder). In this regards, the present volume owes much to the preceding monograph of Boggiatto, Buzano and Rodino. We also discuss the non-commutative residue in $\mathbb{R}^d$ and, in strict connection with Spectral Theory, the Dixmier trace.

The second main line of discourse is devoted to the study of exponential decay and holomorphic extension of the solutions. We refer here to the Gelfand-Shilov spaces $S_\nu^\mu(\mathbb{R}^d)$, replacing successfully the Schwartz space in Applied Mathematics; a self-contained presentation is given in this book. The results, coming from a series of papers by Cappiello, Gramchev and Rodino, refer to semi-linear perturbations of Γ and G equations. Applications, besides to non-linear Quantum Mechanics, are also to travelling waves.

We wish finally to express our gratitude to Ernesto Buzano, of the University of Torino. In addition to reading a large part of the manuscript and suggesting many improvements, he discussed with us the structure of the book and took part in the choice of the contents.

We wish also to express our warmest thanks to Claudia Garetto (University of Innsbruck), Alessandro Morando (University of Brescia), Ubertino Battisti, Paolo Boggiatto, Marco Cappiello, Elena Cordero, Sandro Coriasco, Giuseppe De Donno, Gianluca Garello, Alessandro Oliaro, Patrik Wahlberg (University of Torino) for contributing scientific material, reading the manuscript and suggesting improvements.

Torino, December 2009   Fabio Nicola,   Luigi Rodino

Introduction

To give an introduction to the contents of the book, let us consider initially the basic models to which our pseudo-differential calculus will apply, namely the linear partial differential operators with polynomial coefficients in $\mathbb{R}^d$:

$$P = \sum c_{\alpha\beta} x^\beta D^\alpha, \quad x \in \mathbb{R}^d, \ c_{\alpha\beta} \in \mathbb{C}, \tag{I.1}$$

where in the sum $(\alpha, \beta) \in \mathbb{N}^d \times \mathbb{N}^d$ runs over a finite subset of indices. A natural setting for P is given by the Schwartz space $\mathcal{S}(\mathbb{R}^d)$ and its dual $\mathcal{S}'(\mathbb{R}^d)$. These spaces are invariant under the action of the Fourier transform

$$\mathcal{F}u(\xi) = \widehat{u}(\xi) = \int e^{-ix\xi} u(x) \, dx, \quad \text{with } dx = (2\pi)^{-\frac{d}{2}} dx. \tag{I.2}$$

Note also that the conjugation $\mathcal{F}P\mathcal{F}^{-1}$ gives still an operator of the form (I.1).

The preliminary problem in our analysis will be to establish the global regularity of P by means of the construction of parametrices in the pseudo-differential form, with suitable symbols $a(x, \xi) \in C^\infty(\mathbb{R}^d \times \mathbb{R}^d)$:

$$Au(x) = a(x, D)u(x) = \int e^{ix\xi} a(x, \xi)\widehat{u}(\xi) \, d\xi. \tag{I.3}$$

Namely, a parametrix A of P is a linear map $\mathcal{S}(\mathbb{R}^d) \to \mathcal{S}(\mathbb{R}^d)$, $\mathcal{S}'(\mathbb{R}^d) \to \mathcal{S}'(\mathbb{R}^d)$ such that

$$PA = I + R_1, \quad AP = I + R_2 \tag{I.4}$$

where R_1, R_2 are regularizing, i.e., R_1, R_2: $\mathcal{S}'(\mathbb{R}^d) \to \mathcal{S}(\mathbb{R}^d)$. Starting from the equation $Pu = f$ with $u \in \mathcal{S}'(\mathbb{R}^d)$, $f \in \mathcal{S}(\mathbb{R}^d)$, the second identity in (I.4) implies $APu = u + R_2 u$, hence we conclude $u \in \mathcal{S}(\mathbb{R}^d)$, i.e., in our terminology P is globally regular. In particular all the solutions $u \in \mathcal{S}'(\mathbb{R}^d)$ of $Pu = 0$ belong to $\mathcal{S}(\mathbb{R}^d)$. From the existence of a pseudo-differential parametrix A, we may also deduce the Fredholm property of P in Sobolev spaces with suitable weights.

Define the symbol of P as standard:

$$p(x, \xi) = \sum c_{\alpha\beta} x^\beta \xi^\alpha. \tag{I.5}$$

Which properties of the polynomial $p(x, \xi)$ warrant the construction of the parametrix? The standard (local) ellipticity can be read for $p(x, \xi)$ in (I.5), where we assume $|\alpha| \le m$, $|\beta| \le n$, as well as the condition

$$\sum_{\substack{|\alpha|=m \\ |\beta| \le n}} c_{\alpha\beta} x^\beta \xi^\alpha \ne 0 \quad \text{for } \xi \ne 0. \tag{I.6}$$

This provides the existence of local pseudo-differential parametrices and local regularity, but does not give the required control on the asymptotic behaviour of the solutions for $x \to \infty$.

The basic idea is then to add to (I.6) conditions involving the x-variables in the homogeneous structure. As suggested by the invariance of the class of the operators (I.1) under Fourier conjugation, we expect in these conditions a joint or symmetrical role for the x and ξ variables.

We may identify two different approaches, somewhat in competition with each other from the historical point of view, both based on model equations of Quantum Mechanics.

Γ-*classes.* Our starting model is the harmonic oscillator

$$-\Delta + |x|^2 - \lambda. \tag{I.7}$$

We generalize it by considering

$$P = \sum_{|\alpha|+|\beta| \le m} c_{\alpha\beta} x^\beta D^\alpha \tag{I.8}$$

satisfying the Γ-ellipticity assumption

$$p_m(x, \xi) = \sum_{|\alpha|+|\beta|=m} c_{\alpha\beta} x^\beta \xi^\alpha \ne 0 \quad \text{for } (x, \xi) \ne (0, 0), \tag{I.9}$$

which implies local ellipticity. The operator in (I.7) is Γ-elliptic, independently of $\lambda \in \mathbb{C}$. A parametrix for P in (I.8) is constructed as a pseudo-differential operator with symbol having principal part $1/p(x, \xi)$, which we cut off in the bounded region where possibly $p(x, \xi) = 0$. The natural class containing the symbol of the parametrix is defined by considering $z = (x, \xi) \in \mathbb{R}^{2d}$ and imposing the estimates

$$|\partial_z^\gamma a(z)| \lesssim \langle z \rangle^{m-|\gamma|}, \quad z \in \mathbb{R}^{2d}, \tag{I.10}$$

where now $m \in \mathbb{R}$, $\gamma \in \mathbb{N}^{2d}$ and $\langle z \rangle = (1 + |z|^2)^{1/2} = (1 + |x|^2 + |\xi|^2)^{1/2}$. The corresponding pseudo-differential calculus was first given in Shubin [182], 1971.

G-*classes.* The basic model is the free particle operator in $\mathbb{R}^d$

$$-\Delta - \lambda. \tag{I.11}$$

As a generalization, consider the operator

$$P = \sum_{\substack{|\alpha| \leq m \\ |\beta| \leq n}} c_{\alpha\beta} x^{\beta} D^{\alpha} \tag{I.12}$$

satisfying the following G-ellipticity condition. First, considering the so-called bi-homogeneous principal symbol, we impose

$$p_{m,n}(x, \xi) = \sum_{\substack{|\alpha| = m \\ |\beta| = n}} c_{\alpha\beta} x^{\beta} \xi^{\alpha} \neq 0 \quad \text{for } x \neq 0, \ \xi \neq 0. \tag{I.13}$$

Then we assume the standard local ellipticity (I.6) and the dual property, obtained from (I.6) by interchanging the role of x and ξ:

$$\sum_{\substack{|\alpha| \leq m \\ |\beta| = n}} c_{\alpha\beta} x^{\beta} \xi^{\alpha} \neq 0 \quad \text{for } x \neq 0. \tag{I.14}$$

In the case $n = 0$, i.e., when $P = p(D)$ is an operator with constant coefficients, the local ellipticity (I.6) implies (I.13), whereas (I.14) is satisfied if and only if $p(\xi) \neq 0$ for all $\xi \in \mathbb{R}^d$. So the operator in (I.11) is G-elliptic when $\lambda \notin \mathbb{R}_+ \cup \{0\}$.

Under the assumptions (I.6), (I.13), (I.14), a parametrix can be constructed with symbol in the G-classes defined by the estimates

$$|\partial_{\xi}^{\alpha} \partial_x^{\beta} a(x, \xi)| \lesssim \langle \xi \rangle^{m-|\alpha|} \langle x \rangle^{n-|\beta|}, \quad (x, \xi) \in \mathbb{R}^{2d}, \tag{I.15}$$

for $m \in \mathbb{R}$, $n \in \mathbb{R}$. The corresponding pseudo-differential operators were introduced by Parenti [156], 1972, and then studied in detail by Cordes [59], 1995.

In short: the first aim of this book is to present a simple pseudo-differential calculus, containing both Γ-classes and G-classes as particular cases, and to give in this framework some main results, concerning construction of parametrices, weighted Sobolev spaces, Fredholm property and global regularity. Peculiarities of Γ and G operators, depending on the respective homogeneous structures, are then emphasized. Attention is given to recent results, concerning in particular L^p-boundedness, non-commutative residues, exponential decay and holomorphic extension of solutions of semi-linear Γ and G equations.

A more detailed description of the contents can be found in the Summary preceding each chapter. In the following we illustrate general ideas, list models having importance in the applications and provide some references. A detailed bibliography will be found in the Notes at the end of each chapter.

As for the pseudo-differential calculus in Chapter 1, the symbols are defined by the estimates

$$|\partial_{\xi}^{\alpha} \partial_x^{\beta} a(x, \xi)| \lesssim M(x, \xi) \Psi(x, \xi)^{-|\alpha|} \Phi(x, \xi)^{-|\beta|}, \quad (x, \xi) \in \mathbb{R}^{2d}, \tag{I.16}$$

where M, Φ, Ψ are positive weight functions in $\mathbb{R}^{2d}$ satisfying suitable conditions. In particular, the effectiveness of the symbolic calculus is granted by the strong uncertainty principle

$$(1 + |x|^2 + |\xi|^2)^{\epsilon/2} \lesssim \Phi(x,\xi)\Psi(x,\xi), \quad (x,\xi) \in \mathbb{R}^{2d}, \tag{I.17}$$

for some $\epsilon > 0$. The Γ-classes in (I.10) are recaptured by setting $\Phi(x,\xi) = \Psi(x,\xi) = (1 + |x|^2 + |\xi|^2)^{1/2}$, so that (I.17) is valid with $\epsilon = 2$. The choice $\Phi(x,\xi) = \langle x \rangle$, $\Psi(x,\xi) = \langle \xi \rangle$ gives the G-classes. The strong uncertainty principle is satisfied with $\epsilon = 1$ since $(1 + |x|^2 + |\xi|^2)^{1/2} \le \langle x \rangle \langle \xi \rangle = \Phi(x,\xi)\Psi(x,\xi)$.

In Chapter 2 we fix attention on Γ-classes and their generalizations, considering the case when $p(x,\xi)$ in (I.5) is a H-polynomial in $\mathbb{R}^{2d}$, or in particular a multi-quasi-elliptic polynomial as in Boggiatto, Buzano, Rodino [19]:

$$p(x,\xi) = \sum_{(\beta,\alpha)\in\mathcal{P}} c_{\alpha\beta} x^\beta \xi^\alpha, \tag{I.18}$$

where the Newton polyhedron $\mathcal{P}$ of $p(x,\xi)$ is assumed to be complete (in short: the normals to the faces have strictly positive components), with lower bound

$$|p(x,\xi)| \gtrsim \Lambda_{\mathcal{P}}(x,\xi) = \left(\sum_{(\beta,\alpha)\in\mathcal{P}} x^{2\beta} \xi^{2\alpha} \right)^{1/2} \tag{I.19}$$

for large (x,ξ). Taking $\Psi(x,\xi) = \Phi(x,\xi) = \Lambda_{\mathcal{P}}(x,\xi)^{1/\mu}$, where μ is the so-called formal order of $\mathcal{P}$, we obtain a symbolic calculus which satisfies the strong uncertainty principle. Following the results of Morando [148], Garello and Morando [85], we prove that the corresponding pseudo-differential operators are L^p-bounded, with $1 < p < \infty$. For the operator P with symbol (I.18) we deduce in particular the a priori estimates

$$\sum_{(\beta,\alpha)\in\mathcal{P}} \|x^\beta D^\alpha u\|_{L^p} \lesssim \|Pu\|_{L^p} + \|u\|_{L^p} \tag{I.20}$$

and related Fredholm properties. The results apply for example to the following generalizations of the harmonic oscillator

$$P = -\Delta + V(x) \tag{I.21}$$

where the potential $V(x)$ is a positive multi-quasi-elliptic polynomial with respect to the x-variables.

In Chapter 3, besides G-pseudo-differential operators, we consider extensions of (I.21) to more general potentials $V(x)$, by constructing the parametrix in the classes with weights $\Phi(x,\xi) = 1$, $\Psi(x,\xi) = (1 + |x|^2 + |\xi|^2)^{\rho/2}$, $\rho > 0$, for which (I.17) is still satisfied, cf. Buzano [27].

Chapter 4 is devoted to Spectral Theory for pseudo-differential operators with symbol in the classes defined by (I.16), (I.17). For generic weights Φ, Ψ, we study the complex powers following Buzano and Nicola [30], and compute the trace of the heat kernel. Precise asymptotic expansions are deduced for the counting functions $N(\lambda)$, $\lambda \to +\infty$, of self-adjoint operators in the case of Γ and G classes.

Namely, for a Γ-operator with principal symbol $p_m(x, \xi)$ as in (I.9) we prove the formula of Shubin [182]

$$N(\lambda) \sim \lambda^{\frac{2d}{m}} (2\pi)^{-d} \int_{p_m(x,\xi)\leq 1} dx \, d\xi. \tag{I.22}$$

This gives in particular for the eigenvalues of the harmonic oscillator (I.7) the well-known formula

$$N(\lambda) \sim \frac{\lambda^d}{2^d d!}. \tag{I.23}$$

For a G-operator with bi-homogeneous principal symbol $p_{m,n}(x, \xi)$, $m > 0$, $n > 0$, as in (I.13), we obtain according to Maniccia and Panarese [138]:

$$N(\lambda) \sim \begin{cases} C\lambda^{\frac{d}{m}} \log \lambda, & \text{for } m = n, \\ C'\lambda^{\frac{d}{m}}, & \text{for } m < n, \\ C''\lambda^{\frac{d}{n}}, & \text{for } m > n, \end{cases} \tag{I.24}$$

for constants C, C', C'' which can be computed in terms of $p_{m,n}$, of the symbol in (I.6) and the symbol in (I.14), respectively.

In Chapter 5 we present results on non-commutative residue and Dixmier's trace, from Boggiatto and Nicola [20], Nicola [152], Nicola and Rodino [154]. Let us recall that a linear map on an algebra over $\mathbb{C}$ is called a trace if it vanishes on commutators. For the algebra of the classical pseudo-differential operators on a compact manifold, a trace is given by the so-called Wodzicki's non-commutative residue. For classical Γ-operators in $\mathbb{R}^d$, namely with symbol having asymptotic expansion in homogeneous terms $a(z) \sim \sum_{j=0}^{\infty} a_{m-j}(z)$, $m \in \mathbb{Z}$, we define

$$\operatorname{Res} a(x, D) = \int_{\mathbb{S}^{2d-1}} a_{-2d}(\Theta) \, d\Theta, \tag{I.25}$$

where $z = (x, \xi) \in \mathbb{R}^{2d}$, and $d\Theta$ is the usual surface measure on $\mathbb{S}^{2d-1}$. The map Res in (I.25) turns out to be the unique trace on the algebra of the classical Γ-operators which vanishes on regularizing operators, up to a multiplicative constant. Similarly to the result of Connes in the case of a compact manifold, even in $\mathbb{R}^d$ the map Res coincides with the Dixmier trace Tr_ω, when applied to a Γ-operator of order $-2d$. Namely

$$\operatorname{Res} a(x, D) = 2d(2\pi)^d \operatorname{Tr}_\omega(a(x, D)) \tag{I.26}$$

where, limiting for simplicity to a positive self-adjoint operator $a(x, D)$ with eigenvalues $\lambda_j \to 0^+$, $j = 1, 2, \ldots$,

$$\operatorname{Tr}_\omega(a(x, D)) = \lim_{N \to \infty} \frac{1}{\log N} \sum_{j=1}^{N} \lambda_j. \tag{I.27}$$

We may test (I.26), (I.27) on the inverse of the d-th power of the harmonic oscillator in (I.7):

$$h(x, D) = (-\Delta + |x|^2)^{-d} \tag{I.28}$$

for which we have

$$\operatorname{Res} h(x, D) = \int_{\mathbb{S}^{2d-1}} d\Theta = 2d(2\pi)^d \operatorname{Tr}_\omega(h(x, D)), \tag{I.29}$$

as we easily deduce from (I.23). We obtain similar results for the algebra of G-operators. Concerning operators defined by general weights Φ and Ψ, we content ourselves with giving a sufficient condition for Dixmier traceability. Because of the lack of homogeneity, the problem of the definition of the non-commutative residue is in this case difficult, and challenging.

   In the conclusive Chapter 6 we replace, as a basic space, the Schwartz space $\mathcal{S}(\mathbb{R}^d)$ with the subspaces $S_\nu^\mu(\mathbb{R}^d)$, $\mu > 0$, $\nu > 0$, which are better adapted to the study of the problems of Applied Mathematics. We have $f \in S_\nu^\mu(\mathbb{R}^d)$ if and only if

$$|f(x)| \lesssim e^{-\epsilon|x|^{1/\nu}}, \quad x \in \mathbb{R}^d, \tag{I.30}$$

$$|\widehat{f}(\xi)| \lesssim e^{-\epsilon|\xi|^{1/\mu}}, \quad \xi \in \mathbb{R}^d, \tag{I.31}$$

for some $\epsilon > 0$. If $\mu \le 1$, the function f extends to the complex domain, and the estimates (I.31) imply precise bounds for this extension. After a detailed discussion of the properties of the classes $S_\nu^\mu(\mathbb{R}^d)$ we present the results of Cappiello, Gramchev and Rodino [41], [44], Cappiello and Rodino [45], concerning exponential decay and holomorphic extension of the solutions of Γ-elliptic and G-elliptic equations. In short, the conclusions are the following. As suggested by the behaviour of the eigenfunctions of the harmonic oscillator, i.e., the Hermite functions, all the solutions $u \in \mathcal{S}'(\mathbb{R}^d)$ of a generic Γ-elliptic equation $Pu = 0$ belong to $S_{1/2}^{1/2}(\mathbb{R}^d)$, that implies super-exponential decay and holomorphic extension in $\mathbb{C}^d$. Instead, eigenfunctions of G-elliptic equations are in $S_1^1(\mathbb{R}^d)$; this gives exponential decay and holomorphic extension limited to a strip $\{x + iy \in \mathbb{C}^d : |y| < T\}$.

   Particular attention in the second part of Chapter 6 is reserved for semi-linear equations, because of their importance in applications. Consider for example the semi-linear harmonic oscillator of the Quantum Mechanics:

$$-\Delta + |x|^2 u - \lambda u = G[u], \tag{I.32}$$

where the non-linear term is $G[u] = u^k$, $k \ge 2$, or more generally $G[u] = L(u^k)$ with L a first-order Γ-operator. For the eigenfunctions of (I.32), i.e., homoclinics, we still have super-exponential decay; however, with respect to the linear Γ-case, the entire extension is lost, and analyticity in the complex domain is limited to a strip.

A basic example in the G-case is given by a linear part with constant coefficients

$$\sum_{|\alpha|\leq m} c_\alpha D^\alpha u = F[u]. \tag{I.33}$$

As observed before, we have G-ellipticity if and only if the symbol $p(\xi) = \sum_{|\alpha|\leq m} c_\alpha \xi^\alpha$ is elliptic in the standard sense and $p(\xi) \neq 0$ for all $\xi \in \mathbb{R}^d$. The non-linear term $F[u]$ is an arbitrary polynomial in u and lower order derivatives. The result for (I.33) is the same as in the linear case, namely homoclinics belong to $S_1^1(\mathbb{R}^d)$. Relevant examples are

$$u'' - Cu + u^2 = 0 \quad \text{in } \mathbb{R} \tag{I.34}$$

for the solitary wave $v(t, x) = u(x - Ct)$, $C > 0$, of the Korteweg-de Vries equation, and higher order travelling waves equations; we recapture for all of them the results of exponential decay, expected by the physical intuition. Our result also applies to the d-dimensional extension of (I.34)

$$-\Delta u + u = u^k, \quad k \geq 2, \tag{I.35}$$

appearing in plasma physics and non-linear optics. Finally, as a non-local example, the intermediate-long-wave equation in $\mathbb{R}$

$$\frac{e^D + e^{-D}}{e^D - e^{-D}} Du + \gamma u = u^2, \quad \gamma > -1, \tag{I.36}$$

is contained in our theory, because the symbol $\xi \mathrm{Ctgh}\,\xi + \gamma$ of the Fourier multiplier in the left-hand side of (I.36) is G-elliptic, and its homoclinics belong to $S_1^1(\mathbb{R})$.

Background Material

In this chapter we fix the notation used in the present book and we collect some results from Real Analysis which will be useful in the sequel. For a comprehensive account we refer the reader to Hörmander [119, Vol I]; see also Grubb [102].

0.1 Basic Facts and Notation

We employ standard set-theoretic notation, with one peculiarity: set-theoretic inclusion $A \subset B$ does not exclude equality. Thus proper inclusion has to be stated explicitly: $A \subset B$ and $A \neq B$.

We set $\mathbb{N} = \{0, 1, 2, ...\}$, whereas $\mathbb{Z}$, $\mathbb{R}$ stand for the set of all integer and real numbers, respectively. We define $\mathbb{R}_+ = \{x \in \mathbb{R} : x > 0\}$, $\mathbb{R}_- = \{x \in \mathbb{R} : x < 0\}$.

Given a real number x, we set $x_+ = \max\{x, 0\} = \frac{1}{2}(x + |x|)$ and $x_- = \min\{x, 0\} = \frac{1}{2}(x - |x|)$. Hence x_+ and x_- are *the positive and the negative part of x* respectively. The *integer part* of a real number x is denoted by $[x]$. Then $[x]$ is the unique integer such that

$$[x] \leq x < [x] + 1.$$

If X is a non-empty set and $f, g : X \to [0, +\infty)$, we set

$$f(x) \lesssim g(x), \quad x \in X,$$

if there exists $C > 0$ such that $f(x) \leq Cg(x)$, for all $x \in X$. Moreover, if f and g depend on a further variable $z \in Z$, the statement that, *for all $z \in Z$,*

$$f(x, z) \lesssim g(x, z), \quad x \in X,$$

means that for every $z \in Z$ there exists a real number $C_z > 0$ such that $f(x, z) \leq C_z g(x, z)$ for every $x \in X$. Of course, it may happen that $\sup_{z \in Z} C_z = \infty$. Also, we set

$$f(x) \asymp g(x), \quad x \in X,$$

if $f(x) \lesssim g(x)$ and $g(x) \lesssim f(x)$, $x \in X$, and similarly as above if the functions depend on a further parameter.

We will employ the multi-index notation. Given $\alpha, \beta \in \mathbb{N}^d$ and $x \in \mathbb{R}^d$, we set

$$|\alpha| = \alpha_1 + \cdots + \alpha_d, \qquad \alpha! = \alpha_1! \cdots \alpha_d!,$$
$$\alpha \le \beta \iff \alpha_j \le \beta_j, \quad \text{for } j = 1, \ldots, d,$$
$$\alpha < \beta \iff \alpha \ne \beta \text{ and } \alpha \le \beta,$$
$$\binom{\alpha}{\beta} = \prod_{j=1}^{d} \binom{\alpha_j}{\beta_j} = \frac{\alpha!}{\beta!(\alpha - \beta)!}, \quad \beta \le \alpha,$$
$$x^\alpha = x_1^{\alpha_1} \cdots x_d^{\alpha_d}.$$

Functions are always understood complex-valued, if not stated otherwise. Partial derivatives are denoted by

$$\partial_j = \partial_{x_j} = \frac{\partial}{\partial x_j}, \qquad D_j = D_{x_j} = -i\partial_j, \quad j = 1, \ldots, d,$$

where i is the *imaginary unit*. More generally, we set

$$\partial^\alpha = \partial_1^{\alpha_1} \cdots \partial_d^{\alpha_d} = \partial_x^\alpha = \partial_{x_1}^{\alpha_1} \cdots \partial_{x_d}^{\alpha_d},$$
$$D^\alpha = D_1^{\alpha_1} \cdots D_d^{\alpha_d} = D_x^\alpha = D_{x_1}^{\alpha_1} \cdots D_{x_d}^{\alpha_d}.$$

If $x, \xi \in \mathbb{R}^d$, we set

$$x\xi = x \cdot \xi = \langle x, \xi \rangle = \sum_{j=1}^{d} x_j \xi_j, \qquad |x| = \left(\sum_{j=1}^{d} x_j^2 \right)^{1/2},$$
$$\langle x \rangle = \left(1 + |x|^2 \right)^{1/2}.$$

Observe that $\langle x \rangle$ is a smooth function satisfying $\partial_{x_j} \langle x \rangle = x_j / \langle x \rangle$. In general one verifies that

$$|\partial_x^\alpha \langle x \rangle| \le 2^{|\alpha|+1} |\alpha|! \langle x \rangle^{1-|\alpha|}, \tag{0.1.1}$$

for all $x \in \mathbb{R}^d$ and all $\alpha \in \mathbb{N}^d$.

The following elementary inequality, sometimes called *Peetre's inequality*, will be used throughout the book:

$$\langle x + y \rangle^s \le c_s \langle x \rangle^s \langle y \rangle^{|s|}, \quad x, y \in \mathbb{R}^d, s \in \mathbb{R}, \tag{0.1.2}$$

with a constant $c_s > 0$.

The power of multi-index notation is well explained by the following formulas. For an open subset X of $\mathbb{R}^d$, consider the spaces $C^n(X)$ of functions having continuous partial derivatives of order $\le n$, and $C^\infty(X) = \cap_{n \in \mathbb{N}} C^n(X)$. Given $f, g \in C^n(X)$, we have *Leibniz' formula*:

$$\partial^\alpha(fg) = \sum_{\beta \le \alpha} \binom{\alpha}{\beta} \partial^\beta f \, \partial^{\alpha-\beta} g, \quad |\alpha| \le n. \tag{0.1.3}$$

If we assume furthermore that X is convex, we have *Taylor's formula:*

$$f(y) = \sum_{|\alpha| < n} \frac{1}{\alpha!} \partial^\alpha f(x)(y - x)^\alpha$$

$$+ \sum_{|\alpha| = n} \frac{n}{\alpha!} (y - x)^\alpha \int_0^1 (1 - t)^{n-1} \partial^\alpha f((1 - t)x + ty) \, dt, \quad \text{for } x, y \in X.$$

A *linear differential operator in an open set* $X \subset \mathbb{R}^d$ is defined as

$$A = \sum_{|\alpha| \leq m} a_\alpha(x) D^\alpha, \tag{0.1.4}$$

with $a_\alpha : X \to \mathbb{C}$ for $|\alpha| \leq m$.

The *symbol* of the operator A is the polynomial in the $\xi \in \mathbb{R}^d$ variables

$$a(x, \xi) = \sum_{|\alpha| \leq m} a_\alpha(x) \xi^\alpha; \tag{0.1.5}$$

thus the operator A can be thought of as obtained by substituting $\xi = D$ in (0.1.5). Then it is customary to re-write (0.1.4) as

$$a(x, D) = \sum_{|\alpha| \leq m} a_\alpha(x) D^\alpha.$$

We want to remark that the symbol of $a(x, D)$ can also be computed by the formula

$$a(x, \xi) = e^{-ix \cdot \xi} a(x, D_x) e^{ix \cdot \xi}. \tag{0.1.6}$$

For example the symbol of the Laplace operator $\Delta = \Delta_x = \sum_{j=1}^d \partial_{x_j}^2$ is given by

$$e^{-ix \cdot \xi} \Delta_x e^{ix \cdot \xi} = -|\xi|^2.$$

0.2 Function Spaces and Fourier Transform

As standard, $L^p(\mathbb{R}^d)$, $1 \leq p \leq \infty$, stands for the Banach space of measurable functions $f : \mathbb{R}^d \to \mathbb{C}$ satisfying

$$\|f\|_{L^p(\mathbb{R}^d)} = \left(\int |f(x)|^p \, dx \right)^{1/p} < \infty,$$

if $p < \infty$, or $\|f\|_{L^\infty(\mathbb{R}^d)} = \text{ess} - \sup |f(x)| < \infty$ if $p = \infty$ (where two functions define the same element if they coincide away from a set of Lebesgue measure 0). The inner product in $L^2(\mathbb{R}^d)$ is defined by

$$(f, g) = \int f(x) \overline{g(x)} \, dx,$$

and is hence linear in the first component.

The Schwartz space $\mathcal{S}(\mathbb{R}^d)$ of rapidly decaying functions is defined as the space of all smooth functions f in $\mathbb{R}^d$ such that

$$\sup_{x \in \mathbb{R}^d} |x^\beta \partial^\alpha f(x)| < \infty, \quad \text{for all } \alpha, \beta \in \mathbb{N}^d.$$

It is a Fréchet space with the obvious seminorms. Its topological dual $\mathcal{S}'(\mathbb{R}^d)$ is the space of the temperate distributions. We write $\langle \cdot, \cdot \rangle$ for the duality between Schwartz functions and temperate distributions. In particular, $(u,v) = \langle u, \bar{v} \rangle$ for $u, v \in \mathcal{S}(\mathbb{R}^d)$.

The Fourier transform

$$\mathcal{F}f(\xi) = \widehat{f}(\xi) = \int e^{-ix\xi} f(x)\, đx,$$

with

$$đx = (2\pi)^{-d/2} dx,$$

defines an isomorphism of $\mathcal{S}(\mathbb{R}^d)$ which extends to an isomorphism of $\mathcal{S}'(\mathbb{R}^d)$, and an isometry of $L^2(\mathbb{R}^d)$. The inverse Fourier transform is

$$\mathcal{F}^{-1} f(x) = \int e^{ix\xi} f(\xi)\, đ\xi. \tag{0.2.1}$$

The following identities are valid for functions in $\mathcal{S}(\mathbb{R}^d)$ and distributions in $\mathcal{S}'(\mathbb{R}^d)$:

$$(\widehat{D^\alpha f})(\xi) = \xi^\alpha \widehat{f}(\xi), \tag{0.2.2}$$

$$(\widehat{x^\alpha f}) = (-1)^{|\alpha|} D^\alpha \widehat{f}. \tag{0.2.3}$$

We have also

$$\widehat{f * g} = (2\pi)^{d/2} \widehat{f}\, \widehat{g}, \tag{0.2.4}$$

$$\widehat{fg} = (2\pi)^{d/2} \widehat{f} * \widehat{g}, \tag{0.2.5}$$

for f and g in appropriate subspaces of $\mathcal{S}'(\mathbb{R}^d)$, e.g. $f \in \mathcal{S}(\mathbb{R}^d)$, $g \in \mathcal{S}'(\mathbb{R}^d)$ or vice-versa.

The Sobolev spaces $H^s(\mathbb{R}^d)$, $s \in \mathbb{R}$, are defined by means of the Fourier transform as

$$H^s(\mathbb{R}^d) := \{ f \in \mathcal{S}'(\mathbb{R}^d) : \langle \xi \rangle^s \widehat{f}(\xi) \in L^2(\mathbb{R}^d) \},$$

with $\|f\|_{H^s(\mathbb{R}^d)} = \|\langle \cdot \rangle^s \widehat{f}\|_{L^2(\mathbb{R}^d)}$. When $s = k \in \mathbb{N}$ an equivalent norm is given by

$$\|f\|_{H^k(\mathbb{R}^d)} \asymp \sum_{|\alpha| \le k} \|\partial^\alpha f\|_{L^2(\mathbb{R}^d)}, \quad f \in H^k(\mathbb{R}^d).$$

Moreover, we recall the Sobolev embedding

$$H^s(\mathbb{R}^d) \hookrightarrow L^\infty(\mathbb{R}^d), \quad \text{for } s > d/2. \tag{0.2.6}$$

Finally, we will also use the fact that the space $H^s(\mathbb{R}^d)$, for $s > d/2$, is an algebra with respect to pointwise multiplication, and the following *Schauder estimates* are satisfied:

$$\|fg\|_{H^s(\mathbb{R}^d)} \le C_s \|f\|_{H^s(\mathbb{R}^d)} \|g\|_{H^s(\mathbb{R}^d)}, \quad f,g \in H^s(\mathbb{R}^d). \tag{0.2.7}$$

0.3   Identities and Inequalities for Factorials and Binomial Coefficients

We collect in the sequel some identities and inequalities for factorials and binomial coefficients.

First we recall the *generalized Newton formula:*

$$(t_1 + \ldots + t_d)^N = \sum_{|\alpha|=N} \frac{N!}{\alpha_1! \ldots \alpha_d!} t_1^{\alpha_1} \ldots t_d^{\alpha_d}, \tag{0.3.1}$$

where $N \in \mathbb{N}$ and $t_1, \ldots, t_d$ are real numbers. Fixing $t_1 = \ldots = t_d = 1$ in (0.3.1) we deduce

$$d^N = \sum_{|\alpha|=N} \frac{N!}{\alpha!}. \tag{0.3.2}$$

This implies in particular

$$|\alpha|! \le d^{|\alpha|} \alpha!. \tag{0.3.3}$$

When $d - 2$, we obtain from (0.3.2) and (0.3.3), respectively

$$2^N = \sum_{k+j=N} \frac{N!}{k!\, j!} \tag{0.3.4}$$

and

$$(k+j)! \le 2^{k+j} k!\, j! \tag{0.3.5}$$

for any $k, j \in \mathbb{N}$. Hence

$$(\alpha + \beta)! \le 2^{|\alpha|+|\beta|} \alpha!\, \beta! \tag{0.3.6}$$

for any $\alpha, \beta \in \mathbb{N}^d$, whereas obviously

$$\alpha!\, \beta! \le (\alpha + \beta)!. \tag{0.3.7}$$

From (0.3.4) we have then

$$\sum_{\beta \le \alpha} \binom{\alpha}{\beta} = 2^{|\alpha|}, \quad \alpha \in \mathbb{N}^d, \tag{0.3.8}$$

and in particular

$$\binom{\alpha}{\beta} \le 2^{|\alpha|}, \quad \beta \le \alpha. \tag{0.3.9}$$

Recall now the definition of the *Euler Gamma function*

$$\Gamma(t) = \int_0^{+\infty} e^{-s} s^{t-1} \, ds, \quad t > 0.$$

Observing that

$$\Gamma(t+1) = t\Gamma(t)$$

and $\Gamma(1) = 1$, one obtains

$$n! = \Gamma(n+1), \quad n \in \mathbb{N}. \tag{0.3.10}$$

The asymptotic behaviour of the Gamma function is given by the well-known *Stirling formula*:

$$\Gamma(t+1) = t^t e^{-t} \sqrt{2\pi t} \, e^{\theta(t)/(12t)}, \quad t \ge 1,$$

where $0 < \theta(t) < 1$. In particular we have

$$t^t \le e^t \, \Gamma(t+1),$$

and also $\Gamma(t+1) \le t^t$ for large t. Taking into account (0.3.10), we then obtain for $N = 1, 2, \ldots,$

$$N! = N^N e^{-N} \sqrt{2\pi N} \, e^{\theta_N/(12N)}, \tag{0.3.11}$$

where $0 < \theta_N < 1$; in particular we have

$$N^N \le e^N N!, \tag{0.3.12}$$

whereas obviously $N! \le N^N$ for all $N = 1, 2, \ldots$.

We have from the Taylor expansion of e^t for $t > 0$,

$$t^N \le N! e^t, \quad N = 0, 1, \ldots, \tag{0.3.13}$$

which implies $t^N \le N^N e^t$. This last inequality is improved by

$$t^A \le A^A \, e^{t-A}, \quad A > 0, \ t > 0, \tag{0.3.14}$$

as one gets by taking logarithms. In particular we have

$$t^A e^{-t} \le A^A.$$

Finally it will be useful to observe that

$$\#\{\alpha = (\alpha_1, \ldots, \alpha_d) \in \mathbb{N}^d : |\alpha| \le m\} = \binom{m+d}{m}, \tag{0.3.15}$$

whereas

$$\#\{\alpha = (\alpha_1, \ldots, \alpha_d) \in \mathbb{N}^d : |\alpha| = m\} = \binom{m+d-1}{d-1}. \tag{0.3.16}$$

Chapter 1

Global Pseudo-Differential Calculus

Summary

Pseudo-differential operators, in a broad sense, are linear operators of the type

$$a(x, D)u(x) = \int e^{ix\xi} a(x, \xi) \widehat{u}(\xi) \, đ\xi, \quad u \in S(\mathbb{R}^d). \qquad (1.0.1)$$

The function $a(x, \xi)$, satisfying suitable estimates in $\mathbb{R}^{2d}$, is called the *symbol* of $a(x, D)$. In fact, (1.0.1) makes sense as a temperate distribution even for $a \in S'(\mathbb{R}^{2d})$.

A linear partial differential operator

$$a(x, D) = \sum_{|\alpha| \leq m} c_\alpha(x) D^\alpha, \qquad (1.0.2)$$

with slowly increasing coefficients can be regarded as a pseudo-differential operator with symbol

$$a(x, \xi) = \sum_{|\alpha| \leq m} c_\alpha(x) \xi^\alpha,$$

as one sees by using (0.2.1), (0.2.2). The way of associating to a symbol a the operator $a(x, D)$ in (1.0.1) is called *standard*, or *left, quantization*. The terminology is justified by the special case of the partial differential operators, where this quantization amounts to replacing ξ by D and to putting the coefficients to the left of the derivatives.

Other important examples of pseudo-differential operators are given by convolution operators $u \mapsto T * u$; in fact, assuming the Fourier transform of the distribution T is well defined, we have

$$(T * u)(x) = \int e^{ix\xi} a(\xi) \widehat{u}(\xi) \, đ\xi, \quad u \in S(\mathbb{R}^d), \qquad (1.0.3)$$

where $a(\xi) = (2\pi)^{d/2}\widehat{T}(\xi)$, see (0.2.4). As a particular case we quote the linear partial differential operators with constant coefficients and the convolution with fundamental solutions whose Fourier transform is defined in some sense.

The first task of the theory is to express functional operations between pseudo-differential operators, as transposition and compositions, in terms of analytic operations on the corresponding symbols. For this program to work, the symbols need to satisfy some growth estimates at infinity, together with their derivatives. In Sections 1.1, 1.2, we will consider symbols $a(x, \xi)$ satisfying, for every $\alpha, \beta \in \mathbb{N}^d$, estimates of the type

$$|\partial_\xi^\alpha \partial_x^\beta a(x, \xi)| \lesssim M(x, \xi)\Psi(x, \xi)^{-|\alpha|}\Phi(x, \xi)^{-|\beta|}, \quad \text{for } (x, \xi) \in \mathbb{R}^{2d}, \qquad (1.0.4)$$

where Φ, Ψ, M are positive weight functions in $\mathbb{R}^{2d}$ satisfying some technical conditions. We denote by $S(M; \Phi, \Psi)$ the space of functions $a(x, \xi)$ satisfying estimates (1.0.4). The choice of the functions Φ, Ψ, M depends on the problem under consideration; we will clarify this point later.

Now, the symbol of the composition of two operators of the form (1.0.3) is exactly the product of the corresponding symbols. This is generally not true when dealing with two operators of the form (1.0.1), say $a(x, D)$ and $b(x, D)$; nevertheless, if $a \in S(M_1; \Phi, \Psi)$ and $b \in S(M_2; \Phi, \Psi)$, the composition of $a(x, D)$ and $b(x, D)$ is still a pseudo-differential operator with symbol in $S(M_1 M_2; \Phi, \Psi)$ which can be written as

$$a(x, \xi)b(x, \xi) + r(x, \xi), \qquad (1.0.5)$$

with a remainder $r \in S(M_1 M_2 h; \Phi, \Psi)$, where $h := \Phi^{-1}\Psi^{-1}$ is the so-called Planck function. It is assumed $h(x, \xi) \lesssim 1$, so that $r(x, \xi)$ also belongs to $S(M_1 M_2; \Phi, \Psi)$. Precisely, we have for $r(x, \xi)$ the asymptotic expansion

$$r(x, \xi) \sim \sum_{\alpha \neq 0} \frac{1}{\alpha!}\partial_\xi^\alpha a(x, \xi)D_x^\alpha b(x, \xi), \qquad (1.0.6)$$

which means that, for every $N \geq 2$,

$$r(x, \xi) - \sum_{0 < |\alpha| \leq N-1} \frac{1}{\alpha!}\partial_\xi^\alpha a(x, \xi)D_x^\alpha b(x, \xi) \in S(M_1 M_2 h^N; \Phi, \Psi). \qquad (1.0.7)$$

In fact, in several applications we will assume the so-called strong uncertainty principle $h(x, \xi) \lesssim (1 + |x| + |\xi|)^{-\delta}$ for some $\delta > 0$, which guarantees, for example, that the expression in (1.0.7) has a strong decay at infinity provided N is large enough.

Similarly, we will prove formulas for the symbol of the transpose and the adjoint of a pseudo-differential operator. Here we only remark that the transpose of the differential operator $a(x, D) = \sum_{|\alpha| \leq m} c_\alpha(x)D^\alpha$ is given by

$${}^t a(x, D)u(x) = \sum_{|\alpha| \leq m} (-D)^\alpha \left(c_\alpha(x)u(x)\right),$$

which suggests the study of pseudo-differential operators of the form

$$\mathrm{Op}_1(b)u(x) = \int e^{i(x-y)\xi} b(y,\xi)u(y)\, dy\, d\xi. \tag{1.0.8}$$

In fact, we have ${}^t a(x, D) = \mathrm{Op}_1(b)$, with $b(x,\xi) = \sum_{|\alpha|\le m} c_\alpha(x)(-\xi)^\alpha$. The correspondence $b \mapsto \mathrm{Op}_1(b)$ is called *right quantization*.

More generally, we will develop the whole theory for the so-called τ-*quanti-zation*:

$$\mathrm{Op}_\tau(a)u(x) = \int e^{i(x-y)\xi} a((1-\tau)x + \tau y, \xi)u(y)\, dy\, d\xi, \tag{1.0.9}$$

where $\tau \in \mathbb{R}$. When $\tau = 0$ we recapture the operator $a(x, D)$ in (1.0.1), whereas for $\tau = 1$ we obtain the quantization in (1.0.8). The case $\tau = 1/2$ is also of special interest and yields the *Weyl quantization*

$$a^w u(x) = \int e^{i(x-y)\xi} a\left(\frac{x+y}{2}, \xi\right) u(y)\, dy\, d\xi. \tag{1.0.10}$$

For example, in dimension $d = 1$, if $a(x, \xi) = x\xi$,

$$a^w = \frac{1}{2}(xD_x + D_x x).$$

This quantization has the nice property that real-valued symbols give rise to (formally) self-adjoint operators, which is one reason for its usefulness in Quantum Mechanics.

We will prove in Section 1.2 that every operator of the form (1.0.10), with symbol a in some class $S(M; \Phi, \Psi)$, is continuous on $\mathcal{S}(\mathbb{R}^d)$ and extends to a continuous operator on $\mathcal{S}'(\mathbb{R}^d)$. Moreover symbols in $S(1; \Phi, \Psi)$ yield bounded operators in $L^2(\mathbb{R}^d)$, cf. Section 1.4. This is easy to see by the aid of the symbolic calculus, assuming the strong uncertainty principle. A more general result will be discussed later as an application of Anti-Wick techniques.

Let us now come to the main applications of the symbolic calculus in Section 1.3. The role of (1.0.5) and (1.0.6) is fundamental in this connection. Suppose, for example, that a symbol $a(x, \xi)$ satisfies the estimate $|a(x, \xi)| \gtrsim M(x, \xi)$, for $|x| + |\xi|$ large; we say that a is *elliptic*. Then (1.0.5) and (1.0.6) show that the operator B with symbol $a(x, \xi)^{-1}$ fulfills

$$Ba(x, D) = I + S_1, \quad a(x, D)B = I + S_2, \tag{1.0.11}$$

where S_1 and S_2 have symbols in $S(h; \Phi, \Psi)$. Under the strong uncertainty principle, and by a slightly more refined argument based on (1.0.7), one can in fact construct B as a pseudo-differential operator verifying (1.0.11) with S_1 and S_2 regularizing operators, i.e., continuous as maps $\mathcal{S}'(\mathbb{R}^d) \to \mathcal{S}(\mathbb{R}^d)$. We will refer to B as the *parametrix* of $a(x, D)$. The existence of a parametrix gives at once a global regularity result for any operator with elliptic symbol; namely, if $u \in \mathcal{S}'(\mathbb{R}^d)$

and $a(x, D)u \in S(\mathbb{R}^d)$, then $u \in S(\mathbb{R}^d)$. This also explains the advantage of considering the above quite general symbol classes: dealing with a given symbol $a(x, \xi)$
(typically, of a particular differential operator), one looks for weights Φ, Ψ such
that a satisfies (1.0.4) with $M(x, \xi) = |a(x, \xi)|$ and, if there exist any, $a(x, D)$ is
then globally regular. Many examples and applications of this general principle
will be detailed in the subsequent chapters.

Another application of the parametrix is the theory of Sobolev spaces *tailored*
to the above symbol classes. In Section 1.5 we will define the Sobolev space
$H(M)$ as $A^{-1}(L^2(\mathbb{R}^d))$, endowed with a convenient norm, where A is any pseudo-
differential operator with an elliptic symbol in $S(M; \Phi, \Psi)$. Any pseudo-differential
operator P with symbol in $S(M'; \Phi, \Psi)$ will define a bounded operator from $H(M)$
to $H(M/M')$.

These spaces measure the regularity and decay of functions and temperate
distributions in $\mathbb{R}^d$ and allow us to give precise global regularity results for pseudo-
differential operators with elliptic symbols. For example, if P has an elliptic
symbol in $S(M'; \Phi, \Psi)$ and $u \in S'(\mathbb{R}^d)$ with $Pu \in H(M/M')$, then $u \in H(M)$.
It is worth noticing that when regarded as operators on such tailored Sobolev
spaces, elliptic operators become Fredholm (they have kernel and cokernel of finite
dimension) and an index theory can be developed, cf. Section 1.6.

In Section 1.7 we will consider yet another type of quantization, the so-called
Anti-Wick quantization. To this end, let

$$\mathcal{G}_{y,\eta}(x) = \pi^{-d/4} e^{ix\eta} e^{-\frac{1}{2}|x-y|^2},$$

where $y, \eta \in \mathbb{R}^d$ are regarded as parameters. These functions satisfy $\|\mathcal{G}_{y,\eta}\|_{L^2(\mathbb{R}^d)}$
$= 1$ for every $(y, \eta) \in \mathbb{R}^{2d}$, and have the property to be localized near y and to
have the Fourier transform localized near η, i.e., by denoting the open ball in $\mathbb{R}^d$ of
center x_0 and radius R by $B(x_0, R)$, $\|\mathcal{G}_{y,\eta}\|_{L^2(B(y,1))} \geq c$, and $\|\widehat{\mathcal{G}_{y,\eta}}\|_{L^2(B(\eta,1))} \geq c$,
for some $c > 0$, *uniformly* with respect to y, η. By this reason, the behaviour near
y of a given function $u(x)$, and the behaviour near η of its Fourier transform $\widehat{u}(\xi)$,
are captured together by the *short-time Fourier transform Vu* of u, defined by

$$Vu(y, \eta) = (u, \mathcal{G}_{y,\eta}).$$

It is easy to see that V acts on $S(\mathbb{R}^d) \to S(\mathbb{R}^{2d})$ continuously and extends to
a continuous map $S'(\mathbb{R}^d) \to S'(\mathbb{R}^{2d})$. Moreover $V : L^2(\mathbb{R}^d) \to L^2(\mathbb{R}^{2d})$ is an
isometry, up to a constant factor. Then, the Anti-Wick operator A_a with symbol
$a \in S'(\mathbb{R}^{2d})$ reads

$$A_a u = (2\pi)^{-d} V^*(aVu), \quad u \in S(\mathbb{R}^d),$$

namely as a Fourier multiplier but with the Fourier transform replaced by the
short-time Fourier transform.

It turns out that $A_a : S(\mathbb{R}^d) \to S'(\mathbb{R}^d)$ continuously, and that A_a is in fact a
pseudo-differential operator whose Weyl symbol $b(x, \xi)$ is obtained by convolving
$a(x, \xi)$ with a Gaussian function in $\mathbb{R}^{2d}$. The main feature of the Anti-Wick

quantization is that it preserves positivity, i.e., non-negative symbols yield non-negative operators. This is not true for the Weyl quantization, but the result for Anti-Wick operators can be fruitfully used to prove *lower bounds* for Weyl operators.

Anti-Wick techniques also yield a very elementary proof of the L^2-boundedness of pseudo-differential operators whose symbols $a(x, \xi)$ are bounded together with their derivatives, i.e., satisfying for every $\alpha, \beta \in \mathbb{N}^d$,

$$|\partial_\xi^\alpha \partial_x^\beta a(x, \xi)| \leq C_{\alpha\beta}, \quad (x, \xi) \in \mathbb{R}^{2d}.$$

The functions $\mathcal{G}_{y,\eta}$ can in fact be used to almost diagonalize such operators, in the sense that, after conjugating with the short-time Fourier transform, the operator $Va(x, D)V^*$ has an integral kernel $K(y', \eta'; y, \eta)$ dominated by a *convolution kernel* in L^1. Young's inequality then gives the desired boundedness on L^2.

1.1 Symbol Classes

A positive continuous function $\Phi(x, \xi)$, $(x, \xi) \in \mathbb{R}^{2d}$, is called a *sub-linear weight* if

$$1 \leq \Phi(x, \xi) \lesssim 1 + |x| + |\xi|, \quad \text{for } x, \xi \in \mathbb{R}^d. \tag{1.1.1}$$

It is called a *temperate weight* if, for some $s > 0$,

$$\Phi(x + y, \xi + \eta) \lesssim \Phi(x, \xi)(1 + |y| + |\eta|)^s, \quad \text{for } x, \xi, y, \eta \in \mathbb{R}^d. \tag{1.1.2}$$

Observe that (1.1.2) with $x-y$ and $\xi-\eta$ in place of x and ξ gives, for any temperate weight Φ, the lower bound

$$\Phi(x + y, \xi + \eta) \gtrsim \Phi(x, \xi)(1 + |y| + |\eta|)^{-s}, \quad \text{for } x, \xi, y, \eta \in \mathbb{R}^d, \tag{1.1.3}$$

i.e., Φ^{-1} is temperate as well. Moreover, the product and real powers of temperate weights are still temperate weights. Also, (1.1.2) and (1.1.3) with $x = \xi = 0$ yield

$$(1 + |y| + |\eta|)^{-s} \lesssim \Phi(y, \eta) \lesssim (1 + |y| + |\eta|)^s.$$

Basic examples of temperate weights are the functions $(1 + |x| + |\xi|)^m$, and $(1 + |x|^2 + |\xi|^2)^{m/2}$, $m \in \mathbb{R}$ (by Peetre's inequality (0.1.2) in $\mathbb{R}^{2d}$). Similarly, $\langle x \rangle^m$ and $\langle \xi \rangle^m$, $m \in \mathbb{R}$, regarded as functions in $\mathbb{R}^{2d}$ are temperate weights.

Definition 1.1.1. (*Symbol classes*) Let $\Phi(x, \xi)$, $\Psi(x, \xi)$ be sub-linear and temperate weights. Let $M(x, \xi)$ be a temperate weight. We denote by $S(M; \Phi, \Psi)$, or for short $S(M)$, the space of all smooth functions $a(x, \xi)$, $(x, \xi) \in \mathbb{R}^{2d}$, such that for every $\alpha, \beta \in \mathbb{N}^d$,

$$|\partial_\xi^\alpha \partial_x^\beta a(x, \xi)| \lesssim M(x, \xi)\Psi(x, \xi)^{-|\alpha|}\Phi(x, \xi)^{-|\beta|}, \quad \text{for } (x, \xi) \in \mathbb{R}^{2d}. \tag{1.1.4}$$

The following remarks follow at once from Definition 1.1.1.

If $a \in S(M)$, then $\partial_\xi^\alpha \partial_x^\beta a \in S(M\Psi^{-|\alpha|}\Phi^{-|\beta|})$. By Leibniz' rule, if $a \in S(M_1)$ and $b \in S(M_2)$, then $ab \in S(M_1 M_2)$. Moreover, if $M_1 \lesssim M_2$, then $S(M_1) \subset S(M_2)$.

The following family of seminorms

$$\|a\|_{k,S(M;\Phi,\Psi)}$$
$$= \sup_{|\alpha|+|\beta|\leq k} \sup_{(x,\xi)\in\mathbb{R}^{2d}} |\partial_\xi^\alpha \partial_x^\beta a(x,\xi)| M(x,\xi)^{-1} \Psi(x,\xi)^{|\alpha|} \Phi(x,\xi)^{|\beta|}, \quad (1.1.5)$$

with $k \in \mathbb{N}$, defines a Fréchet topology on $S(M;\Phi,\Psi)$ (notice that these seminorms are actually norms).

When $M(x,\xi) = \langle\xi\rangle^m$, $m \in \mathbb{R}$, $\Phi(x,\xi) = 1$, $\Psi(x,\xi) = \langle\xi\rangle^\rho$, $0 \leq \rho \leq 1$, the class $S(M;\Phi,\Psi)$ is usually denoted by $S_{\rho,0}^m$ in the literature.

Now, according to the general theory of Fréchet spaces, a subset $B \subset S(M;\Phi, \Psi)$ is bounded if $\sup_{a\in B} \|a\|_{k,S(M;\Phi,\Psi)} < \infty$ for every $k \in \mathbb{N}$. Hence, a bounded subset of $S(M;\Phi,\Psi)$ is also bounded in $C^\infty(\mathbb{R}^{2d})$. Since every bounded sequence in $C^\infty(\mathbb{R}^{2d})$ has a convergent subsequence in $C^\infty(\mathbb{R}^{2d})$ (see, e.g., Treves [190, Theorem 14.4, page 146]), we deduce the following fact.

Proposition 1.1.2. *For any bounded sequence $a_n \in S(M;\Phi,\Psi)$ the following conditions are equivalent:*

1) *a_n converges in $\mathcal{S}'(\mathbb{R}^{2d})$;*

2) *a_n converges pointwise;*

3) *a_n converges in $C^\infty(\mathbb{R}^{2d})$.*

Notice that the Schwartz space $\mathcal{S}(\mathbb{R}^{2d})$ is contained in every symbol class $S(M;\Phi,\Psi)$. This is due to the fact that Φ and Ψ are sub-linear, and M satisfies a lower bound of the type (1.1.3).

Example 1.1.3. Let $p(x,\xi)$ be a polynomial of degree N. Then $|p(x,\xi)| \lesssim (1 + |x| + |\xi|)^N$. Since the derivatives $\partial_\xi^\alpha \partial_x^\beta p(x,\xi)$, $|\alpha| + |\beta| \leq N$, are polynomials of degree $N - |\alpha| - |\beta|$, we see that $p \in S(M;\Phi,\Psi)$, with $M(x,\xi) = (1 + |x| + |\xi|)^N$, and $\Phi(x,\xi) = \Psi(x,\xi) = 1 + |x| + |\xi|$. Actually, since all the weights Φ and Ψ we are considering have sub-linear growth, for any couple of such weights we have $p \in S(M;\Phi,\Psi)$ with $M(x,\xi) = (1 + |x| + |\xi|)^N$.

The following proposition shows a way to approximate symbols in $S(M;\Phi,\Psi)$ by Schwartz functions.

Lemma 1.1.4. *Let $\chi \in C_0^\infty(\mathbb{R}^{2d})$ and let Φ and Ψ be any sub-linear weights. Then $\chi_\epsilon(x,\xi) := \chi(\epsilon x, \epsilon\xi)$, $0 \leq \epsilon \leq 1$, is bounded in $S(1;\Phi,\Psi)$. If moreover $\chi(x,\xi) = 1$ for (x,ξ) in a neighbourhood of the origin, χ_ϵ tends to 1 in $S(M_0;\Phi,\Psi)$ as $\epsilon \to 0$, where $M_0(x,\xi)$ is any (temperate) weight which tends to infinity at infinity.*

Proof. We have

$$\partial_\xi^\alpha \partial_x^\beta \chi_\epsilon(x,\xi) = \epsilon^{|\alpha|+|\beta|}(\partial_\xi^\alpha \partial_x^\beta \chi)(\epsilon x, \epsilon \xi).$$

Now there exists a constant $C > 0$, independent of ϵ, such that on the support of χ_ϵ we have $\epsilon(|x| + |\xi|) \leq C$, and therefore $\epsilon \leq (C+1)(1 + |x| + |\xi|)^{-1}$. Since Φ and Ψ are sub-linear weights, it follows that, for every $\alpha, \beta \in \mathbb{N}^d$,

$$|\partial_\xi^\alpha \partial_x^\beta \chi_\epsilon(x,\xi)| \lesssim \Psi(x,\xi)^{-|\alpha|}\Phi(x,\xi)^{-|\beta|}, \quad (x,\xi) \in \mathbb{R}^{2d}, \ 0 \leq \epsilon \leq 1, \qquad (1.1.6)$$

which is the first part of the statement.

If $\chi(x,\xi) = 1$ for (x,ξ) in a neighbourhood of the origin, we see that $\epsilon \geq C'(1 + |x| + |\xi|)^{-1}$ on the support of $\chi_\epsilon - 1$, for a constant $C' > 0$ independent of ϵ. Hence, for every $L > 0$ there exists $\epsilon_0 > 0$ such that $M_0(x,\xi) > L$ on the support of $\chi_\epsilon - 1$, if $\epsilon < \epsilon_0$. As a consequence,

$$|\partial_\xi^\alpha \partial_x^\beta (\chi_\epsilon(x,\xi) - 1)| \lesssim L^{-1}M_0(x,\xi)\Psi(x,\xi)^{-|\alpha|}\Phi(x,\xi)^{-|\beta|}, \quad (x,\xi) \in \mathbb{R}^{2d},$$

if $\epsilon < \epsilon_0$, which is the second part of the statement. $\qquad\square$

Proposition 1.1.5. *Let $a \in S(M; \Phi, \Psi)$ and let M_0 be any (temperate) weight which tends to infinity at infinity. Then there exists a sequence a_n of Schwartz functions, bounded in $S(M; \Phi, \Psi)$ and convergent to a in $S(MM_0; \Phi, \Psi)$.*

Proof. Let $\chi \in C_0^\infty(\mathbb{R}^{2d})$, $\chi = 1$ in a neighbourhood of the origin. It suffices to set

$$a_n(x,\xi) = \chi(x/n, \xi/n)a(x,\xi)$$

and apply Lemma 1.1.4. $\qquad\square$

A key role in the development of the symbolic calculus will be played by the so-called *uncertainty principle*, i.e., the lower bound

$$\Phi(x,\xi)\Psi(x,\xi) \gtrsim 1, \quad (x,\xi) \in \mathbb{R}^{2d}, \qquad (1.1.7)$$

which follows from the above hypotheses $\Phi(x,\xi) \geq 1$ and $\Psi(x,\xi) \geq 1$. The function

$$h(x,\xi) := \Phi(x,\xi)^{-1}\Psi(x,\xi)^{-1} \qquad (1.1.8)$$

will be called the *Planck function*. As a consequence of the uncertainty principle we have the inclusions

$$S(Mh^N; \Phi, \Psi) \subset S(M; \Phi, \Psi), \quad N \in \mathbb{N}. \qquad (1.1.9)$$

Often in the following we will make use of the *strong uncertainty principle*, that is

$$\Phi(x,\xi)\Psi(x,\xi) \gtrsim (1 + |x| + |\xi|)^\delta, \quad (x,\xi) \in \mathbb{R}^{2d}, \qquad (1.1.10)$$

for some $\delta > 0$. The importance of the strong uncertainty principle is clear when dealing with operators of the form $D_\xi^\alpha D_x^\alpha$, $|\alpha| = N$, which map $S(M; \Phi, \Psi)$ into

$S(Mh^N; \Phi, \Psi)$. If the strong uncertainty principle is satisfied we see that symbols in $S(Mh^N; \Phi, \Psi)$ decay at infinity, together with their derivatives, as much as we want if N is large enough, and in fact

$$\bigcap_{N \in \mathbb{N}} S(Mh^N; \Phi, \Psi) = \mathcal{S}(\mathbb{R}^{2d}). \tag{1.1.11}$$

For any given sequence of symbols $a_n \in S(Mh^n; \Phi, \Psi)$, $n \in \mathbb{N}$, we write

$$a(x, \xi) \sim \sum_n a_n(x, \xi) \tag{1.1.12}$$

if, for every $N \geq 1$,

$$a - \sum_{j=0}^{N-1} a_j \in S(Mh^N; \Phi, \Psi). \tag{1.1.13}$$

The right-hand side of (1.1.12) is called *asymptotic expansion* of a. Because of (1.1.9), from (1.1.12) we have that $a \in S(M; \Phi, \Psi)$ and, generally speaking, nothing more. Instead, if the strong uncertainty principle is satisfied, the above asymptotic expansion determines a modulo Schwartz functions. More precisely, the following result holds.

Proposition 1.1.6. *Assume the strong uncertainty principle (1.1.10). Let $a_n \in S(Mh^n; \Phi, \Psi)$, $n \in \mathbb{N}$. Then there exists a symbol $a(x, \xi) \in S(M; \Phi, \Psi)$ verifying (1.1.12). Moreover a is uniquely determined modulo Schwartz functions.*

Proof. The uniqueness of a modulo Schwartz functions is a consequence of (1.1.11).

Let now $\chi \in C_0^\infty(\mathbb{R}^{2d})$, $\chi = 1$ in a neighbourhood of the origin. Let ϵ_j be a positive sequence converging to 0, and set $A_j(x, \xi) = (1 - \chi(\epsilon_j x, \epsilon_j \xi))a_j(x, \xi)$. Hence $A_j \in S(Mh^j)$. By Lemma 1.1.4, $1 - \chi(\epsilon_j x, \epsilon_j \xi)$ tends to zero in $S(M_0)$ with, say, $M_0(x, \xi) = 1 + |x| + |\xi|$. Hence, we can choose ϵ_j converging to zero so rapidly that

$$\|A_j\|_{k, S(MM_0 h^j)} \leq 2^{-j}, \quad \text{for } k \leq j.$$

Set $a(x, \xi) := \sum_{j=0}^\infty A_j(x, \xi)$. Since this sum is locally finite, $a(x, \xi)$ is a well defined smooth function. Moreover, by the strong uncertainty principle, there is $N' \in \mathbb{N}$ such that $h^{-N'} \gtrsim M_0$. Hence, if $k \leq N$,

$$\left\| a - \sum_{j=0}^{N-1} A_j \right\|_{k; S(Mh^{N-N'})} \lesssim \left\| \sum_{j=N}^\infty A_j \right\|_{k; S(MM_0 h^N)}$$

$$\leq \sum_{j=N}^\infty \|A_j\|_{k; S(MM_0 h^N)}$$

$$\lesssim \sum_{j=N}^\infty \|A_j\|_{k; S(MM_0 h^j)} < \infty.$$

Then, for every fixed k, N, we choose N'' such that $N'' - N' \geq N$, and we obtain

$$\left\| a - \sum_{j=0}^{N''-1} A_j \right\|_{k,S(Mh^N)} \lesssim \left\| a - \sum_{j=0}^{N''-1} A_j \right\|_{k,S(Mh^{N''-N'})} < \infty.$$

Since, for $j \geq N$, $A_j \in S(Mh^j) \subset S(Mh^N)$, we see that

$$\left\| a - \sum_{j=0}^{N-1} A_j \right\|_{k,S(Mh^N)} < \infty.$$

The same holds with a_j in place of A_j, because $A_j - a_j \in S(\mathbb{R}^{2d})$.

  This concludes the proof. $\qquad\qquad\qquad\qquad\qquad\qquad\qquad\qquad\qquad\qquad\qquad$ □

Remark 1.1.7. It is worth noticing that (1.1.13) could give significant information from a micro-local point of view, even if the strong uncertainty principle is not satisfied (e.g. looking at sub-regions, as cones, where the strong uncertainty principal holds). Hence in the next section we will deal with asymptotic expansions without assuming the strong uncertainty principle.

1.2 Basic Calculus

1.2.1 Action on S

Let $a \in S'(\mathbb{R}^{2d})$, $\tau \in \mathbb{R}$, and consider the operator $\mathrm{Op}_\tau(a)$ defined formally by

$$\mathrm{Op}_\tau(a)u(x) = \int e^{i(x-y)\xi} a((1-\tau)x + \tau y, \xi) u(y)\, dy\, d\xi, \quad u \in S(\mathbb{R}^d). \qquad (1.2.1)$$

We will call a the *τ-symbol* of the *pseudo-differential operator* $\mathrm{Op}_\tau(a)$. Important special cases are obtained when $\tau = 0$, $\tau = 1$ and $\tau = 1/2$, which correspond to the so-called *left quantization* $a(x, D)$ in (1.0.1), the *right quantization* in (1.0.8) and the *Weyl quantization* a^w in (1.0.10), respectively.

  The rigorous meaning we give to (1.2.1) is the following one: $\mathrm{Op}_\tau(a)$ is the continuous operator $S(\mathbb{R}^d) \to S'(\mathbb{R}^d)$ with distribution kernel $K_\tau \in S'(\mathbb{R}^{2d})$ given by

$$K_\tau(x, y) = (2\pi)^{-d/2} \mathcal{F}^{-1}_{\xi \to x-y} a((1-\tau)x + \tau y, \xi). \qquad (1.2.2)$$

This means that

$$\langle \mathrm{Op}_\tau(a)u, v \rangle = \langle K_\tau, v \otimes u \rangle, \quad u, v \in S(\mathbb{R}^d), \qquad (1.2.3)$$

where $\langle \cdot, \cdot \rangle$ stands for the usual duality between temperate distributions and Schwartz functions.

  Observe that when $a \in S(\mathbb{R}^{2d})$ the integral in (1.2.1) is absolutely convergent and the action just defined reduces to (1.2.1). Other interpretations are given in Proposition 1.2.3 below.

Proposition 1.2.1. *The correspondence $a \mapsto K_\tau$ is an isomorphism of $\mathcal{S}(\mathbb{R}^{2d})$, of $\mathcal{S}'(\mathbb{R}^{2d})$, and also of $L^2(\mathbb{R}^{2d})$. The inverse map is given by*

$$a(x, \xi) = (2\pi)^{d/2} \mathcal{F}_{y \to \xi} K_\tau(x + \tau y, x - (1 - \tau)y). \tag{1.2.4}$$

Proof. The partial Fourier transform as well as the composition with the change of variable $\mathcal{J}(x, y) = ((1 - \tau)x + \tau y, x - y)$ are isomorphisms of $\mathcal{S}(\mathbb{R}^{2d})$, of $\mathcal{S}'(\mathbb{R}^{2d})$, and also of $L^2(\mathbb{R}^{2d})$. This gives the first part of the statement. The second part is just an easy computation.   □

In particular, any pseudo-differential operator corresponds to only one τ-symbol.

Operators with Schwartz symbols, namely Schwartz kernels, are called (globally) *regularizing* operators. They are characterized by the property that they extend to continuous operators $\mathcal{S}'(\mathbb{R}^d) \to \mathcal{S}(\mathbb{R}^d)$.

The following simple result will be used often in the following.

Proposition 1.2.2. *For any given sequence $a_n \in \mathcal{S}'(\mathbb{R}^{2d})$, convergent to $a \in \mathcal{S}'(\mathbb{R}^{2d})$, it turns out that $\operatorname{Op}_\tau(a_n)u \to \operatorname{Op}_\tau(a)u$ in $\mathcal{S}'(\mathbb{R}^d)$ for every $u \in \mathcal{S}(\mathbb{R}^d)$.*

Proof. The kernels of $\operatorname{Op}_\tau(a_n)$ converge to that of $\operatorname{Op}_\tau(a)$ in $\mathcal{S}'(\mathbb{R}^{2d})$, since the map $a \mapsto K_\tau$ is continuous on $\mathcal{S}'(\mathbb{R}^{2d})$.   □

Proposition 1.2.3. *Let $a \in S(M; \Phi, \Psi)$. Then the integral in (1.2.1) is well defined as an iterated integral. Moreover, $\operatorname{Op}_\tau(a)u$, $u \in \mathcal{S}(\mathbb{R}^d)$, defined as a distribution in (1.2.3), coincides with the function expressed by that integral.*

Proof. Since $u \in \mathcal{S}(\mathbb{R}^d)$ and all the derivatives of a are dominated by the same function $M(x, \xi)$, having polynomial growth, we see by repeated integrations by parts that the integral

$$\int e^{i(x-y)\xi} a((1 - \tau)x + \tau y, \xi) u(y)\, dy$$

defines a function $b(x, \xi)$ satisfying the estimates $|b(x, \xi)| \lesssim \langle x \rangle^s \langle \xi \rangle^{-N}$, for some $s \in \mathbb{R}$ and every N. This gives the first part of the statement. This estimate also implies that we can apply Fubini's theorem and interchange the integrals with respect to x and ξ in

$$\iiint e^{i(x-y)\xi} a((1 - \tau)x + \tau y, \xi) u(y) v(x)\, dy\, d\xi\, dx.$$

The change of variable $\mathcal{J}(x, y) = ((1 - \tau)x + \tau y, x - y)$ then reduces that expression to $(2\pi)^{-d/2} \langle a, \mathcal{F}_2^{-1}((v \otimes u) \circ \mathcal{J}^{-1}) \rangle$, ($\mathcal{F}_2$ being the partial Fourier transform in the second variable), which is exactly the meaning of the right-hand side of (1.2.3).   □

Now we show that every pseudo-differential operator $\mathrm{Op}_{\tau_1}(a)$ can be written as $\mathrm{Op}_{\tau_2}(b)$ for another symbol b, with a precise dependence of a. To study this problem we introduce the following family of Fourier multipliers in $\mathbb{R}^{2d}$:

$$e^{i\tau D_x \cdot D_\xi} a(x,\xi) := \int e^{i(xy+\xi\eta)} e^{i\tau y\eta} \widehat{a}(y,\eta)\,dy\,d\eta, \quad a \in \mathcal{S}'(\mathbb{R}^{2d}),\ \tau \in \mathbb{R}. \quad (1.2.5)$$

In particular, for $\tau = 0$ we have the identity operator. These operators define isomorphisms of $\mathcal{S}(\mathbb{R}^{2d})$ and $\mathcal{S}'(\mathbb{R}^{2d})$, because such are the Fourier transform in $\mathbb{R}^{2d}$ and the multiplication by $e^{i\tau y\eta}$. Moreover, as operators on $\mathcal{S}(\mathbb{R}^{2d})$ and $\mathcal{S}'(\mathbb{R}^{2d})$ they satisfy the group property

$$e^{i\tau_1 D_x \cdot D_\xi} e^{i\tau_2 D_x \cdot D_\xi} = e^{i(\tau_1+\tau_2) D_x \cdot D_\xi}, \quad \tau_1, \tau_2 \in \mathbb{R}, \quad (1.2.6)$$

as one sees at once from their definition. Also, by Parseval's formula, $e^{i\tau D_x \cdot D_\xi}$ is in fact a unitary operator on $L^2(\mathbb{R}^{2d})$.

Now we show other useful expressions for $e^{i\tau D_x \cdot D_\xi} a$, when $a \in \mathcal{S}(\mathbb{R}^{2d})$. Since the Fourier transform of $(2\pi|\tau|)^{-d} e^{-iy\eta/\tau}$, $\tau \neq 0$, is $(2\pi)^{-d} e^{i\tau x\xi}$ (x and ξ being the variables dual to y and η respectively), cf. Hörmander [119, Vol. I, Theorem 7.6.1], and using the formula $\widehat{a * b} = (2\pi)^d \widehat{a}\,\widehat{b}$, for $a \in \mathcal{S}(\mathbb{R}^{2d})$, $b \in \mathcal{S}'(\mathbb{R}^{2d})$, cf. (0.2.4), we see that, for $\tau \neq 0$,

$$e^{i\tau D_x \cdot D_\xi} a = |\tau|^{-d} \int e^{-iy\eta/\tau} a(x+y,\xi+\eta)\,dy\,d\eta, \quad a \in \mathcal{S}(\mathbb{R}^{2d}). \quad (1.2.7)$$

By changes of variables we obtain from this formula the following two expressions, as absolutely convergent integrals ($\tau \neq 0$):

$$e^{i\tau D_x \cdot D_\xi} a = \int e^{-iy\eta} a(x+y,\xi+\tau\eta)\,dy\,d\eta, \quad a \in \mathcal{S}(\mathbb{R}^{2d}), \quad (1.2.8)$$

$$e^{i\tau D_x \cdot D_\xi} a = \int e^{-i(y-x)(\eta-\xi)} a((1-\tau)x + \tau y, \eta)\,dy\,d\eta, \quad a \in \mathcal{S}(\mathbb{R}^{2d}). \quad (1.2.9)$$

We now study the action of the operator $e^{i\tau D_x \cdot D_\xi}$ on the class of symbols $S(M)$. According to Remark 1.1.7, we do not assume the strong uncertainty principle.

Theorem 1.2.4. *Let* $a \in S(M; \Phi, \Psi)$. *Then* $b(x,\xi) := e^{i\tau D_x \cdot D_\xi} a \in S(M; \Phi, \Psi)$ *and the following asymptotic expansion holds:*

$$b(x,\xi) \sim \sum_\alpha (\alpha!)^{-1} \tau^{|\alpha|} \partial_\xi^\alpha D_x^\alpha a(x,\xi). \quad (1.2.10)$$

Moreover the map $e^{i\tau D_x \cdot D_\xi}$ *is continuous on* $S(M; \Phi, \Psi)$.

Formula (1.2.10) means that

$$b(x,\xi) - \sum_{|\alpha|<N} (\alpha!)^{-1} \tau^{|\alpha|} \partial_\xi^\alpha D_x^\alpha a(x,\xi) \in S(Mh^N; \Phi, \Psi),$$

where h is the Planck function in (1.1.8). Notice that the asymptotic expansion (1.2.10) could be obtained formally by considering the Taylor expansion of the exponential $e^{i\tau D_x \cdot D_\xi}$.

Proof. Suppose first $a \in \mathcal{S}(\mathbb{R}^{2d})$ and, for every $N \in \mathbb{N}$, $N \geq 1$, consider the Schwartz function

$$b_N(x, \xi) := e^{i\tau D_x \cdot D_\xi} a(x, \xi) - \sum_{|\alpha| < N} (\alpha!)^{-1} \tau^{|\alpha|} \partial_\xi^\alpha D_x^\alpha a(x, \xi). \qquad (1.2.11)$$

We claim that each seminorm of b_N in $S(Mh^N)$ is estimated by a seminorm of a in $S(M)$. To prove this, we use the formula (1.2.8) for $e^{i\tau D_x \cdot D_\xi} a$. That integral is absolutely convergent (we can suppose $\tau \neq 0$); hence we can regard it as an iterated integral, in the indicated order. We then Taylor expand the function $a(x + y, \xi + \tau\eta)$ at $\eta = 0$:

$$a(x + y, \xi + \tau\eta) = \sum_{|\alpha| < N} (\alpha!)^{-1} \partial_\xi^\alpha a(y + x, \xi)(\tau\eta)^\alpha + r_N(x, \xi, y, \eta),$$

where

$$r_N(x, \xi, y, \eta) = N \sum_{|\alpha| = N} (\alpha!)^{-1} \int_0^1 (1 - t)^{N-1} \partial_\xi^\alpha a(x + y, \xi + t\tau\eta)(\tau\eta)^\alpha \, dt.$$

As an iterated integral, we have

$$\iint e^{-iy\eta} \partial_\xi^\alpha a(x + y, \xi)(\tau\eta)^\alpha dy \, d\eta = \tau^{|\alpha|} \partial_\xi^\alpha D_x^\alpha a(x, \xi),$$

where we have used the formula $(\tau\eta)^\alpha e^{-iy\eta} = (-\tau D_y)^\alpha e^{-iy\eta}$, integrated by parts, and finally used $\int \hat{u}(\eta) \, d\eta = u(0)$, if $u \in \mathcal{S}(\mathbb{R}^d)$.

Hence, we have $b_N(x, \xi) = \int e^{-iy\eta} r_N(x, \xi, y, \eta) dy \, d\eta$. By Fubini's theorem it suffices to prove the desired estimates, uniformly with respect to $t \in [0, 1]$, for any term

$$\iint e^{-iy\eta} \partial_\xi^\alpha a(x + y, \xi + t\tau\eta)(\tau\eta)^\alpha \, dy \, d\eta$$

$$= \tau^N \iint e^{-iy\eta} \partial_\xi^\alpha D_y^\alpha a(x + y, \xi + t\tau\eta) \, dy \, d\eta, \qquad |\alpha| = N.$$

To this end, we use the identity $e^{-iy\eta} = \langle y \rangle^{-2N'} (1 - \Delta_\eta)^{N'} \langle \eta \rangle^{-2N'} (1 - \Delta_y)^{N'} e^{-iy\eta}$, valid for every $N' \in \mathbb{N}$, and we integrate by parts in the last integral, which is then dominated by

$$|\tau|^N \iint \langle \eta \rangle^{-2N'} |(1 - \Delta_y)^{N'} [\langle y \rangle^{-2N'} (1 - \Delta_\eta)^{N'} \partial_\xi^\alpha D_y^\alpha a(x + y, \xi + t\tau\eta)]| \, dy \, d\eta.$$

By (0.1.1) and Leibniz' rule, the expression under the integral sign can be estimated by (see (1.1.5))

$$\|a\|_{2N+4N',S(M)}\langle\eta\rangle^{-2N'}\langle y\rangle^{-2N'}(Mh^N)(x+y,\xi+t\tau\eta)$$
$$\lesssim \|a\|_{2N+4N',S(M)}\langle\eta\rangle^{-2N'}\langle y\rangle^{-2N'}(Mh^N)(x,\xi)(1+|y|+|\eta|)^{s+2Ns},$$

since M, Φ and Ψ are temperate weights (without loss of generality we assumed the constant s in the temperance estimates to be the same for the three weights). Hence if we choose N' satisfying $-2N'+(1+2N)s < -2d$, we obtain

$$|b_N(x,\xi)| \lesssim \|a\|_{2N+4N',S(M)}M(x,\xi)h(x,\xi)^N, \quad (x,\xi)\in\mathbb{R}^{2d}.$$

The estimates for the derivatives of b_N follow similarly. Indeed,

$$\partial_x^\gamma\partial_\xi^\beta b_N(x,\xi) := e^{i\tau D_x\cdot D_\xi}\partial_x^\gamma\partial_\xi^\beta a(x,\xi) - \sum_{|\alpha|<N}(\alpha!)^{-1}\tau^{|\alpha|}\partial_\xi^\alpha D_x^\alpha\partial_x^\gamma\partial_\xi^\beta a(x,\xi),$$

so that one can repeat the above arguments with $\partial_x^\gamma\partial_\xi^\beta a(x,\xi)$ in place of a. This gives the claim at the beginning of the present proof.

Let now $a \in S(M)$. By Proposition 1.1.5 there is a sequence a_n of Schwartz functions convergent to a in $\mathcal{S}'(\mathbb{R}^{2d})$ and bounded in $S(M)$. It follows from what we already proved that the sequence

$$b_N^{(n)}(x,\xi) := e^{i\tau D_x\cdot D_\xi}a_n(x,\xi) - \sum_{|\alpha|<N}(\alpha!)^{-1}\tau^{|\alpha|}\partial_\xi^\alpha D_x^\alpha a_n(x,\xi)$$

is bounded in $S(Mh^N)$. Moreover, it is convergent in $\mathcal{S}'(\mathbb{R}^{2d})$ to the distribution $b_N(x,\xi)$ given in (1.2.11), since the operators $e^{i\tau D_x\cdot D_\xi}$ and $\partial_\xi^\alpha D_x^\alpha$ are continuous on $\mathcal{S}'(\mathbb{R}^{2d})$. It follows from Proposition 1.1.2 that $b_N^{(n)}$ is convergent in $C^\infty(\mathbb{R}^{2d})$, so that b_N is in fact a smooth function in $S(Mh^N)$.

The continuity of the map $e^{i\tau D_x\cdot D_\xi}$ on $S(M)$ also follows from these arguments. $\square$

The following formula (1.2.12) expresses the τ_2-symbol of a pseudo-differential operator in terms of its τ_1-symbol.

Proposition 1.2.5. *Let* $a \in \mathcal{S}'(\mathbb{R}^{2d})$, $\tau_1,\tau_2 \in \mathbb{R}$. *We have*

$$\mathrm{Op}_{\tau_1}(a) = \mathrm{Op}_{\tau_2}\left(e^{i(\tau_1-\tau_2)D_x\cdot D_\xi}a\right). \tag{1.2.12}$$

Proof. First we consider the case $\tau_2 = 0$. By Proposition 1.2.2, it suffices to consider Schwartz symbols, because $\mathcal{S}(\mathbb{R}^{2d})$ is dense in $\mathcal{S}'(\mathbb{R}^{2d})$ and $e^{i(\tau_1-\tau_2)D_x\cdot D_\xi}$ is continuous on $\mathcal{S}'(\mathbb{R}^{2d})$. Now, it follows from (1.2.9) that, for $a \in \mathcal{S}(\mathbb{R}^{2d})$, $u \in \mathcal{S}(\mathbb{R}^d)$,

$$\mathrm{Op}_0\left(e^{i\tau D_x\cdot D_\xi}a\right)u(x)$$
$$= \iint e^{i(x-z)\xi}\left(\int e^{-i(y-x)(\eta-\xi)}a((1-\tau)x+\tau y,\eta)u(z)\,dy\,d\eta\right)dz\,d\xi$$

$$= \iint e^{i(x-y)\eta} a((1-\tau)x + \tau y, \eta) \left(\iint e^{i(y-z)\xi} u(z)\, dz\, d\xi \right) dy\, d\eta,$$

where we repeatedly applied Fubini's theorem. By the inversion formula for the Fourier transform this last expression is just $\mathrm{Op}_\tau(a)u$. This proves the desired result for $\tau_2 = 0$.

In the general case, we observe that by what we have just proved,

$$\mathrm{Op}_0\left(e^{i\tau_1 D_x \cdot D_\xi} a\right) = \mathrm{Op}_{\tau_1}(a), \text{ and } \mathrm{Op}_0\left(e^{i\tau_2 D_x \cdot D_\xi} b\right) = \mathrm{Op}_{\tau_2}(b).$$

If we choose $b = e^{i(\tau_1 - \tau_2)D_x \cdot D_\xi} a$ in this last formula and apply (1.2.6) we see that (1.2.12) is verified. $\square$

Remark 1.2.6. It follows from Proposition 1.2.5 and Theorem 1.2.4 that if the τ_1-symbol a_{τ_1} of a pseudo-differential operator belongs to a class $S(M; \Phi, \Psi)$, then its τ_2-symbol a_{τ_2} belongs to the same class as well. Moreover, one has the asymptotic expansion

$$a_{\tau_2}(x, \xi) \sim \sum_\alpha (\alpha!)^{-1} (\tau_1 - \tau_2)^{|\alpha|} \partial_\xi^\alpha D_x^\alpha a_{\tau_1}(x, \xi). \qquad (1.2.13)$$

We now study the action on $\mathcal{S}(\mathbb{R}^d)$ of pseudo-differential operators with symbol in $S(M; \Phi, \Psi)$.

Proposition 1.2.7. *Let $a \in S(M; \Phi, \Psi)$, $\tau \in \mathbb{R}$. Then the operator $\mathrm{Op}_\tau(a)$ is continuous on $\mathcal{S}(\mathbb{R}^d)$.*

Proof. By Proposition 1.2.5 and Theorem 1.2.4 it suffices to consider the case $\tau = 0$. Now, by Proposition 1.2.3 the distribution $a(x, D)u$, $u \in \mathcal{S}(\mathbb{R}^d)$, is in fact a function given by the absolutely convergent integral

$$a(x, D)u(x) = \int e^{ix\xi} a(x, \xi)\widehat{u}(\xi)\, d\xi.$$

The estimates

$$\left|\partial_x^\alpha \left(e^{ix\xi} a(x, \xi)\right)\right| \lesssim M(x, \xi)\langle\xi\rangle^{|\alpha|} \lesssim \langle x\rangle^s \langle\xi\rangle^{s+|\alpha|},$$

valid for some $s \geq 0$, show that it is possible to differentiate under the integral sign, and we deduce that $a(x, D)u$ is in fact a smooth function. We also see that every derivative $\partial^\alpha a(x, D)u$ is a finite sum of terms of the same form as $a(x, D)u$, with a different symbol in $S(M; \Phi, \Psi)$ and for a new Schwartz function u. Hence to conclude the proof it is sufficient to prove that $\langle x\rangle^{-2N} a(x, D)u(x)$ is dominated, for every $N \in \mathbb{N}$, by a fixed power of $\langle x\rangle$. The identity $\langle x\rangle^{2N} e^{ix\xi} = (1 - \Delta_\xi)^N e^{ix\xi}$ and an integration by parts yield

$$\langle x\rangle^{2N} |a(x, D)u(x)| \leq \int |(1 - \Delta_\xi)^N (a(x, \xi)\widehat{u}(\xi))|\, d\xi.$$

Since, for every $N' \in \mathbb{N}$ and some $s \geq 0$, we have

$$|(1 - \Delta_\xi)^N (a(x, \xi)\widehat{u}(\xi))| \lesssim M(x, \xi)\langle\xi\rangle^{-N'} \lesssim \langle x\rangle^s \langle\xi\rangle^{s-N'},$$

we obtain $\langle x\rangle^{2N}|a(x, D)u(x)| \lesssim \langle x\rangle^s$, as desired. $\qquad\square$

Remark 1.2.8. It follows from the proof of Proposition 1.2.7 that the map $(a, u) \mapsto a(x, D)u$ from $S(M; \Phi, \Psi) \times \mathcal{S}(\mathbb{R}^d)$ to $\mathcal{S}(\mathbb{R}^d)$ is in fact continuous.

1.2.2 Adjoint and Transposed Operator. Action on $\mathcal{S}'$

Let $a \in \mathcal{S}'(\mathbb{R}^{2d})$. The formal adjoint of $\mathrm{Op}_\tau(a)$ is defined as the unique operator $\mathrm{Op}_\tau(a)^* : \mathcal{S}(\mathbb{R}^d) \to \mathcal{S}'(\mathbb{R}^d)$ satisfying

$$(\mathrm{Op}_\tau(a)u, v) = (u, \mathrm{Op}_\tau(a)^*v), \quad \text{for every } u, v \in \mathcal{S}(\mathbb{R}^d).$$

Here we set $(w, v) = \langle w, \overline{v}\rangle$, for $w \in \mathcal{S}'(\mathbb{R}^d)$ $v \in \mathcal{S}(\mathbb{R}^d)$, or $w \in \mathcal{S}(\mathbb{R}^d)$ $v \in \mathcal{S}'(\mathbb{R}^d)$, where $\langle\cdot, \cdot\rangle$ stands for the usual duality between temperate distributions and Schwartz functions.

It is easy to write $\mathrm{Op}_\tau(a)^*$ in the pseudo-differential form. Indeed, if $K(x, y)$ denotes the distribution kernel of $\mathrm{Op}_\tau(a)$, that of $\mathrm{Op}_\tau(a)^*$ is $\overline{K(y, x)}$. On the other hand, by (1.2.2),

$$\overline{K(y, x)} = (2\pi)^{-d/2}\mathcal{F}^{-1}_{\xi \to x-y}\,\overline{a}((\tau x + (1 - \tau)y, \xi),$$

which is the kernel of $\mathrm{Op}_{1-\tau}(\overline{a})$. Hence

$$\mathrm{Op}_\tau(a)^* = \mathrm{Op}_{1-\tau}(\overline{a}), \quad a \in \mathcal{S}'(\mathbb{R}^{2d}). \tag{1.2.14}$$

Let us now point out two important consequences of this formula. First, the formal adjoint of an operator $a(x, D)$, with $a \in S(M; \Phi, \Psi)$, is a pseudo-differential operator with symbol in the same class. Second, real-valued Weyl symbols give rise to (formally) self-adjoint operators.

Proposition 1.2.9. Let $a \in S(M; \Phi, \Psi)$. Then the formal adjoint of $\mathrm{Op}_\tau(a)$ is the pseudo-differential operator with τ-symbol

$$b(x, \xi) = e^{i(1-2\tau)D_x \cdot D_\xi}\overline{a}(x, \xi) \in S(M; \Phi, \Psi).$$

Moreover we have the asymptotic expansion

$$b(x, \xi) \sim \sum_\alpha (\alpha!)^{-1}(1 - 2\tau)^{|\alpha|}\partial^\alpha_\xi D^\alpha_x \overline{a}(x, \xi).$$

Proof. The desired result follows at once from (1.2.14), Proposition 1.2.5 and Theorem 1.2.4. $\qquad\square$

Consider now the Weyl operator $a^w = \mathrm{Op}_{1/2}(a)$ (see (1.0.10)).

Proposition 1.2.10. *Let $a \in \mathcal{S}'(\mathbb{R}^{2d})$. Then*

$$\bar{a} = a \iff (a^{\mathrm{w}})^* = a^{\mathrm{w}}.$$

Proof. This is a consequence of (1.2.14) with $\tau = 1/2$.   □

Similarly, one can consider the transposed operator of $\mathrm{Op}_\tau(a)$, defined as the unique operator ${}^t\mathrm{Op}_\tau(a) : \mathcal{S}(\mathbb{R}^d) \to \mathcal{S}'(\mathbb{R}^d)$ satisfying

$$\langle {}^t\mathrm{Op}_\tau(a)u, v \rangle = \langle u, \mathrm{Op}_\tau(a)v \rangle, \quad \text{for every } u, v \in \mathcal{S}(\mathbb{R}^d). \tag{1.2.15}$$

Remark 1.2.11. It follows at once from this definition that ${}^t({}^t\mathrm{Op}_\tau(a)) = \mathrm{Op}_\tau(a)$. Similarly, $(\mathrm{Op}_\tau(a)^*)^* = \mathrm{Op}_\tau(a)$. Indeed, $\mathrm{Op}_\tau(a)^*u = {}^t\mathrm{Op}_\tau(a)\bar{u}$.

If $K(x, y)$ denotes the distribution kernel of $\mathrm{Op}_\tau(a)$, that of ${}^t\mathrm{Op}_\tau(a)$ is $K(y, x)$. Hence, by arguing as above we obtain the formula

$$ {}^t\mathrm{Op}_\tau(a(x, \xi)) = \mathrm{Op}_{1-\tau}(a(x, -\xi)), \quad a \in \mathcal{S}'(\mathbb{R}^{2d}). \tag{1.2.16}$$

Proposition 1.2.12. *Let $a \in S(M; \Phi, \Psi)$. Then the transpose of $\mathrm{Op}_\tau(a)$ is the pseudo-differential operator with τ-symbol*

$$b(x, \xi) = e^{i(1-2\tau)D_x \cdot D_\xi} a(x, -\xi) \in S(M; \Phi, \Psi).$$

Moreover we have the asymptotic expansion

$$b(x, \xi) \sim \sum_\alpha (\alpha!)^{-1}(1 - 2\tau)^\alpha \partial_\xi^\alpha D_x^\alpha a(x, -\xi).$$

Proof. The desired result follows at once from (1.2.16), Proposition 1.2.5 and Theorem 1.2.4.   □

Proposition 1.2.13. *Let $a \in S(M; \Phi, \Psi)$, $\tau \in \mathbb{R}$. Then the operator $\mathrm{Op}_\tau(a)$ extends to a continuous operator $\mathcal{S}'(\mathbb{R}^d) \to \mathcal{S}'(\mathbb{R}^d)$.*

Proof. We know from Proposition 1.2.12 that the transposed operator ${}^t\mathrm{Op}_\tau(a)$ is still a pseudo-differential operator with τ-symbol in the same class $S(M; \Phi, \Psi)$. Hence by Proposition 1.2.7 it maps $\mathcal{S}(\mathbb{R}^d) \to \mathcal{S}(\mathbb{R}^d)$. The desired extension of $\mathrm{Op}_\tau(a)$ is therefore given by

$$\langle \mathrm{Op}_\tau(a)u, v \rangle = \langle u, {}^t\mathrm{Op}_\tau(a)v \rangle, \quad u \in \mathcal{S}'(\mathbb{R}^d), \ v \in \mathcal{S}(\mathbb{R}^d). \qquad \square$$

Remark 1.2.14. The action on $\mathcal{S}'(\mathbb{R}^d)$ of a pseudo-differential operator is defined in Proposition 1.2.13 exactly to make the relation (1.2.15) true for every $v \in \mathcal{S}'(\mathbb{R}^d)$, $u \in \mathcal{S}(\mathbb{R}^d)$.

1.2.3 Composition of Operators

We now consider the composition of two pseudo-differential operators with symbols in the above classes. Notice that the composition is well defined on $S(\mathbb{R}^d)$ and on $S'(\mathbb{R}^d)$ by Propositions 1.2.7 and 1.2.13.

Remark 1.2.15. As a consequence of Propositions 1.2.7 and 1.2.13, if $a \in S(M)$ and $b \in S(\mathbb{R}^{2d})$ (i.e., $b(x, D)$ is regularizing), we see that both the compositions $a(x, D)b(x, D)$ and $b(x, D)a(x, D)$ are regularizing operators.

Theorem 1.2.16. *Let $a_1 \in S(M_1; \Phi, \Psi)$, $a_2 \in S(M_2; \Phi, \Psi)$. Then, as operators on $S(\mathbb{R}^d)$ or on $S'(\mathbb{R}^d)$, we have $a_1(x, D)a_2(x, D) = b(x, D)$, where*

$$b(x, \xi) = e^{iD_y \cdot D_\eta} a_1(x, \eta) a_2(y, \xi) \Big|_{\eta = \xi, y = x} \in S(M_1 M_2; \Phi, \Psi), \qquad (1.2.17)$$

with the asymptotic expansion

$$b(x, \xi) \sim \sum_\alpha (\alpha!)^{-1} \partial_\xi^\alpha a_1(x, \xi) D_x^\alpha a_2(x, \xi). \qquad (1.2.18)$$

Moreover the map $(a_1, a_2) \mapsto b$ is continuous from $S(M_1; \Phi, \Psi) \times S(M_2; \Phi, \Psi)$ into $S(M_1 M_2; \Phi, \Psi)$.

Proof. Consider first the case in which $a_1, a_2 \in S(\mathbb{R}^{2d})$. Then, for $u \in S(\mathbb{R}^d)$, we have

$$\mathcal{F}\left(a_2(x, D)u\right)(\eta) = \int e^{iy(\xi - \eta)} a_2(y, \xi) \hat{u}(\xi) \, d\xi \, dy.$$

Hence $a_1(x, D)a_2(x, D) = b(x, D)$, with

$$b(x, \xi) = \int e^{-iy\eta} a_1(x, \xi + \eta) a_2(x + y, \xi) \, dy \, d\eta. \qquad (1.2.19)$$

By (1.2.8) with $\tau = 1$, this proves the equality in (1.2.17) when $a_1, a_2 \in S(\mathbb{R}^{2d})$.
Let still $a_1, a_2 \in S(\mathbb{R}^{2d})$ and set

$$b_N(x, \xi) = e^{iD_y \cdot D_\eta} a_1(x, \eta) a_2(y, \xi) \Big|_{\eta = \xi, y = x}$$

$$- \sum_{|\alpha| < N} (\alpha!)^{-1} \partial_\xi^\alpha a_1(x, \xi) D_x^\alpha a_2(x, \xi). \qquad (1.2.20)$$

We claim that every seminorm of b_N in $S(M_1 M_2 h^N)$ is estimated by the product of a seminorm of a_1 in $S(M_1)$ and a seminorm of a_2 in $S(M_2)$. This can be proved by repeating the same arguments as in the first part of the proof of Theorem 1.2.4. Namely, in (1.2.19) one considers the Taylor expansion at $\eta = 0$ of the function $a_1(x, \xi + \eta) a_2(x + y, \xi)$:

$$a_1(x, \xi + \eta) a_2(x + y, \xi) = \sum_{|\alpha| < N} (\alpha!)^{-1} \partial_\xi^\alpha a_1(x, \xi) a_2(x + y, \xi) \eta^\alpha + r_N(x, \xi, y, \eta),$$

where

$$r_N(x,\xi,y,\eta) = N \sum_{|\alpha|=N} (\alpha!)^{-1} \int_0^1 (1-t)^{N-1} (\partial_\xi^\alpha a_1)(x,\xi+t\eta) a_2(x+y,\xi)\eta^\alpha \, dt.$$

Since, as iterated integrals,

$$\iint e^{-iy\eta} \partial_\xi^\alpha a_1(x,\xi) a_2(x+y,\xi)\eta^\alpha dy\, d\eta = \partial_\xi^\alpha a_1(x,\xi) D_x^\alpha a_2(x,\xi),$$

it suffices to prove the desired estimate, uniformly with respect to $t \in [0,1]$, for every term

$$\iint e^{-iy\eta} \left(\partial_\xi^\alpha a_1\right)(x,\xi+t\eta) a_2(x+y,\xi)\eta^\alpha\, dy\, d\eta$$

$$= \iint e^{-iy\eta} \left(\partial_\xi^\alpha a_1\right)(x,\xi+t\eta) D_y^\alpha a_2(x+y,\xi)\, dy\, d\eta, \quad |\alpha| = N.$$

This can be achieved by repeated integration by parts, the relevant estimates being the following ones, with $k = |\delta| + |\beta| + |\gamma|$:

$$|\partial_x^\delta \partial_\xi^\beta \partial_\eta^\gamma a_1(x,\xi+t\eta)|$$

$$\leq \|a_1\|_{k,S(M_1)} M_1(x,\xi+t\eta) \Phi(x,\xi+t\eta)^{-|\delta|} \Psi(x,\xi+t\eta)^{-|\beta|}$$

$$\lesssim \|a_1\|_{k,S(M_1)} M_1(x,\xi) \Phi(x,\xi)^{-|\delta|} \Psi(x,\xi)^{-|\beta|} \langle \eta \rangle^{s(1+|\delta|+|\beta|)},$$

and

$$|\partial_x^\delta \partial_\xi^\beta \partial_y^\gamma a_2(x+y,\xi)|$$

$$\leq \|a_2\|_{k,S(M_2)} M_2(x+y,\xi) \Phi(x+y,\xi)^{-|\delta|} \Psi(x+y,\xi)^{-|\beta|}$$

$$\lesssim \|a_2\|_{k,S(M_2)} M_2(x,\xi) \Phi(x,\xi)^{-|\delta|} \Psi(x,\xi)^{-|\beta|} \langle y \rangle^{s(1+|\delta|+|\beta|)},$$

where we used the temperance property of M_1, M_2, Φ, Ψ (and s is the constant in such temperance estimates). This gives the desired claim.

Suppose now $a_1 \in S(M_1)$ and $a_2 \in S(M_2)$. Observe that, for fixed x,ξ, the function $a_1(x,\eta)a_2(y,\xi)$ belongs to $S(M_1(x,\eta)M_2(y,\xi); \Phi(y,\xi), \Psi(x,\eta))$, as symbol class in $\mathbb{R}^{2d}_{y,\eta}$. Hence, by Theorem 1.2.4 we see that the distribution

$$e^{iD_y \cdot D_\eta} a_1(x,\eta) a_2(y,\xi)$$

is in fact a smooth function of y,η and it makes sense to evaluate it at $\eta = \xi, y = x$. Let therefore $b(x,\xi)$ be given by (1.2.17).

For $j = 1,2$ consider, according to Proposition 1.1.5, a sequence of symbols $a_j^{(n)}$, bounded in $S(M_j)$ and converging to a_j in $S(M_j M_0)$ with, say, $M_0(x,\xi) = 1 + |x| + |\xi|$. It follows from Remark 1.2.8 that $a_2^{(n)}(x,D)u$, $u \in \mathcal{S}(\mathbb{R}^d)$, converges to $a_2(x, D)u$ in $S(\mathbb{R}^d)$ and $a_1^{(n)}(x, D)a_2^{(n)}(x, D)u$ converges to $a_1(x, D)a_2(x, D)u$ in $S(\mathbb{R}^d)$.

On the other hand, it follows from Theorem 1.2.4 that the sequence of functions $e^{iD_y \cdot D_\eta} a_1^{(n)}(x, \eta)a_2^{(n)}(y, \xi)$, for every fixed x, ξ, converges to

$$e^{iD_y \cdot D_\eta} a_1(x, \eta)a_2(y, \xi)$$

in

$$S((M_1 M_0)(x, \eta)\,(M_2 M_0)(y, \xi);\, \Phi(y, \xi),\, \Psi(x, \eta))$$

(as symbol class in $\mathbb{R}^{2d}_{y,\eta}$), hence pointwise in $\mathbb{R}^{2d}_{y,\eta}$. The sequence

$$b^{(n)}(x, \xi) := e^{iD_y \cdot D_\eta} a_1^{(n)}(x, \eta)a_2^{(n)}(y, \xi)|_{\eta=\xi, y=x},$$

converges therefore pointwise to $b(x, \xi)$. Since, for the first part of the proof, $b^{(n)}$ is bounded in $S(M_1 M_2)$, by Proposition 1.1.2 $b^{(n)} \to b$ in $C^\infty(\mathbb{R}^{2d})$, so that $b \in S(M_1 M_2)$. Again by Proposition 1.1.2, $b^{(n)} \to b$ in $S'(\mathbb{R}^{2d})$ as well, and therefore $b^{(n)}(x, D)u \to b(x, D)u$ in $S'(\mathbb{R}^d)$ by Proposition 1.2.2. This shows that $a_1(x, D)a_2(x, D) = b(x, D)$.

Finally, to prove the desired asymptotic expansion, we consider the sequence

$$b_N^{(n)}(x, \xi) := e^{iD_y \cdot D_\eta} a_1^{(n)}(x, \eta)a_2^{(n)}(y, \xi)\Big|_{\eta=\xi, y=x}$$
$$- \sum_{|\alpha|<N} (\alpha!)^{-1} \partial_\xi^\alpha a_1^{(n)}(x, \xi)D_x^\alpha a_2^{(n)}(x, \xi),$$

which is bounded in $S(M_1 M_2 h^N)$ for the first part of the proof, and converges in $C^\infty(\mathbb{R}^{2d})$ to the function $b_N(x, \xi)$ given in (1.2.20), by what we just observed. Hence $b_N \in S(M_1 M_2 h^N)$.

Also, the continuity of the bilinear map $(a_1, a_2) \mapsto b$ from $S(M_1) \times S(M_2)$ into $S(M_1 M_2)$ follows. $\qquad\square$

By combining Theorem 1.2.16 with the Remark 1.2.6 it follows that for every $\tau_1, \tau_2, \tau \in \mathbb{R}$, $a \in S(M_1)$, $b \in S(M_2)$ we have

$$\mathrm{Op}_{\tau_1}(a)\mathrm{Op}_{\tau_2}(b) = \mathrm{Op}_\tau(c)$$

for a suitable $c \in S(M_1 M_2)$. An asymptotic expansion of the symbol can also be obtained from (1.2.18) after writing $\mathrm{Op}_{\tau_1}(a)$ and $\mathrm{Op}_{\tau_2}(b)$ as pseudo-differential operators with 0-symbols and then coming back to the τ-symbol, hence using repeatedly (1.2.13). This is however quite a long task involving identities of binomial coefficients and which is detailed, e.g., in Boggiatto, Buzano and Rodino [19] and Shubin [183]. Instead, here we will present a fast heuristic approach which is useful to *find* the desired asymptotic expansions. We consider, for simplicity, just the case of the Weyl quantization ($\tau_1 = \tau_2 = \tau = 1/2$), which is the most important one after the standard quantization.

Theorem 1.2.17. Let $a \in S(M_1; \Phi, \Psi)$, $b \in S(M_2; \Phi, \Psi)$. Then $a^w b^w = c^w$ for a symbol $c \in S(M_1 M_2; \Phi, \Psi)$. Moreover $c(x, \xi)$ has the asymptotic expansion

$$c(x, \xi) \sim \sum_{\alpha, \beta} (-1)^{|\beta|} (\alpha! \beta!)^{-1} 2^{-|\alpha+\beta|} \partial_\xi^\alpha D_x^\beta a(x, \xi) \partial_\xi^\beta D_x^\alpha b(x, \xi).$$

Proof. The first part of the statement was already observed above. For the asymptotic expansion we argue heuristically as follows. We know from Theorem 1.2.16 and Proposition 1.2.5 that c is given by

$$c(x, \xi) = e^{-\frac{i}{2} D_x \cdot D_\xi} \left\{ \left[e^{iD_y \cdot D_\eta} \left(e^{\frac{i}{2} D_x \cdot D_\eta} a(x, \eta) e^{\frac{i}{2} D_y \cdot D_\xi} b(y, \xi) \right) \right] \Big|_{\eta = \xi, y = x} \right\}.$$

Now, the key observation is that we can bring the operator $e^{-\frac{i}{2} D_x \cdot D_\xi}$ before the evaluation at $y = x, \eta = \xi$, after modifying it as $e^{-\frac{i}{2}(D_x + D_y) \cdot (D_\xi + D_\eta)}$. This is seen by expanding this exponential in Taylor series and applying the chain rule[1]. After that, we can rearrange the exponentials together. We obtain

$$c(x, \xi) = e^{\frac{i}{2}(D_y \cdot D_\eta - D_x \cdot D_\xi)} a(x, \eta) b(y, \xi) \Big|_{\eta = \xi, y = x}. \tag{1.2.21}$$

A further Taylor expansion of this exponential gives the desired asymptotic expansion, after observing that

$$\frac{1}{j!} \left(\frac{i}{2} (D_y \cdot D_\eta - D_x \cdot D_\xi) \right)^j a(x, \eta) b(y, \xi) \Big|_{\eta = \xi, y = x}$$
$$= \sum_{|\alpha+\beta|=j} (-1)^{|\beta|} (\alpha! \beta!)^{-1} 2^{-j} \partial_\xi^\alpha D_x^\beta a(x, \xi) \partial_\xi^\beta D_x^\alpha b(x, \xi). \quad \square$$

Remark 1.2.18. If we denote by $\sigma(x, y; \xi, \eta) = \xi y - x \eta$ the standard symplectic form in $\mathbb{R}^{2d}$, we see that (1.2.21) can be rewritten as

$$c(x, \xi) = e^{\frac{i}{2} \sigma(D_x, D_y; D_\eta, D_\xi)} a(x, \eta) b(y, \xi) \Big|_{y = x, \eta = \xi}, \tag{1.2.22}$$

which suggests a relationship between Weyl quantization and symplectic geometry. Indeed, the Weyl calculus enjoys a property of invariance with respect to symplectic changes of variables. However, we will not discuss this in the present book.

1.3 Global Regularity

In this section we discuss one of the main motivations and applications of the symbolic calculus just developed, namely the construction of inverses of elliptic operators modulo regularizing operators.

---

[1] Basically, this argument relies on the observation that for a differentiable function $f(x, y)$ of two real variables, it turns out $\frac{d}{dx} f(x, x) = \left(\frac{\partial f}{\partial x} + \frac{\partial f}{\partial y} \right) \Big|_{y=x}$.

1.3.1 Hypoellipticity and Construction of the Parametrix

We introduce two particular symbol classes.

Definition 1.3.1. A symbol a is called (globally) *elliptic* in the class $S(M; \Phi, \Psi)$ if $a \in S(M; \Phi, \Psi)$ and for some $R > 0$,

$$|a(x, \xi)| \gtrsim M(x, \xi), \quad \text{for } |x| + |\xi| \geq R. \tag{1.3.1}$$

In the sequel we shall also refer to the following more general definition.

Definition 1.3.2. A symbol $a \in S(M; \Phi, \Psi)$ is called (globally) *hypoelliptic* if there exist a temperate weight $M_0(x, \xi)$ and $R > 0$ such that

$$|a(x, \xi)| \gtrsim M_0(x, \xi), \quad \text{for } |x| + |\xi| \geq R, \tag{1.3.2}$$

and, for every $\alpha, \beta \in \mathbb{N}^d$,

$$|\partial_\xi^\alpha \partial_x^\beta a(x, \xi)| \lesssim |a(x, \xi)| \Psi(x, \xi)^{-|\alpha|} \Phi(x, \xi)^{-|\beta|}, \quad \text{for } |x| + |\xi| \geq R. \tag{1.3.3}$$

We will denote by $\mathrm{Hypo}(M, M_0; \Phi, \Psi)$, or also for short by $\mathrm{Hypo}(M, M_0)$, the class of such symbols.

Remark 1.3.3. In particular, if (1.3.2) holds with $M_0(x, \xi) = M(x, \xi)$, then a is elliptic. Vice versa, every elliptic symbol $a \in S(M; \Phi, \Psi)$ is hypoelliptic, since (1.3.2) holds with $M_0(x, \xi) = M(x, \xi)$, whereas (1.3.3) is true because $|a(x, \xi)| \asymp M(x, \xi)$, for $|x| + |\xi| \geq R$.

It is important to observe that we can talk about pseudo-differential operators with a hypoelliptic symbol without specifying the type of quantization, as one sees by combining (1.2.12) and the following result.

Proposition 1.3.4. *Assume the strong uncertainty principle* (1.1.10). *Let* $a \in S(M; \Phi, \Psi)$ *be a hypoelliptic symbol and* $\tau \in \mathbb{R}$. *Then the symbol* $e^{i\tau D_x \cdot D_\xi} a$ *is hypoelliptic as well.*

Proof. Because of (1.2.10), we have, for every $N > 1$,

$$e^{i\tau D_x \cdot D_\xi} a - a = \sum_{0 < |\alpha| < N} (\alpha!)^{-1} \tau^{|\alpha|} \partial_\xi^\alpha D_x^\alpha a + r_N,$$

where $r_N \in S(Mh^N; \Phi, \Psi)$, with h as in (1.1.8). Hence, let a satisfy (1.3.2) and (1.3.3). Since $h(x, \xi) \to 0$ as $(x, \xi) \to \infty$ by the strong uncertainty principle, we see that $e^{i\tau D_x \cdot D_\xi} a$ will satisfy the same inequalities, possibly for a larger R, if we prove that

$$|\partial_\xi^\beta \partial_x^\gamma b(x, \xi)| \lesssim |a(x, \xi)| h(x, \xi) \Psi(x, \xi)^{-|\beta|} \Phi(x, \xi)^{-|\gamma|}, \quad |x| + |\xi| \geq R,$$

for $b(x, \xi) = r_N(x, \xi)$, with N large enough, or $b(x, \xi) = \partial_\xi^\alpha D_x^\alpha a(x, \xi)$, $|\alpha| > 0$. This is clear when $b(x, \xi) = r_N(x, \xi)$, if N is chosen so that $Mh^N \lesssim M_0 h$ (again, this choice is possible by the strong uncertainty principle). On the other hand, for $b(x, \xi) = \partial_\xi^\alpha D_x^\alpha a(x, \xi)$ we deduce from (1.3.3) that

$$|\partial_\xi^\beta \partial_x^\gamma b(x, \xi)| \lesssim |a(x, \xi)| h(x, \xi)|^{|\alpha|} \Psi(x, \xi)^{-|\beta|} \Phi(x, \xi)^{-|\gamma|}.$$

Since $|\alpha| > 0$, this gives the desired estimate.   □

Lemma 1.3.5. *Let* $a \in \mathrm{Hypo}(M, M_0; \Phi, \Psi)$. *Then, if* p *is an integer, or if* p *is any complex number and* $a(x, \xi)$ *does not take values in* $\mathbb{R}_- \cup \{0\}$ *for* $|x| + |\xi| \geq R$, *it turns out, for every* $\alpha, \beta \in \mathbb{N}^d$,

$$|\partial_\xi^\alpha \partial_x^\beta a(x, \xi)^p| \lesssim |a(x, \xi)|^p \Psi(x, \xi)^{-|\alpha|} \Phi(x, \xi)^{-|\beta|}, \quad \text{for } |x| + |\xi| \geq R. \tag{1.3.4}$$

In particular, $a^p \in S(M^p; \Phi, \Psi)$ *if* $p \geq 0$ *and* $a^p \in S(M_0^p; \Phi, \Psi)$ *if* $p < 0$ *(the power* a^p *is defined by taking the principal branch of the logarithm).*

Proof. First of all we observe that, by (1.3.2), $a(x, \xi) \neq 0$ for $|x| + |\xi| \geq R$. Now, let p be an integer, or let p be any complex number and assume that $a(x, \xi)$ does not take values in $\mathbb{R}_- \cup \{0\}$ for $|x| + |\xi| \geq R$. Then every derivative $\partial_\xi^\alpha \partial_x^\beta a(x, \xi)^p$, $|\alpha| + |\beta| > 0$, is well defined there, and by induction one sees that it is a linear combination of terms of the form

$$a(x, \xi)^{p-k} \prod_{j=1}^k \partial_\xi^{\gamma_j} \partial_x^{\delta_j} a(x, \xi),$$

with $1 \leq k \leq |\alpha| + |\beta|$, $\gamma_j, \delta_j \in \mathbb{N}^d$, $j = 1, \ldots, k$, $\gamma_1 + \ldots + \gamma_k = \alpha$, $\delta_1 + \ldots + \delta_k = \beta$. The desired estimate then follows from (1.3.3).

The last part of the statement is immediate.   □

Theorem 1.3.6. *Assume the strong uncertainty principle and let* $a \in \mathrm{Hypo}(M, M_0; \Phi, \Psi)$. *Then there exist* $b \in S(M_0^{-1}; \Phi, \Psi)$ *and regularizing operators* S_1, S_2 *such that*

$$b(x, D)a(x, D) = I + S_1, \quad a(x, D)b(x, D) = I + S_2, \tag{1.3.5}$$

as operators on $\mathcal{S}(\mathbb{R}^d)$ *and on* $\mathcal{S}'(\mathbb{R}^d)$. *Such a map* $b(x, D)$ *is called a parametrix of* $a(x, D)$.

Proof. Let $b_1(x, \xi) = a^{-1}(x, \xi)$ for $|x| + |\xi| \geq R$ and $b_1 \in C^\infty(\mathbb{R}^{2d})$. We claim that

$$b_1(x, D)a(x, D) = I - r_1(x, D), \tag{1.3.6}$$

where $r_1(x, \xi) \in S(h)$. Indeed, since $b_1 \in S(M_0^{-1})$ by Lemma 1.3.5, we deduce from (1.2.18) that

$$r_1(x, \xi) - \sum_{0 < |\alpha| < N} (\alpha!)^{-1} \partial_\xi^\alpha a(x, \xi)^{-1} D_x^\alpha a(x, \xi) \in S(M M_0^{-1} h^N).$$

By the strong uncertainty principle we have $MM_0^{-1}h^N \lesssim h$ for h large enough, so that $S(MM_0^{-1}h^N) \subset S(h)$. On the other hand, by (1.3.3) and (1.3.4) we see that, in the above sum, $\partial_\xi^\alpha a(x,\xi)^{-1} D_x^\alpha a(x,\xi) \in S(h^{|\alpha|})$. This gives the claim.

Now, let $r_n(x,\xi) \in S(h^n)$, $n \geq 0$, be the symbol of $r_1(x,D)^n$ and, according to Proposition 1.1.6, let $p(x,\xi) \in S(1)$ satisfy $p(x,\xi) \sim \sum_n r_n(x,\xi)$. By using

$$\left(\sum_{n=0}^N r_1(x,D)^n \right) (1 - r_1(x,D)) = I - r_1(x,D)^{N+1},$$

we see from (1.3.6) that the symbol of the operator

$$p(x,D)b_1(x,D)a(x,D) - I$$

$$= -r(x,D)^{N+1} + \left(p(x,D) - \sum_{n=0}^N r_1(x,D)^n \right) b_1(x,D)a(x,D) \quad (1.3.7)$$

belongs to $S(h^{N+1})$ for every N (indeed, we know from (1.3.6) that the operator $b_1(x,D)a(x,D)$ has symbol in $S(1)$). So the first equation in (1.3.5) holds for a suitable S_1, $b(x,\xi)$ being the symbol of $p(x,D)b_1(x,D)$.

By arguing similarly one finds an operator $c(x,D)$ such that $a(x,D)c(x,D) = I + T$, for some regularizing operator T. Let us now prove that $b(x,D) - c(x,D)$ is a regularizing operator. This will give the second equation in (1.3.5). To this end, we observe that

$$b(x,D) = b(x,D)I = b(x,D)\left(a(x,D)c(x,D) - T\right) = b(x,D)a(x,D)c(x,D) + T_1$$

and

$$c(x,D) = Ic(x,D) = (b(x,D)a(x,D) - S_1)c(x,D) = b(x,D)a(x,D)c(x,D) + T_2$$

with T_1 and T_2 regularizing. By subtracting these two equations we see that $b(x,D) - c(x,D) = T_1 - T_2$ is a regularizing operator. $\qquad\square$

Remark 1.3.7. Observe that the arguments at the end of the proof of Theorem 1.3.6 show that the parametrix, independently of the pseudo-differential representation, is unique modulo regularizing operators; namely, if $B_j : S(\mathbb{R}^d) \to S(\mathbb{R}^d)$, $j = 1,2$, extends to a continuous operator $S'(\mathbb{R}^d) \to S'(\mathbb{R}^d)$ and moreover $B_j a(x,D) = I + S'_j$ and $a(x,D)B_j = I + S''_j$, with S'_j and S''_j regularizing, then $B_1 - B_2$ is regularizing.

Definition 1.3.8. We say that a linear continuous map $A : S(\mathbb{R}^d) \to S(\mathbb{R}^d)$, $S'(\mathbb{R}^d) \to S'(\mathbb{R}^d)$ is *globally regular* if $u \in S'(\mathbb{R}^d)$ and $Au \in S(\mathbb{R}^d)$ imply $u \in S(\mathbb{R}^d)$.

Corollary 1.3.9. *Assume the strong uncertainty principle and let a be a symbol in* $\mathrm{Hypo}(M, M_0; \Phi, \Psi)$. *Then $a(x,D)$ is globally regular.*

Proof. By Theorem 1.3.6 there exists $b \in S(M_0^{-1}; \Phi, \Psi)$ such that $b(x, D)a(x, D) = I + R$ on $\mathcal{S}'(\mathbb{R}^d)$, where R is a regularizing operator. If $u \in \mathcal{S}'(\mathbb{R}^d)$ and $Au \in \mathcal{S}(\mathbb{R}^d)$,

$$u = b(x, D)a(x, D)u - Ru = b(x, D)f - Ru \in \mathcal{S}(\mathbb{R}^d),$$

for R maps $\mathcal{S}'(\mathbb{R}^d)$ into $\mathcal{S}(\mathbb{R}^d)$ and $b(x, D)$ maps $\mathcal{S}(\mathbb{R}^d)$ into itself (Proposition 1.2.7). $\qquad\square$

1.3.2   Slow Variation and Construction of Elliptic Symbols

We now address ourselves to the question of the existence of elliptic symbols in any given class $S(M; \Phi, \Psi)$. This question is equivalent to asking whether there exists a weight M', *equivalent* to M, which is also a symbol in $S(M; \Phi, \Psi)(= S(M'; \Phi, \Psi))$. We show that this is in fact the case if the following *slow variation conditions* are satisfied:

There exist constants $c, C > 0$ such that,

$$\begin{cases} |x - y| \le c\Phi(y, \eta) \\ |\xi - \eta| \le c\Psi(y, \eta) \end{cases} \implies \begin{cases} C^{-1}\Phi(y, \eta) \le \Phi(x, \xi) \le C\Phi(y, \eta), \\ C^{-1}\Psi(y, \eta) \le \Psi(x, \xi) \le C\Psi(y, \eta), \\ C^{-1}M(y, \eta) \le M(x, \xi) \le CM(y, \eta). \end{cases} \tag{1.3.8}$$

Indeed, consider a smooth function $\chi(x, \xi)$ with $0 \le \chi(x, \xi) \le 1$, supported in the box $|x| \le c$, $|\xi| \le c$, and equal to 1 in the box $|x| \le c/2$, $|\xi| \le c/2$, where c is the constant in (1.3.8). Set

$$\chi_{y,\eta}(x, \xi) = \chi((x - y)/\Phi(y, \eta), (\xi - \eta)/\Psi(y, \eta)).$$

Observe that $\chi_{y,\eta}(x, \xi)$ is supported in the box $|x - y| \le c\Phi(y, \eta)$, $|\xi - \eta| \le c\Psi(y, \eta)$. Hence, if $\chi_{y,\eta}(x, \xi) \ne 0$, then the conclusions in (1.3.8) hold. Using this fact one deduces that the family of functions $\chi_{y,\eta}$ is bounded in $S(1; \Phi, \Psi)$, i.e., for every $\alpha, \beta \in \mathbb{N}^d$ there exist constants $C_{\alpha\beta}$ such that

$$|\partial_\xi^\alpha \partial_x^\beta \chi_{y,\eta}(x, \xi)| \le C_{\alpha\beta} \Psi(x, \xi)^{-|\alpha|} \Phi(x, \xi)^{-|\beta|}, \quad x, \xi, y, \eta \in \mathbb{R}^d. \tag{1.3.9}$$

Moreover, we have

$$\int \chi_{y,\eta}(x, \xi)dy\, d\eta \asymp h(x, \xi)^d, \quad (x, \xi) \in \mathbb{R}^{2d}. \tag{1.3.10}$$

To verify the upper bound in (1.3.10), notice that the support of $\chi_{y,\eta}$ is contained in the box $|x - y| \le cC\Phi(x, \xi)$, $|\xi - \eta| \le cC\Psi(x, \xi)$. Integrating on this larger set yields

$$\int \chi_{y,\eta}(x, \xi)dy\, d\eta \le (cC)^{2d} \|\chi\|_{L^\infty} h(x, \xi)^d.$$

For the lower bound in (1.3.10), we observe that $\chi_{y,\eta}(x,\xi) = 1$ on the set $|x - y| \leq cC^{-1}\Phi(x,\xi)/2$, $|\xi - \eta| \leq cC^{-1}\Psi(x,\xi)/2$. An integration on this set gives the estimate

$$\int \chi_{y,\eta}(x,\xi)\,dy\,d\eta \geq (cC^{-1}/2)^{2d}h(x,\xi)^d.$$

Now, for any weight $M(x,\xi)$, we set

$$M'(x,\xi) = \int M(y,\eta)\chi_{y,\eta}(x,\xi)h(y,\eta)^{-d}dy\,d\eta. \qquad (1.3.11)$$

Let us verify that $M' \in S(M; \Phi, \Psi)$ and $M'(x,\xi) \gtrsim M(x,\xi)$, $(x,\xi) \in \mathbb{R}^{2d}$.
   The equivalence

$$M'(x,\xi) \asymp M(x,\xi), \quad (x,\xi) \in \mathbb{R}^{2d},$$

follows at once from the above observation that, on the support of $\chi_{y,\eta}(x,\xi)$, $M(x,\xi)$ is equivalent to $M(y,\eta)$ and $h(y,\eta)$ is equivalent to $h(x,\xi)$, together with (1.3.10). The estimates for the derivatives of $M'(x,\xi)$ follow similarly, because we can differentiate (1.3.11) under the integral sign, apply (1.3.9) and argue as above.

Remark 1.3.10. The above discussion also shows the existence of a continuous (weighted) partition of unity of symbols $\chi'_{y,\eta}(x,\xi) \geq 0$ which belong to a bounded subset of $S(1; \Phi, \Psi)$, namely

$$\int \chi'_{y,\eta}(x,\xi)h(y,\eta)^{-d}\,dy\,d\eta = 1, \quad (x,\xi) \in \mathbb{R}^{2d}.$$

Indeed, the above construction with $M(x,\xi) = 1$ yields an elliptic symbol $M'(x,\xi) \in S(1; \Phi, \Psi)$. Then the functions

$$\chi'_{y,\eta}(x,\xi) = \chi_{y,\eta}(x,\xi)/M'(x,\xi)$$

have the desired property.

Example 1.3.11. Set $\Phi(x,\xi) = (1 + |x|^2 + |\xi|^2)^{\rho_1/2}$, $\Psi(x,\xi) = (1 + |x|^2 + |\xi|^2)^{\rho_2/2}$ and $M(x,\xi) = (1+|x|^2+|\xi|^2)^{s/2}$, $\rho_1, \rho_2, s \in \mathbb{R}$. Then we see that the slow variation conditions (1.3.8) are satisfied for some $c, C > 0$ if $0 \leq \rho_1, \rho_2 \leq 1$. Indeed, since $(1 + |x|^2 + |\xi|^2)^{1/2} \asymp (1 + |x| + |\xi|)$, it suffices to verify that, for some constants $c, C > 0$,

$$\begin{cases} |x - y| \leq c(1 + |y| + |\eta|)^{\rho_1} \\ |\xi - \eta| \leq c(1 + |y| + |\eta|)^{\rho_2} \end{cases} \implies C^{-1}(1+|y|+|\eta|) \leq 1+|x|+|\xi| \leq C(1+|y|+|\eta|).$$

To this end we observe that by the triangle inequality,

$$|(1 + |x| + |\xi|) - (1 + |y| + |\eta|)| \leq |x - y| + |\xi - \eta|$$

so the desired estimate follows, if $\rho_1, \rho_2 \leq 1$, for example with $c = 1/4$, $C = 2$. In view of (0.1.1) we have actually $M \in S(M; \Phi, \Psi)$.

Example 1.3.12. Let $\Phi(x,\xi) = \langle x \rangle^{\rho_1}$, $\Psi(x,\xi) = \langle \xi \rangle^{\rho_2}$, $M(x,\xi) = \langle x \rangle^{s_1} \langle \xi \rangle^{s_2}$, $\rho_1, \rho_2, s_1, s_2 \in \mathbb{R}$. Essentially the same arguments as in the previous example show that the slow variation conditions are satisfied if $0 \leq \rho_1 \leq 1$, $0 \leq \rho_2 \leq 1$.

1.4   Boundedness on L^2

The basic boundedness result we are going to prove is the continuity on $L^2(\mathbb{R}^d)$ of pseudo-differential operators "of order 0", i.e., with symbols in $S(1) = S(1; \Phi, \Psi)$. Here we assume the strong uncertainty principle (1.1.10). A more general result without that assumption will be proved in Theorem 1.7.14 below.

Because of Remark 1.2.6 we can limit ourselves to considering the standard (left) quantization.

Here we regard $a(x, D)$ as a continuous map $S'(\mathbb{R}^d) \to S'(\mathbb{R}^d)$ (cf. Proposition 1.2.13). The boundedness of $a(x, D)$ on $L^2(\mathbb{R}^d)$ is then easily seen to be equivalent to the estimate

$$\|a(x, D)u\| \lesssim \|u\|, \quad u \in S(\mathbb{R}^d),$$

where $\|u\| = \|u\|_{L^2(\mathbb{R}^d)}$. Indeed, since $S(\mathbb{R}^d)$ is dense in $L^2(\mathbb{R}^d)$, this estimate tells us that the restriction of $a(x, D)$ to $S(\mathbb{R}^d)$ (which maps $S(\mathbb{R}^d)$ into itself) extends to a bounded operator on $L^2(\mathbb{R}^d)$. This extension coincides with the restriction of $a(x, D)$ to $L^2(\mathbb{R}^d)$ by the uniqueness of the limit in $S'(\mathbb{R}^d)$.

Theorem 1.4.1. *Assume the strong uncertainty principle. Let* $a \in S(1; \Phi, \Psi)$. *Then the operator* $a(x, D)$ *is bounded on* $L^2(\mathbb{R}^d)$. *Moreover we have the uniform estimate*

$$\|a(x, D)u\| \lesssim \|a\|_{k, S(1; \Phi, \Psi)} \|u\|, \quad u \in S(\mathbb{R}^d), \ a \in S(1; \Phi, \Psi) \qquad (1.4.1)$$

for some $k \in \mathbb{N}$ *(see* (1.1.5)).

Proof. The first observation is that the boundedness of $a(x, D)$ follows from that of $A := a(x, D)^* a(x, D)$, where $a(x, D)^*$ is the formal adjoint of $a(x, D)$. Indeed, $\|a(x, D)u\|^2 = (Au, u)$, for $u \in S(\mathbb{R}^d)$.

Now, let $C > 0$ satisfy $|a(x, \xi)|^2 \leq C$, $(x, \xi) \in \mathbb{R}^{2d}$. Then the symbol $C + 1 - |a(x, \xi)|^2$ is elliptic in $S(1)$, and does not vanish anywhere in $\mathbb{R}^{2d}$. Hence, by Lemma 1.3.5 we have

$$b(x, \xi) := (C + 1 - |a(x, \xi)|^2)^{1/2} \in S(1).$$

As a consequence of Proposition 1.2.9 and Theorem 1.2.16, it turns out that

$$b(x, D)^* b(x, D) = (C + 1)I - A + r(x, D),$$

for some $r \in S(h)$. Hence,

$$\|a(x, D)u\|^2 \leq (C + 1)\|u\|^2 + (r(x, D)u, u).$$

Therefore it remains to prove that symbols in $S(h)$ give rise to bounded operators on $L^2(\mathbb{R}^d)$. The remark at the beginning of the proof shows that this follows from a similar result for symbols in $S(h^2)$. By a finite induction we see that it suffices to prove boundedness of operators with symbols in $S(h^N)$, with N large enough. To this end, let N be such that $h(x,\xi)^N \lesssim (1+|x|+|\xi|)^{-d-1}$, which is possible by the strong uncertainty principle. Then symbols in $S(h^N)$ are in $L^2(\mathbb{R}^{2d})$. By Proposition 1.2.1 the kernels of the corresponding operators belong to $L^2(\mathbb{R}^{2d})$, so that such operators are clearly bounded (see for details the subsequent Proposition 4.4.19 in Section 4.4 concerning the abstract theory of Hilbert-Schmidt operators). The uniform bound in (1.4.1) follows from this proof and the continuity properties established in Proposition 1.2.9, cf. Theorem 1.2.4, and Theorem 1.2.16. $\quad\square$

We now exhibit a sufficient condition for a pseudo-differential operator to be compact on $L^2(\mathbb{R}^d)$.

Theorem 1.4.2. *Assume the strong uncertainty principle. Let $a \in S(M;\Phi,\Psi)$, where M tends to 0 at infinity. Then the operator $a(x,D)$ is compact on $L^2(\mathbb{R}^d)$.*

Proof. According to Proposition 1.1.5 with $M_0 = M^{-1}$, let $a_n(x,\xi)$ be a sequence of Schwartz symbols, converging to $a(x,\xi)$ in $S(1)$. It follows from (1.4.1) that the corresponding operators $a_n(x,D)$ converge to $a(x,D)$ in the space of bounded operators on $L^2(\mathbb{R}^d)$. Hence it suffices to prove that every operator with Schwartz symbol is compact on $L^2(\mathbb{R}^d)$. Now, such an operator maps $L^2(\mathbb{R}^d)$ into $\mathcal{S}(\mathbb{R}^d)$ continuously, and the embedding $\mathcal{S}(\mathbb{R}^d) \hookrightarrow L^2(\mathbb{R}^d)$ is compact. Indeed, every bounded sequence in $\mathcal{S}(\mathbb{R}^d)$ has a convergent subsequence in $\mathcal{S}(\mathbb{R}^d)$ (see for example Treves [190]), hence in $L^2(\mathbb{R}^d)$. Alternatively, operators with symbols in $\mathcal{S}(\mathbb{R}^{2d}) \subset L^2(\mathbb{R}^{2d})$ are Hilbert-Schmidt, therefore compact; see Proposition 4.4.19 below. $\quad\square$

1.5 Sobolev Spaces

We now introduce a Sobolev space associated to a symbol class $S(M;\Phi,\Psi)$, to measure both the local regularity and decay at infinity of functions or temperate distributions in $\mathbb{R}^{2d}$. We assume in this section the strong uncertainty principle.

Moreover, we will always work with symbol classes $S(M;\Phi,\Psi)$ for which M is a regular weight, in the following sense.

Definition 1.5.1. A temperate weight M is called *regular* (with respect to Φ,Ψ) if it is a symbol in its own class, i.e., $M \in S(M;\Phi,\Psi)$.

All the weights we will consider in the applications of the subsequent chapters are regular. The assumption that M is a regular weight guarantees the existence of an elliptic symbol in $S(M;\Phi,\Psi)$, namely M itself. We saw in Section 1.3.2 that, under slow variation assumptions, we can always replace M by an equivalent weight which is regular.

Let $a \in S(M;\Phi,\Psi)$ be an elliptic symbol and set $A = a(x,D)$. According to Theorem 1.3.6, let $B = b(x,D)$, $b \in S(M^{-1};\Phi,\Psi)$ be a (left) parametrix of A, i.e.,

$$BA = I + R,$$ (1.5.1)

where R is a regularizing operator.

Definition 1.5.2. (*Sobolev spaces*) Let Φ, Ψ be fixed, and let A, R be as above. We define

$$H(M) = \{u \in \mathcal{S}'(\mathbb{R}^d) : \ Au \in L^2(\mathbb{R}^d)\},$$

endowed with the norm

$$\|u\|_{H(M)} = \|Au\|_{L^2(\mathbb{R}^d)} + \|Ru\|_{L^2(\mathbb{R}^d)}.$$ (1.5.2)

The unpleasant term $\|Ru\|_{L^2(\mathbb{R}^d)}$ is essential to obtain a norm. Indeed, if $Au = 0$ and $Ru = 0$ it follows from (1.5.1) that $u = 0$ (see however Proposition 1.7.12 below). The space $H(M)$ could a priori depend on the weights Φ, Ψ, but we will show in Section 1.7.4 below that actually this is *not* the case.

Proposition 1.5.3. *We have:*

(a) *Different choices of A and R in* (1.5.2) *produce equivalent norms.*

(b) *The scalar product*

$$(u, v)_{H(M)} = (Au, Au)_{L^2(\mathbb{R}^d)} + (Ru, Ru)_{L^2(\mathbb{R}^d)}$$

defines a Hilbert space structure on $H(M)$, and induces a norm equivalent to that in (1.5.2).

Proof. (a) Let A' be another pseudo-differential operator with elliptic symbol in $S(M)$, and B' a corresponding parametrix. Hence, $B'A' = I + R'$, with R' regularizing. Of course, it suffices to verify that

$$\|Au\|_{L^2(\mathbb{R}^d)} + \|Ru\|_{L^2(\mathbb{R}^d)} \lesssim \|A'u\|_{L^2(\mathbb{R}^d)} + \|R'u\|_{L^2(\mathbb{R}^d)}, \quad u \in \mathcal{S}(\mathbb{R}^d).$$ (1.5.3)

We substitute $u = B'A'u - R'u$ in the left-hand side of (1.5.3). The operator AB' has symbol in $S(1)$ and is therefore bounded on $L^2(\mathbb{R}^d)$ by Theorem 1.4.1. Since R and RB' are regularizing operators, and therefore bounded on $L^2(\mathbb{R}^d)$, we obtain

$$\|Au\|_{L^2(\mathbb{R}^d)} + \|Ru\|_{L^2(\mathbb{R}^d)} \lesssim \|A'u\|_{L^2(\mathbb{R}^d)} + \|R'u\|_{L^2(\mathbb{R}^d)} + \|AR'u\|_{L^2(\mathbb{R}^d)}.$$

If we replace again $u = B'A'u - R'u$ in the last term of this expression and argue similarly, we obtain the desired estimate.

(b) The only non-trivial fact is the completeness of $H(M)$. To prove it, let u_n be a Cauchy sequence in $H(M)$. Then, since $L^2(\mathbb{R}^d)$ is complete, Au_n and Ru_n converge in $L^2(\mathbb{R}^d)$. As a consequence, $u_n = BAu_n - Ru_n$ converges in $\mathcal{S}'(\mathbb{R}^d)$ to, say, $u \in \mathcal{S}'(\mathbb{R}^d)$. Since R is regularizing, $Ru \in \mathcal{S}(\mathbb{R}^d)$ and $Ru_n \to Ru$ in $\mathcal{S}(\mathbb{R}^d)$, therefore in $L^2(\mathbb{R}^d)$. It remains to prove that $Au \in L^2(\mathbb{R}^d)$ and $Au_n \to Au$ in $L^2(\mathbb{R}^d)$. This follows from the uniqueness of the limit in $\mathcal{S}'(\mathbb{R}^d)$ by combining the fact that Au_n converges in $L^2(\mathbb{R}^d)$ and tends to Au in $\mathcal{S}'(\mathbb{R}^d)$. □

Here are some basic properties of Sobolev spaces.

Proposition 1.5.4.

(a) $S(\mathbb{R}^d)$ *is a dense subspace of* $H(M)$, *and the embedding* $S(\mathbb{R}^d) \hookrightarrow H(M)$ *is continuous.*

(b) *The embedding* $H(M) \hookrightarrow S'(\mathbb{R}^d)$ *is continuous.*

Proof. (a) The continuity of the embedding $S(\mathbb{R}^d) \hookrightarrow H(M)$ is clear, because A and R are continuous on $S(\mathbb{R}^d)$ and $S(\mathbb{R}^d) \hookrightarrow L^2(\mathbb{R}^d)$ continuously. To prove the density property, let $u \in H(M)$ and let v_n be a sequence of Schwartz functions converging to Au in $L^2(\mathbb{R}^d)$. Set $u_n = Bv_n - Ru \in S(\mathbb{R}^d)$. Let us verify that $u_n \to u$ in $H(M)$. Since AB is bounded on $L^2(\mathbb{R}^d)$, we have $Au_n - Au = ABv_n - ARu - Au \to ABAu - ARu - Au = 0$ in $L^2(\mathbb{R}^d)$. On the other hand, since $u_n \to u$ in $S'(\mathbb{R}^d)$, $Ru_n \to Ru$ in $S(\mathbb{R}^d)$, and therefore in $L^2(\mathbb{R}^d)$.

(b) Let $u_n \in H(M)$ be a sequence converging to u in $H(M)$. Then $Au_n \to Au$ and $Ru_n \to Ru$ in $L^2(\mathbb{R}^d)$, and therefore in $S'(\mathbb{R}^d)$. Hence, for every $\phi \in S(\mathbb{R}^d)$, $\langle u_n, \phi \rangle = \langle Au_n, {}^tB\phi \rangle - \langle Ru_n, \phi \rangle$ converges to $\langle Au, {}^tB\phi \rangle - \langle Ru, \phi \rangle = \langle u, \phi \rangle$, which is the desired conclusion. Here we applied to B the Remark 1.2.14. $\qquad\square$

Proposition 1.5.5. *Let* M, M_1, M_2, M' *be regular weights and let* P *be a pseudo-differential operator with symbol in* $S(M'; \Phi, \Psi)$.

(a) P *defines a bounded operator from* $H(M)$ *into* $H(M/M')$.

(b) *If moreover* $M'(x, \xi) M_2(x, \xi)/M_1(x, \xi)$ *tends to zero at infinity,* P *defines a compact operator* $H(M_1) \to H(M_2)$.

Proof. (a) Let A, B, R be as at the beginning of this section, so that the norm in $H(M)$ is defined in terms of A and R. Let A' be a pseudo-differential operator with elliptic symbol in $S(M/M')$, with parametrix B' having symbol in $S(M'/M)$, and $R' = B'A' - I$. We use A', R' to define the norm of $H(M/M')$. Then we consider the expression

$$\|A'Pu\|_{L^2(\mathbb{R}^d)} + \|R'Pu\|_{L^2(\mathbb{R}^d)},$$

and we substitute $u = BAu - Ru$. Since $A'PB$ is a pseudo-differential operator with symbol in $S(1)$, it is bounded on $L^2(\mathbb{R}^d)$. On the other hand, $R'PB$ and $R'B$ are regularizing. Hence we obtain

$$\|A'Pu\|_{L^2(\mathbb{R}^d)} + \|R'Pu\|_{L^2(\mathbb{R}^d)} \lesssim \|Au\|_{L^2(\mathbb{R}^d)} + \|Ru\|_{L^2(\mathbb{R}^d)} + \|A'PRu\|_{L^2(\mathbb{R}^d)}.$$

Substituting $u = BAu - Ru$ again in the last term of this estimate and arguing similarly one obtains the desired boundedness result.

(b) Let A_j, $j = 1, 2$, be a pseudo-differential operator with elliptic symbol in $S(M_j)$, with parametrix B_j having symbol in $S(M_j^{-1})$, and $R_j = B_jA_j - I$. We use A_j, R_j to define the norm of $H(M_j)$. Write

$$P = IPI = (B_2A_2 - R_2)P(B_1A_1 - R_1).$$

We obtain $P = B_2(A_2PB_1)A_1 + S$, where S is a regularizing operator. By the point (a), $A_1 : H(M_1) \to L^2(\mathbb{R}^d)$ and $B_2 : L^2(\mathbb{R}^d) \to H(M_2)$ continuously. Since A_2PB_1 has a symbol in $S(M_2M'/M_1)$, it is compact on $L^2(\mathbb{R}^d)$ by Theorem 1.4.2. Hence it remains to prove that S is compact from $H(M_1)$ to $H(M_2)$. To see this, let u_n be a bounded sequence in $H(M_1)$. It follows that Su_n is bounded in $\mathcal{S}(\mathbb{R}^d)$. Hence, for a suitable subsequence u_{n_k}, it turns out that Su_{n_k} converges in $\mathcal{S}(\mathbb{R}^d)$, therefore in $H(M_2)$. $\square$

Remark 1.5.6. In Proposition 1.5.5 (a), the norm of P as a bounded operator $H(M) \to H(M/M')$ can be in fact estimated by a seminorm of its symbol in $S(M'; \Phi, \Psi)$.

Corollary 1.5.7. *Let M_1, M_2 be regular weights.*

(a) *If $M_2 \lesssim M_1$, then $H(M_1) \hookrightarrow H(M_2)$ continuously.*

(b) *If $M_2(x, \xi)/M_1(x, \xi)$ tends to 0 at infinity, then the embedding $H(M_1) \hookrightarrow H(M_2)$ is compact.*

Proof. This result is a special case of Proposition 1.5.5, with $P = I$ (hence $M'(x, \xi) = 1$). $\square$

The continuity result in Proposition 1.5.5 (a) allows one to improve the global regularity result in Corollary 1.3.9, and also gives an a priori estimate for pseudo-differential operators with hypoelliptic symbols.

Proposition 1.5.8. *Let $M', M_0, M, \tilde{M}$ be arbitrary regular weights and let P be a pseudo-differential operator with a symbol in $\mathrm{Hypo}(M', M_0; \Phi, \Psi)$. We have the a priori estimate*

$$\|u\|_{H(M)} \lesssim \|Pu\|_{H(M/M_0)} + \|u\|_{H(\tilde{M})}, \quad u \in \mathcal{S}(\mathbb{R}^d). \tag{1.5.4}$$

Moreover, if $u \in \mathcal{S}'(\mathbb{R}^d)$ and $Pu \in H(M/M_0)$, it turns out $u \in H(M)$.

Proof. By Theorem 1.3.6 there exists a pseudo-differential operator Q with symbol in $S(M_0^{-1}; \Phi, \Psi)$, such that $I = QP - R$, where R is regularizing. As a consequence of Proposition 1.5.5 (a), B maps continuously $H(M/M_0)$ into $H(M)$, whereas by Proposition 1.5.4 R maps continuously $H(\tilde{M})$ into $H(M)$. This proves the statement. $\square$

We now come to the representation of the dual of a Sobolev space as a Sobolev space itself.

Proposition 1.5.9. *Let M be a regular weight. The pairing*

$$\langle u, v \rangle = \int u(x)v(x)\, dx, \quad u, v \in \mathcal{S}(\mathbb{R}^d),$$

extends to a continuous bilinear map

$$H(M) \times H(M^{-1}) \to \mathbb{C}.$$

Moreover, the dual $H(M)'$ is isomorphic to $H(M^{-1})$ via this extended pairing.

Proof. Let A be a pseudo-differential operator with an elliptic symbol in $S(M)$ and B a parametrix, so that $R = BA - I$ is regularizing. Since $u = BAu - Ru$ we have

$$\langle u, v \rangle = \langle BAu - Ru, v \rangle = \langle Au, {}^t Bv \rangle - \langle Ru, v \rangle, \quad u, v \in \mathcal{S}(\mathbb{R}^d).$$

In the last term of this expression we can replace v by ${}^t A^t Bv - {}^t Rv$. We obtain

$$\langle u, v \rangle = \langle Au, {}^t Bv \rangle - \langle ARu, {}^t Bv \rangle + \langle Ru, {}^t Rv \rangle.$$

We now apply Cauchy-Schwarz' inequality in $L^2(\mathbb{R}^d)$ to each of these terms. Since by Propositions 1.2.12 and 1.5.5 (a) the operators A, AR are bounded $H(M) \to L^2(\mathbb{R}^d)$ and ${}^t B$ and ${}^t R$ are bounded $H(M^{-1}) \to L^2(\mathbb{R}^d)$, we deduce the first part of the statement.

Let now $v \in H(M)'$. Since $\mathcal{S}(\mathbb{R}^d)$ is dense in $H(M)$ and the embedding $\mathcal{S}(\mathbb{R}^d) \hookrightarrow H(M)$ is continuous, we have $H(M)' \hookrightarrow \mathcal{S}'(\mathbb{R}^d)$. Hence $v \in \mathcal{S}'(\mathbb{R}^d)$ and we have to prove that $v \in H(M^{-1})$, i.e., $A'v \in L^2(\mathbb{R}^d)$, where A' is any pseudo-differential operator with an elliptic symbol in $S(M^{-1})$. By the Riesz representation theorem and since $\mathcal{S}(\mathbb{R}^d)$ is dense in $L^2(\mathbb{R}^d)$, this is equivalent to the estimate

$$|\langle A'v, u \rangle| \le C\|u\|_{L^2(\mathbb{R}^d)}, \quad u \in \mathcal{S}(\mathbb{R}^d).$$

Now, this is true because $\langle A'v, u \rangle = \langle v, {}^t A'u \rangle$, and ${}^t A'$ is bounded $L^2(\mathbb{R}^d) \to H(M)$ by Propositions 1.2.12 and 1.5.5 (a).

Hence, we verified that the map

$$H(M^{-1}) \to H(M)', \quad v \mapsto \langle \cdot, v \rangle$$

is a continuous bijection, and therefore a homeomorphism by the Open Mapping Theorem. $\qquad \square$

Example 1.5.10. Particular cases of special interest are given by the Sobolev spaces $H(\Phi^t \Psi^s)$, for $s, t \in \mathbb{R}$ (when Φ and Ψ are regular weights). It is easy to see that, because of the strong uncertainty principle,

$$\bigcap_{s, t \in \mathbb{R}} H(\Phi^t \Psi^s) = \mathcal{S}(\mathbb{R}^d), \quad \bigcup_{s, t \in \mathbb{R}} H(\Phi^t \Psi^s) = \mathcal{S}'(\mathbb{R}^d). \tag{1.5.5}$$

By further specializing Φ and Ψ, we find spaces for which other specific properties can be proved. This will be detailed in the subsequent chapters.

1.6   Fredholm Properties

We discuss in this section some applications of the continuity and compactness results just proved.

First we present the needed results from the abstract theory of Fredholm operators in Hilbert spaces.

1.6.1　Abstract Theory

Let H_1, H_2 be complex Hilbert spaces and $\mathcal{B}(H_1, H_2)$ be the space of bounded operators $H_1 \rightarrow H_2$. We define the kernel and cokernel of an operator $T \in \mathcal{B}(H_1, H_2)$ respectively as

$$\operatorname{Ker} T = \{x \in H_1 : Tx = 0\}, \qquad \operatorname{Coker} T = H_2/T(H_1).$$

Since T is bounded, $\operatorname{Ker} T$ is a closed subspace of H_1. However, $T(H_1)$ need not be closed. Observe that if $W \subset H_2$ is an (algebraic) supplementary space, i.e., $W \cap T(H_1) = \{0\}$ and $W + T(H_1) = H_2$, then W is isomorphic to $H_2/T(II_1)$ via the map $x \mapsto x + T(H_1)$. Clearly, if $T(H_1)$ is closed, this holds with $W = T(H_1)^{\perp}$.

If $T \in \mathcal{B}(H_1, H_2)$ we define the adjoint operator $T^* \in \mathcal{B}(H_2, H_1)$ by the formula

$$(T^*x, y)_{H_1} = (x, Ty)_{H_2}, \quad x \in H_2,\ y \in H_1, \tag{1.6.1}$$

where $(\cdot, \cdot)_{H_j}$, $j = 1, 2$, denotes the inner product in H_j. It follows from this formula that $H_2 = \overline{T(H_1)} \oplus \operatorname{Ker} T^*$. Hence, if $T(H_1)$ is closed, $\operatorname{Ker} T^*$ is a supplementary space of $T(H_1)$.

We also recall that, for an operator $T \in \mathcal{B}(H_1, H_2)$, the transposed ${}^t T \in \mathcal{B}(H_2', H_1')$ is defined by the formula

$$\langle {}^t Tx, y \rangle_1 = \langle x, Ty \rangle_2, \quad x \in H_2',\ y \in H_1, \tag{1.6.2}$$

where $\langle \cdot, \cdot \rangle_j$, $j = 1, 2$, denotes the duality pairing between H_j' and H_j. Let us observe that the conjugate-linear map $x \mapsto (\cdot, x)_{H_2}$ defines an isomorphism of $\operatorname{Ker} T^*$ into $\operatorname{Ker} {}^t T$.

Definition 1.6.1. $T \in \mathcal{B}(H_1, H_2)$ is called a *Fredholm operator* if $\dim \operatorname{Ker} T$ is finite and $T(H_1)$ is closed and has finite codimension; one then defines the *index* of T as the integer number

$$\operatorname{ind} T = \dim \operatorname{Ker} T - \dim \operatorname{Coker} T. \tag{1.6.3}$$

The space of Fredholm operators $H_1 \rightarrow H_2$ is denoted by $\operatorname{Fred}(H_1, H_2)$.

Fredholm operators have therefore a small kernel and a large range, so that one could think to them as almost invertible operators.

It follows from the above discussion that if $T \in \operatorname{Fred}(H_1, H_2)$, then

$$\operatorname{ind} T = \dim \operatorname{Ker} T - \dim \operatorname{Ker} T^*, \tag{1.6.4}$$

and also

$$\operatorname{ind} T = \dim \operatorname{Ker} T - \dim \operatorname{Ker} {}^t T. \tag{1.6.5}$$

Observe that in the above definition the assumption that $T(H_1)$ is closed is in fact superfluous:

Proposition 1.6.2. *Let $T \in \mathcal{B}(H_1, H_2)$, with $T(H_1)$ of finite codimension. Then $T(H_1)$ is closed.*

Proof. T induces a map $H_1/\operatorname{Ker} T \to H_2$ (still denoted by T in the sequel). Let n be the codimension of $T(H_1)$ in H_2, and consider a map

$$S: \ \mathbb{C}^n \to H_2$$

such that $S(\mathbb{C}^n)$ is a supplementary space for $T(H_1)$. Then the map

$$T_1: \ H_1/\operatorname{Ker} T \oplus \mathbb{C}^n \to H_2, \quad (x, y) \mapsto Tx + Sy$$

is bijective. By the Open Mapping Theorem it is a homeomorphism, which proves that $T(H_1) = T_1(H_1/\operatorname{Ker} T \oplus \{0\})$ is closed in H_2. $\qquad \square$

A basic but important observation is that, if H_1 and H_2 have finite dimension, then every linear operator $H_1 \to H_2$ is Fredholm and

$$\operatorname{ind} T = \dim H_1 - \dim H_2, \tag{1.6.6}$$

as a consequence of elementary linear algebra. In particular, in that case the index is in fact independent of T. Although this is not longer true in the infinite dimensional setting, one still has the following stability result.

Theorem 1.6.3. *Let $T \in \operatorname{Fred}(H_1, H_2)$, and $S \in \mathcal{B}(H_1, H_2)$ with $\|S\|$ sufficiently small. Then $\dim \operatorname{Ker}(T + S) \leq \dim \operatorname{Ker} T$, $T + S$ is Fredholm and $\operatorname{ind}(T + S) = \operatorname{ind} T$.*

Proof. Let $V \subset H_1$ be any closed subspace of finite codimension, such that $V \cap \operatorname{Ker} T = \{0\}$. Since $T(V)$ has finite codimension in $T(H_2)$, which has finite codimension in H_2, we see that $T(V)$ has finite codimension in H_2. In particular $T(V)$ is closed as a consequence of Proposition 1.6.2. Moreover, the operator T induces an operator $\tilde{T} : H_1/V \to H_2/T(V)$, acting therefore between finite dimensional spaces. A direct inspection shows that the natural map $\operatorname{Ker} T \hookrightarrow H_1 \to H_1/V$ defines an isomorphism $\operatorname{Ker} T \to \operatorname{Ker} \tilde{T}$, whereas the natural map $H_2/T(H_1) \hookrightarrow H_2/T(V) \to (H_2/T(V)) / \tilde{T}(H_1/V)$ defines an isomorphism $\operatorname{Coker} T \to \operatorname{Coker} \tilde{T}$. Hence

$$\operatorname{ind} T = \operatorname{ind} \tilde{T} = \dim H_1/V - \dim H_2/T(V), \tag{1.6.7}$$

where we used (1.6.6).

Now, consider the linear map

$$T(V)^\perp \oplus V \to H_2, \quad (x, y) \mapsto x + (T + S)y.$$

It depends continuously on S and is a homeomorphism for $S = 0$, hence also for $\|S\|$ small enough. Therefore for such S we see that $V \cap \operatorname{Ker}(T + S) = \{0\}$ and $T(V)^\perp$ is isomorphic to $H_2/(T + S)(V)$. In particular, $(T + S)(V)$ has finite codimension. The inequality $\dim \operatorname{Ker}(T + S) \leq \dim \operatorname{Ker} T$ also follows by this argument by choosing $V = (\operatorname{Ker} T)^{\perp}$, for then the orthogonal projection of $\operatorname{Ker}(T + S)$ on $\operatorname{Ker} T$ is one-to-one. Hence $T + S$ is Fredholm too.

Finally, to prove the equality of the indices, observe that (1.6.7) applied to $T + S$ in place of T yields

$$\begin{aligned}
\operatorname{ind}(T + S) &= \dim H_1/V - \dim H_2/(T + S)(V) \\
&= \dim H_1/V - \dim T(V)^{\perp} \\
&= \operatorname{ind} T,
\end{aligned}$$

which concludes the proof.   □

Corollary 1.6.4. $\operatorname{Fred}(B_1, B_2)$ *is open in* $\mathcal{B}(H_1, H_2)$, $\dim \operatorname{Ker} T$ *is upper semicontinuous on* $\operatorname{Fred}(B_1, B_2)$ *and* $\operatorname{ind} \colon \operatorname{Fred}(H_1, H_2) \to \mathbb{Z}$ *is constant on the connected components of* $\operatorname{Fred}(H_1, H_2)$.

Corollary 1.6.5. *If* $T_1 \in \operatorname{Fred}(H_1, H_2)$ *and* $T_2 \in \operatorname{Fred}(H_2, H_3)$, *then we have* $T_2 T_1 \in \operatorname{Fred}(H_1, H_3)$ *and the following "logarithmic law" holds:*

$$\operatorname{ind} T_2 T_1 = \operatorname{ind} T_1 + \operatorname{ind} T_2.$$

Proof. Since T_1 maps $\operatorname{Ker} T_2 T_1$ in $\operatorname{Ker} T_2$ with kernel $\operatorname{Ker} T_1$, we have

$$\dim \operatorname{Ker} T_2 T_1 \leq \dim \operatorname{Ker} T_1 + \dim \operatorname{Ker} T_2.$$

Similarly,

$$\dim \operatorname{Coker} T_2 T_1 \leq \dim \operatorname{Coker} T_1 + \dim \operatorname{Coker} T_2,$$

so that $T_2 T_1 \in \operatorname{Fred}(H_1, H_3)$.
If I_2 is the identity operator in B_2, it follows from what we have just proved that, in block matrix notation,

$$\begin{pmatrix} I_2 & 0 \\ 0 & T_2 \end{pmatrix} \begin{pmatrix} I_2 \cos t & I_2 \sin t \\ -I_2 \sin t & I_2 \cos t \end{pmatrix} \begin{pmatrix} T_1 & 0 \\ 0 & I_2 \end{pmatrix}$$

is for every t a Fredholm operator from $H_1 \oplus H_2$ to $H_2 \oplus H_3$. For $t = 0$ it is the direct sum of the operators T_1 and T_2 whereas for $t = -\pi/2$ it is the operator $(f_1, f_2) \mapsto (-f_2, T_2 T_1 f_1)$. The corresponding indices are $\operatorname{ind} T_1 + \operatorname{ind} T_2$ and $\operatorname{ind} T_2 T_1$ respectively. Hence the result follows from the stability of the index in Theorem 1.6.3.   □

Theorem 1.6.6. *If* $T \in \operatorname{Fred}(H_1, H_2)$ *and* $K \in \mathcal{B}(H_1, H_2)$ *is compact, then* $T + K \in \operatorname{Fred}(H_1, H_2)$ *and* $\operatorname{ind}(T + K) = \operatorname{ind} T$.

Proof. Once we prove that $T + K$ is Fredholm, the equality of the indices follows by applying the stability result in Theorem 1.6.3 to the homotopy of operators $T + tK$, $0 \leq t \leq 1$.

First we verify that $T + K$ is Fredholm when T is invertible. To prove that $\mathrm{Ker}(T + K)$ has finite dimension, we show that every bounded sequence x_n in $\mathrm{Ker}(T + K)$ has a convergent subsequence, i.e., that the closed unit ball in $\mathrm{Ker}(T + K)$ is compact. In fact, $Tx_n = -Kx_n$ is then convergent, for K is compact. Since T is invertible, the claim follows.

To prove that $\mathrm{Coker}(T + K)$ has finite dimension, we use the splitting

$$H_2 = \overline{(T + K)(H_1)} \oplus \mathrm{Ker}(T + K)^*.$$

Since the above argument applied to $T^* + K^*$ shows that $\mathrm{Ker}(T + K)^*$ has finite dimension, it suffices to prove that $(T + K)(H_1)$ is closed, i.e.,

$$\|x\| \leq C\|(T + K)x\|, \quad x \in \mathrm{Ker}(T + K)^\perp,$$

for some constant $C > 0$. This is seen by contradiction: if some sequence $x_n \in \mathrm{Ker}(T + K)^\perp$ satisfies $\|x_n\| = 1$, $(T + K)x_n \to 0$, then essentially the same argument as above shows that x_n must be convergent, say to $y \in \mathrm{Ker}(T + K)^\perp$, $\|y\| = 1$, because $\mathrm{Ker}(T + K)^\perp$ is closed. On the other hand, by continuity one would obtain $(T + K)y = 0$, which is a contradiction.

Consider now the general case when T is just Fredholm. We then split $H_1 = (\mathrm{Ker}\, T)^\perp \oplus \mathrm{Ker}\, T$, $H_2 = T(H_1) \oplus T(H_1)^\perp$. Correspondingly the operator $T + K$ is represented as

$$T + K = \begin{pmatrix} T_{11} + K_{11} & K_{12} \\ K_{21} & K_{22} \end{pmatrix},$$

where $T_{11} : (\mathrm{Ker}\, T)^\perp \to T(H_1)$ is invertible, and $K_{11}, K_{12}, K_{21}, K_{22}$ are compact. Hence, from the first part of the present proof we get that $T_{11} + K_{11}$ is Fredholm. Therefore

$$\mathrm{Ker}(T + K) \subset \{x \oplus y \in (\mathrm{Ker}\, T)^\perp \oplus \mathrm{Ker}\, T, \ (T_{11} + K_{11})x = -K_{12}y\}$$

has finite dimension, since $K_{12}y$, with $y \in \mathrm{Ker}\, T$, varies in a finite dimensional space; moreover

$$(T + K)(H_1) \supset \{(T_{11} + K_{11})x + K_{21}x, \ x \in (\mathrm{Ker}\, T)^\perp\}$$

has finite codimension, because K_{21} takes values in the finite dimensional space $T(H_1)^\perp$.   □

When $H_1 = H_2$, Theorem 1.6.6 applies, in particular, to operators of the form $I + K$ with K compact, which have index 0. The conclusion reads as the familiar *Fredholm alternative*: there exist finite dimensional subspaces $W_1, W_2 \subset H_1$, having the same dimension, such that the equation $x + Kx = y$ is solvable precisely when $y \perp W_2$ and the solution is then unique modulo W_1. This generalizes what happens for an $n \times n$ linear system.

Corollary 1.6.7. *If $T \in \mathcal{B}(H_1, H_2)$ and $S_j \in \mathcal{B}(H_2, H_1)$, $j = 1, 2$, satisfy*

$$\begin{cases} TS_2 = I_2 + K_2, \\ S_1 T = I_1 + K_1 \end{cases}$$

with K_j compact, then T, S_1, S_2 are Fredholm operators and $\operatorname{ind} T = -\operatorname{ind} S_j$, $j = 1, 2$.

Proof. Since

$$\dim \operatorname{Ker} \ T \le \dim \operatorname{Ker}(I_1 + K_1),$$

and

$$\dim \operatorname{Coker} \ T \le \dim \operatorname{Coker}(I_2 + K_2),$$

we get $T \in \operatorname{Fred}(H_1, H_2)$ by Theorem 1.6.6. Moreover

$$S_1 T S_2 = S_1 + S_1 K_2 = S_2 + K_1 S_2,$$

so that $S_2 - S_1$ is compact. Hence $S_2 T - I_1$ and $T S_1 - I_2$ are compact as well, and therefore $S_1, S_2 \in \operatorname{Fred}(H_2, H_1)$ by the first part of the proof. Finally $\operatorname{ind} T + \operatorname{ind} S_j = \operatorname{ind}(I_j + K_j) = 0$ by Corollary 1.6.5 and Theorem 1.6.6. $\square$

Remark 1.6.8. The definition of Fredholm operator also applies to operators acting on Banach spaces, and the results in this section extend to that more general framework. Here we considered only the case of Hilbert spaces for the sake of simplicity.

1.6.2 Pseudo-Differential Operators

We come now to the case of pseudo-differential operators with symbols in the classes $S(M'; \Phi, \Psi)$, acting on the Sobolev spaces $H(M)$ introduced in the previous section. We continue to assume the strong uncertainty principle and we suppose that the implied weights M, M' are regular, according to Definition 1.5.1 (cf. also Section 1.3.2).

Given a pseudo-differential operator A with symbol in $S(M')$, denote by $A|_{H(M)}$ the restriction of $A : \mathcal{S}'(\mathbb{R}^d) \to \mathcal{S}'(\mathbb{R}^d)$ to $H(M)$. We know from Proposition 1.5.5 (a) that $A|_{H(M)} : H(M) \to H(M/M')$ is bounded. Assume in addition that A is elliptic, and let B be a parametrix. Then B has symbol in $S(M'^{-1})$, so that it defines a bounded operator $H(M/M') \to H(M)$. Moreover the operators $BA - I$ and $AB - I$ have symbols in $S(h)$ and therefore they define compact operators on $H(M)$ and $H(M/M')$, respectively, by Proposition 1.5.5 (b). We can then apply Corollary 1.6.7 with $H_1 = H(M)$, $H_2 = H(M/M')$ and we deduce that $A|_{H(M)} \in \operatorname{Fred}(H_1, H_2)$.

An interesting fact is that the index of $A|_{H(M)}$ is actually independent of the weight M. Indeed, the global regularity result in Corollary 1.3.9 shows that the kernel $\operatorname{Ker} A|_{H(M)}$ consists of Schwartz functions, so that it does not depend on M.

Similarly, we claim that $\dim \operatorname{Coker} A|_{H(M)}$ is independent of M. To see this, let us observe that by the discussion before Definition 1.6.1, $\operatorname{Coker} A|_{H(M)}$ is isomorphic to $\operatorname{Ker}{}^t(A|_{H(M)})$. According to (1.6.2), the transposed ${}^t(A|_{H(M)})$: $H(M/M')' \to H(M)'$ is defined by

$$\langle {}^t(A|_{H(M)})\, u, v \rangle = \langle u, Av \rangle, \quad u \in H(M/M')', \ v \in H(M),$$

where we may read $\langle \cdot, \cdot \rangle$ as the usual duality in Schwartz spaces. In fact, without invoking Proposition 1.5.9 explicitly, observe that, since $\mathcal{S}(\mathbb{R}^d)$ is a dense subspace of $H(M/M')$, we have $H(M/M')' \hookrightarrow \mathcal{S}'(\mathbb{R}^d)$. Hence the above operator coincides with the restriction to $H(M/M')'$ of the transposed ${}^t A : \mathcal{S}'(\mathbb{R}^d) \to \mathcal{S}'(\mathbb{R}^d)$ (see Remarks 1.2.11 and 1.2.14). We deduce that $\operatorname{Ker}{}^t(A|_{H(M)}) = \operatorname{Ker}{}^t A \cap H(M/M')'$. Since ${}^t A$ is elliptic as well, $\operatorname{Ker}{}^t A$ is in fact a subspace of $\mathcal{S}(\mathbb{R}^d)$, so that $\operatorname{Ker}{}^t(A|_{H(M)}) = \operatorname{Ker}{}^t A$ is independent of M. This gives the claim.

Notice that, since $A^* u = {}^t\overline{A}\overline{u}$, $\operatorname{Ker}{}^t A$ is also isomorphic to the kernel of the formal adjoint $A^* : \mathcal{S}'(\mathbb{R}^d) \to \mathcal{S}'(\mathbb{R}^d)$.

Finally we observe that, by Propositions 1.5.5 (b) and Theorem 1.6.6, the index of $A|_{H(M)}$ is invariant by perturbations with operators having symbols in $S(M'M_0)$, where M_0 is any regular weight which tends to 0 at infinity.

We can summarize this discussion in the following theorem.

Theorem 1.6.9. *Let M, M' be regular weights, and let A be a pseudo-differential operator with an elliptic symbol in $S(M'; \Phi, \Psi)$.*

(i) $A|_{H(M)} \in \operatorname{Fred}(H(M), H(M/M'))$;

(ii) $\operatorname{ind} A|_{H(M)} = \dim \operatorname{Ker} A - \dim \operatorname{Ker} A^*$, $\operatorname{ind} A|_{H(M)} = \dim \operatorname{Ker} A - \dim \operatorname{Ker}{}^t A$, *where A^* is the formal adjoint and ${}^t A$ the transposed operator, so that the index is independent of M;*

(iii) *If B is a pseudo-differential operator with symbol in $S(M'M_0; \Phi, \Psi)$, where M_0 is any regular weight which tends to 0 at infinity, then*

$$A|_{H(M)} + B|_{H(M)} \in \operatorname{Fred}(H(M), H(M/M')),$$

and $\operatorname{ind}\left(A|_{H(M)} + B|_{H(M)}\right) = \operatorname{ind} A|_{H(M)}$.

1.7 Anti-Wick Quantization

We now introduce yet another quantization $a \mapsto A_a$, which has the advantage to preserve *positivity*. Then we compare it with the Weyl one.

1.7.1   Short-Time Fourier Transform and Anti-Wick Operators

Let
$$\mathcal{G}_0(x) = \pi^{-d/4} e^{-\frac{1}{2}|x|^2},$$
and
$$\mathcal{G}_{y,\eta}(x) = \pi^{-d/4} e^{ix\eta} e^{-\frac{1}{2}|x-y|^2}, \tag{1.7.1}$$

where (y, η) are parameters in $\mathbb{R}^{2d}$. Observe that

$$\|\mathcal{G}_0\|^2_{L^2(\mathbb{R}^d)} = \|\mathcal{G}_{y,\eta}\|^2_{L^2(\mathbb{R}^d)} = \pi^{-d/2} \int e^{-|x|^2}\, dx = 1.$$

Let moreover
$$T_y \mathcal{G}_0(t) = \mathcal{G}_0(t - y), \quad t, y \in \mathbb{R}^d.$$

Definition 1.7.1. (*Short-time Fourier transform*) For $u \in \mathcal{S}'(\mathbb{R}^d)$, we define the short-time Fourier transform Vu of u as the temperate distribution in $\mathbb{R}^{2d}$ given by

$$Vu(y, \eta) = (u, \mathcal{G}_{y,\eta}) = (2\pi)^{d/2} \mathcal{F}_{t \to \eta}(u(t)(T_y \mathcal{G}_0)(t)), \tag{1.7.2}$$

where $(\cdot, \cdot)$ denotes the sesqui-linear form on $\mathcal{S}'(\mathbb{R}^d) \times \mathcal{S}(\mathbb{R}^d)$ which agrees with the inner product in $L^2(\mathbb{R}^d)$ on Schwartz functions.

Proposition 1.7.2. *The short-time Fourier transform acts continuously* $\mathcal{S}'(\mathbb{R}^d) \to \mathcal{S}'(\mathbb{R}^{2d})$, $\mathcal{S}(\mathbb{R}^d) \to \mathcal{S}(\mathbb{R}^{2d})$ *and* $L^2(\mathbb{R}^d) \to L^2(\mathbb{R}^{2d})$. *Moreover,*

$$\|Vu\|_{L^2(\mathbb{R}^{2d})} = (2\pi)^{d/2} \|u\|_{L^2(\mathbb{R}^d)}, \quad u \in L^2(\mathbb{R}^d). \tag{1.7.3}$$

Proof. The first part of the statement follows by observing that the map $u(t) \mapsto u(t) T_y \mathcal{G}_0(t)$ acts continuously $\mathcal{S}'(\mathbb{R}^d) \to \mathcal{S}'(\mathbb{R}^{2d})$, $\mathcal{S}(\mathbb{R}^d) \to \mathcal{S}(\mathbb{R}^{2d})$ and $L^2(\mathbb{R}^d) \to L^2(\mathbb{R}^{2d})$, whereas the partial Fourier transform is a continuous map on $\mathcal{S}'(\mathbb{R}^{2d})$, on $\mathcal{S}(\mathbb{R}^{2d})$ and on $L^2(\mathbb{R}^{2d})$. The equality (1.7.3) is a consequence of Parseval's formula and the fact that $\|\mathcal{G}_0\|_{L^2(\mathbb{R}^d)} = 1$. $\square$

The adjoint map $\mathcal{S}(\mathbb{R}^{2d}) \to \mathcal{S}(\mathbb{R}^d)$,

$$(V^* F)(t) = (2\pi)^{d/2} \int \mathcal{F}^{-1}_{\eta \to t} F(y, \eta) T_y \mathcal{G}_0(t)\, dy, \quad F \in \mathcal{S}(\mathbb{R}^{2d}), \tag{1.7.4}$$

extends to a well defined continuous map $\mathcal{S}'(\mathbb{R}^{2d}) \to \mathcal{S}'(\mathbb{R}^d)$, and $L^2(\mathbb{R}^{2d}) \to L^2(\mathbb{R}^d)$. Moreover,

$$\|V^* F\|_{L^2(\mathbb{R}^d)} = (2\pi)^{d/2} \|F\|_{L^2(\mathbb{R}^{2d})}, \quad F \in L^2(\mathbb{R}^{2d}). \tag{1.7.5}$$

Also, it follows at once from the expressions in (1.7.2) and (1.7.4) that

$$V^* V = (2\pi)^d I. \tag{1.7.6}$$

Definition 1.7.3. (*Anti-Wick operator*) Let $a \in S'(\mathbb{R}^{2d})$. We define the *Anti-Wick operator with symbol a* as the map $S(\mathbb{R}^d) \to S'(\mathbb{R}^d)$ given by

$$A_a u = (2\pi)^{-d} V^* (aVu), \quad u \in S(\mathbb{R}^d). \tag{1.7.7}$$

The factor $(2\pi)^{-d}$ is introduced for the symbol $a = 1$ to give rise to the identity operator, due to (1.7.6).

Formula (1.7.7) is equivalent to saying

$$\langle A_a u, \overline{v} \rangle = (2\pi)^{-d} \langle a, Vu \overline{Vv} \rangle, \quad u, v \in S(\mathbb{R}^d). \tag{1.7.8}$$

As an immediate consequence of this definition we have the following results.

Proposition 1.7.4. *Let $a_n \in S'(\mathbb{R}^{2d})$ be a sequence convergent to a in $S'(\mathbb{R}^{2d})$. Then $A_{a_n} u \to A_a u$ in $S'(\mathbb{R}^d)$, for every $u \in S(\mathbb{R}^d)$.*

Proposition 1.7.5. *Let $a \in S'(\mathbb{R}^{2d})$ be real-valued. Then A_a is formally self-adjoint.*

Since Vu is a Schwartz function if u is, we deduce that if a is, say, a locally integrable function with polynomial growth (in particular, a symbol in $S(M; \Phi, \Psi)$), then by (1.7.8) and Fubini's theorem we can write the action of the operator A_a as

$$A_a u(x) = \int a(y, \eta)(u, \mathcal{G}_{y,\eta}) \mathcal{G}_{y,\eta}(x) \, dy \, d\eta$$

$$= \int a(y, \eta) \left(P_{y,\eta} u \right)(x) \, dy \, d\eta, \quad u \in S(\mathbb{R}^d), \tag{1.7.9}$$

where $P_{y,\eta}$ is the rank-one orthogonal projection on the line spanned by $\mathcal{G}_{y,\eta}$, i.e.,

$$P_{y,\eta} u(x) = \left(\int \overline{\mathcal{G}_{y,\eta}(t)} u(t) \, dt \right) \mathcal{G}_{y,\eta}(x). \tag{1.7.10}$$

Also, observe that if a is smooth and has polynomial growth, together with its derivatives, then it is a multiplier for $S(\mathbb{R}^{2d})$, and A_a maps $S(\mathbb{R}^d)$ into itself continuously.

Proposition 1.7.6. *Let $a(x, \xi)$ be a locally integrable function with polynomial growth. If $a(x, \xi) \geq 0$ for almost every $(x, \xi) \in \mathbb{R}^{2d}$, then $A_a \geq 0$, that is*

$$(A_a u, u)_{L^2(\mathbb{R}^d)} \geq 0, \quad \text{for all } u \in S(\mathbb{R}^d). \tag{1.7.11}$$

Moreover if $a(x, \xi) > 0$ for almost every $(x, \xi) \in \mathbb{R}^{2d}$, then $A_a > 0$, that is $(A_a u, u)_{L^2(\mathbb{R}^d)} > 0$ for $u \neq 0$.

Proof. We have, from (1.7.8),

$$(A_a u, u) = \int a(y, \eta) |(u, \mathcal{G}_{y,\eta})|^2 \, dy \, d\eta.$$

Under the assumption $a(y,\eta) \geq 0$ we conclude $(A_a u, u) \geq 0$. If $a(y,\eta) > 0$ for almost every $(y,\eta) \in \mathbb{R}^{2d}$, then $(A_a u, u) = 0$ would imply $(u, \mathcal{G}_{y,\eta}) = 0$ for almost every $(y,\eta) \in \mathbb{R}^{2d}$. Since

$$(u, \mathcal{G}_{y,\eta}) = \pi^{-\frac{d}{4}}(2\pi)^{d/2}\mathcal{F}_{x \to \eta}\left[u(x)e^{-\frac{|x-y|^2}{2}}\right](\eta),$$

this gives $u = 0$.   □

Proposition 1.7.7. *Let $a \in L^\infty(\mathbb{R}^{2d})$. Then A_a extends to a bounded operator on $L^2(\mathbb{R}^d)$, with*

$$\|A_a\|_{\mathcal{B}(L^2(\mathbb{R}^d))} \leq \|a\|_{L^\infty(\mathbb{R}^{2d})}.$$

Proof. The result is clear from the definition in (1.7.7), taking into account (1.7.3) and (1.7.5). Alternatively, one can use (1.7.3), (1.7.8) and duality.   □

1.7.2 Relationship with the Weyl Quantization

Anti-Wick operators, as continuous operators $\mathcal{S}(\mathbb{R}^d) \to \mathcal{S}'(\mathbb{R}^d)$, have a distribution kernel in $\mathcal{S}'(\mathbb{R}^{2d})$, and therefore can be regarded as pseudo-differential operators with distribution symbols. We now study in detail the relationship between the Anti-Wick symbol and the Weyl one.

Observe, first of all, that the integral kernel of $P_{y,\eta}$ in (1.7.10) is the Schwartz function

$$K_{y,\eta}(x,t) = \mathcal{G}_{y,\eta}(x)\overline{\mathcal{G}_{y,\eta}(t)}. \tag{1.7.12}$$

As a consequence, $P_{y,\eta}$ can be regarded as a pseudo-differential operator with Schwartz symbol. For example, considering the Weyl quantization, we can write

$$P_{y,\eta}u(x) = \int e^{i(x-t)\xi}\sigma_{y,\eta}\left(\frac{x+t}{2},\xi\right)u(t)\,dt\,d\xi, \tag{1.7.13}$$

where the Weyl symbol $\sigma_{y,\eta}$ of $P_{y,\eta}$ can be computed from the above expression for its kernel by means of (1.2.4) with $\tau = 1/2$. Namely,

$$\sigma_{y,\eta}(x,\xi) = (2\pi)^{d/2}\mathcal{F}_{t\to\xi}K_{y,\eta}\left(x + \frac{1}{2}t, x - \frac{1}{2}t\right) = 2^d e^{-(|x-y|^2 + |\xi-\eta|^2)}. \tag{1.7.14}$$

We have therefore the following basic result.

Proposition 1.7.8. *The orthogonal projection operator $P_{y,\eta}$ on the unitary vector (1.7.1) can be written as a formally self-adjoint pseudo-differential operator, with real Weyl symbol $\sigma_{y,\eta} \in \mathcal{S}(\mathbb{R}^{2d})$ given by (1.7.14).*   □

Consider now $a \in \mathcal{S}(\mathbb{R}^{2d})$. Inserting (1.7.13) into (1.7.9) we obtain, for $u \in \mathcal{S}(\mathbb{R}^d)$,

$$A_a u(x) = \int e^{i(x-t)\xi}b\left(\frac{x+t}{2},\xi\right)u(t)\,dt\,d\xi, \tag{1.7.15}$$

with

$$b(x,\xi) = 2^d \int a(y,\eta)e^{-(|x-y|^2+|\xi-\eta|^2)}\, dy\, d\eta$$

$$= (2\pi)^{-d}(a * \sigma_{0,0})(x,\xi). \tag{1.7.16}$$

Proposition 1.7.9. *Let* $a \in S'(\mathbb{R}^{2d})$. *Then* $A_a = b^w$, *where* $b \in S'(\mathbb{R}^{2d})$ *is given in* (1.7.16).

Proof. The result was proved above for Schwartz symbols. The general case follows by a limiting argument, taking into account Propositions 1.7.4 and 1.2.2 and the fact that the map $a \mapsto a * \sigma_{0,0}$ is continuous on $S'(\mathbb{R}^{2d})$. $\square$

We now take symbols a in the classes $S(M; \Phi, \Psi)$. We consider, for simplicity, only the case in which $\Phi \equiv \Psi$. This implies $\Psi \equiv h^{-1/2}$ (h being the Planck function in (1.1.8)).

Theorem 1.7.10. *Let* $a \in S(M; \Psi, \Psi)$. *Then* A_a *is a pseudo-differential operator with symbol in the same class. More precisely, its Weyl symbol* $b(x,\xi)$ *defined in* (1.7.16) *belongs to* $S(M; \Psi, \Psi)$ *and has asymptotic expansion*

$$b(x,\xi) \sim \sum_{\alpha,\beta} \frac{c_{\alpha\beta}}{\alpha!\beta!}\partial_\xi^\alpha \partial_x^\beta a(x,\xi), \tag{1.7.17}$$

in the sense that, for every $N \geq 1$,

$$R_N = b - \sum_{|\alpha+\beta|<N} \frac{c_{\alpha\beta}}{\alpha!\beta!}\partial_\xi^\alpha \partial_x^\beta a \in S(Mh^{N/2}; \Psi, \Psi), \tag{1.7.18}$$

where

$$c_{\alpha\beta} = 2^d \int \eta^\alpha y^\beta e^{-(|y|^2+|\eta|^2)}\, dy\, d\eta. \tag{1.7.19}$$

In particular $c_{00} = 1$ *and* $c_{\alpha\beta} = 0$ *for odd* $|\alpha + \beta|$.

Proof. It will be sufficient to prove that (1.7.18) holds true for $b(x,\xi)$ defined as in (1.7.16).

We may Taylor expand

$$a(y,\eta) = \sum_{|\alpha+\beta|<N} \frac{1}{\alpha!\beta!}\partial_\xi^\alpha \partial_x^\beta a(x,\xi)(\eta-\xi)^\alpha(y-x)^\beta + r_N(x,y,\xi,\eta), \tag{1.7.20}$$

with

$$r_N(x,y,\xi,\eta) = N \sum_{|\alpha+\beta|=N} \frac{1}{\alpha!\beta!}(\eta-\xi)^\alpha(y-x)^\beta$$

$$\times \int_0^1 (1-t)^{N-1}\left(\partial_\xi^\alpha \partial_x^\beta a\right)(x+t(y-x),\xi+t(\eta-\xi))\, dt. \tag{1.7.21}$$

Inserting this in the expression

$$b(x, \xi) = 2^d \int a(y, \eta) e^{-(|x-y|^2 + |\xi-\eta|^2)} \, dy \, d\eta,$$

we obtain the equality in (1.7.18) with $c_{\alpha\beta}$ given by (1.7.19) and

$$R_N(x, \xi) = 2^d \int r_N(x, y, \xi, \eta) e^{-(|x-y|^2 + |\xi-\eta|^2)} \, dy \, d\eta. \tag{1.7.22}$$

To prove that $R_N \in S(Mh^{N/2}; \Psi, \Psi)$ we estimate $\partial_\xi^\gamma \partial_x^\delta R_N$; in view of (1.7.21) and (1.7.22) it will be sufficient to obtain bounds independent of t for the terms

$$I(t, x, \xi) := \int \eta^\alpha y^\beta e^{-(|y|^2 + |\eta|^2)} \left(\partial_\xi^{\alpha+\gamma} \partial_x^{\beta+\delta} a \right) (x + ty, \xi + t\eta) \, dy \, d\eta, \tag{1.7.23}$$

where $|\alpha + \beta| = N$. Appealing once again to the temperance property of M and Ψ we have, for some $s \geq 0$,

$$\left| \left(\partial_\xi^{\alpha+\gamma} \partial_x^{\beta+\delta} a \right) (x + ty, \xi + t\eta) \right| \lesssim M(x + ty, \xi + t\eta) \Psi(x + ty, \xi + t\eta)^{-|\alpha+\beta+\gamma+\delta|}$$

$$\lesssim M(x, \xi) \Psi(x, \xi)^{-|\alpha+\beta+\gamma+\delta|} (1 + |y| + |\eta|)^{s(1+|\alpha+\beta+\gamma+\delta|)},$$

which gives for (1.7.23) the expected estimates

$$|I(t, x, \xi)| \lesssim M(x, \xi) h(x, \xi)^{N/2} \Psi(x, \xi)^{-|\gamma+\delta|},$$

(recall that here $\Psi \equiv h^{-1/2}$). The theorem is therefore proved.   □

Not every pseudo-differential operator with symbol in $S(M; \Psi, \Psi)$ admits an Anti-Wick symbol a in the same class; observe in fact that the Weyl symbol $b(x, \xi)$ in (1.7.16) must be an analytic function of $(x, \xi) \in \mathbb{R}^{2d}$. We have however a converse of Theorem 1.7.10 modulo regularizing operators.

Proposition 1.7.11. *Let a be a symbol in $S(M; \Psi, \Psi)$. Then there exists a symbol $s \in S(Mh; \Psi, \Psi)$ such that $A_a = a^w + s^w$.*

Assume the strong uncertainty principle (1.1.10) is satisfied. Then for every $b \in S(M; \Psi, \Psi)$ there exist $a \in S(M; \Psi, \Psi)$ and $r \in S(\mathbb{R}^{2d})$ such that $b^w = A_a + r^w$.

Proof. The first part of the statement follows at once from Theorem 1.7.10, with $N = 2$, for $c_{0,0} = 1$ and $c_{\alpha,\beta} = 0$ if $|\alpha + \beta| = 1$.

To prove the second part, we observe that, by the result just proved, $b^w - A_b$ is a pseudo-differential operator with Weyl symbol $b^{(1)} \in S(Mh; \Psi, \Psi)$. The same arguments applied to $b^{(1)}(x, \xi)$ in place of $b(x, \xi)$ shows that $b^w - A_{b+b^{(1)}}$ is a pseudo-differential operator with symbol in $S(Mh^2; \Psi, \Psi)$. By induction one shows the existence of a sequence of symbols $b^{(n)} \in S(Mh^n; \Psi, \Psi)$, $b^{(0)} = b$, such that, after setting $c^{(N)} = \sum_{n=0}^N b^{(n)}$, the difference $b^w - A_{c^{(N-1)}}$ is a pseudo-differential operator with symbol in $S(Mh^N; \Psi, \Psi)$. According to Proposition 1.1.6 there exists a symbol $a \sim \sum_{n=0}^{\infty} b^{(n)}$ in $S(M; \Psi, \Psi)$. We obtain that $b^{\mathrm{w}} - A_a$ is a pseudo-differential operator with symbol in $S(Mh^N; \Psi, \Psi)$ for every N, i.e., in $\mathcal{S}(\mathbb{R}^{2d})$ by (1.1.11).

This concludes the proof. $\qquad \square$

For polynomial symbols there is in fact an exact converse of Theorem 1.7.10. We will discuss it in Section 1.8.

We conclude this section with an application to the theory of Sobolev spaces. More precisely, one can exploit the injectivity of an Anti-Wick operator with strictly positive symbol (cf. Proposition 1.7.6) to define an equivalent norm in the Sobolev spaces $S(M; \Psi, \Psi)$, hence avoiding the use of a parametrix.

Proposition 1.7.12. *Assume the strong uncertainty principle. Let a be an elliptic symbol in $S(M; \Psi, \Psi)$, with $a(x, \xi) > 0$ for $(x, \xi) \in \mathbb{R}^{2d}$. Then A_a defines an isomorphism of $\mathcal{S}(\mathbb{R}^d)$ and of $\mathcal{S}'(\mathbb{R}^d)$, and also of $H(M)$ into $L^2(\mathbb{R}^d)$. Its inverse A_a^{-1} is still a pseudo-differential operator with Weyl symbol in $S(M^{-1}; \Psi, \Psi)$.*

As a consequence, an equivalent norm in $H(M)$ is given by $\|A_a u\|_{L^2(\mathbb{R}^d)}$, $u \in H(M)$.

Proof. We start by proving that A_a is bijective and continuous as a map on $\mathcal{S}(\mathbb{R}^d)$, on $\mathcal{S}'(\mathbb{R}^d)$, and $H(M) \to L^2(\mathbb{R}^d)$. By Theorem 1.7.10, A_a is a pseudo-differential operator with Weyl symbol $b \in S(M; \Psi, \Psi)$ so that the continuity follows from Propositions 1.2.7, 1.2.13 and 1.5.5 (a). Since $b - a \in S(Mh; \Psi, \Psi)$, and $h(x, \xi)$ tends to zero at infinity by the strong uncertainty principle, we see that b is elliptic as well. As a consequence, Corollary 1.3.9 tells us that the kernel of $A_a : \mathcal{S}'(\mathbb{R}^d) \to \mathcal{S}'(\mathbb{R}^d)$ consists of Schwartz functions and therefore reduces to the trivial space by Proposition 1.7.6. This proves the injectivity of the map on all the above spaces.

Now, since A_a is formally self-adjoint, by Theorem 1.6.9 its index is 0 as a map $H(M) \to L^2(\mathbb{R}^d)$, so that this map is also surjective. Again by Corollary 1.3.9, A_a is surjective as a map on $\mathcal{S}(\mathbb{R}^d)$ as well. By the Open Mapping Theorem it follows that A_a defines an isomorphism of $\mathcal{S}(\mathbb{R}^d)$ and also of $H(M)$ into $L^2(\mathbb{R}^d)$. Let us now prove that the same is true as a map on $\mathcal{S}'(\mathbb{R}^d)$ and that the inverse map is pseudo-differential. To this end, let B denote a parametrix of A_a, with Weyl symbol in $S(M^{-1}; \Psi, \Psi)$, and A_a^{-1} the inverse of A_a, as an operator on $\mathcal{S}(\mathbb{R}^d)$. We have $A_a^{-1} = A_a^{-1}(A_a B - R) = B - R'$, as operators on $\mathcal{S}(\mathbb{R}^d)$, with R, R' regularizing. It follows that A_a^{-1} is a pseudo-differential operator with Weyl symbol in $S(M^{-1}; \Psi, \Psi)$ and therefore extends to a continuous operator on $\mathcal{S}'(\mathbb{R}^d)$. The identities $A_a^{-1} A_a = I$, $A_a A_a^{-1} = I$, valid on $\mathcal{S}(\mathbb{R}^d)$, extend therefore to $\mathcal{S}'(\mathbb{R}^d)$.

This concludes the proof. $\qquad \square$

Remark 1.7.13. The proof of Proposition 1.7.12 shows that, if $a \in S(M; \Phi, \Psi)$ is real and elliptic, and a^{w} is injective on $\mathcal{S}(\mathbb{R}^d)$ then a^{w} is an isomorphism of $\mathcal{S}(\mathbb{R}^d)$ and of $\mathcal{S}'(\mathbb{R}^d)$ and its inverse is a pseudo-differential operator with (real elliptic) Weyl symbol in $S(M^{-1}; \Phi, \Psi)$.

1.7.3 Applications to Boundedness on L^2 and Almost Positivity of Pseudo-Differential Operators

Here we prove the boundedness on $L^2(\mathbb{R}^d)$ of pseudo-differential operators with symbols in $S(1;1,1)$ ($M(x,\xi) = \Phi(x,\xi) = \Psi(x,\xi) \equiv 1$), i.e., satisfying for every $k \in \mathbb{N}$

$$\|a\|_{k,S(1;1,1)} = \sup_{|\alpha|+|\beta| \leq k} \sup_{(x,\xi) \in \mathbb{R}^{2d}} |\partial_\xi^\alpha \partial_x^\beta a(x,\xi)| < \infty. \tag{1.7.24}$$

Since the weights Φ and Ψ we are considering satisfy (1.1.1), every class $S(1;\Phi,\Psi)$ is contained in $S(1;1,1)$. Hence we obtain, in particular, the boundedness of pseudo-differential operators with symbols in $S(1;\Phi,\Psi)$, without assuming the strong uncertainty principle.

The idea of the proof is to use the family of functions $\mathcal{G}_{y,\eta}$ in (1.7.1) to reach an almost diagonalization of such pseudo-differential operators.

Theorem 1.7.14. *Let a be a smooth function in $\mathbb{R}^{2d}$ satisfying the estimates in (1.7.24). The corresponding operator $a(x,D)$ is bounded on $L^2(\mathbb{R}^d)$, with the uniform estimate*

$$\|a(x,D)\|_{\mathcal{B}(L^2(\mathbb{R}^d))} \lesssim \|a\|_{N,S(1;1,1)}, \quad a \in S(1;1,1), \tag{1.7.25}$$

where N depends only on the dimension d.

Proof. It suffices to prove (1.7.25) for Schwartz symbols, since the general case follows by a limiting argument which uses Proposition 1.1.5. Hence, let $a \in \mathcal{S}(\mathbb{R}^{2d})$. In view of (1.7.6), we write

$$a(x,D) = (2\pi)^{-4d} V^* V a(x,D) V^* V,$$

where V is the short time Fourier transform in (1.7.2) and V^* its adjoint in (1.7.4). Because of (1.7.3) and (1.7.5), it suffices to prove that

$$\|V a(x,D) V^* F\|_{L^2(\mathbb{R}^{2d})} \lesssim \|a\|_{N,S(1;1,1)} \|F\|_{L^2(\mathbb{R}^{2d})}, \quad F \in \mathcal{S}(\mathbb{R}^{2d}). \tag{1.7.26}$$

To this end we look at the integral kernel $K(y',\eta';y,\eta)$ in $\mathbb{R}^{4d}$ of the operator $V a(x,D) V^*$. Since for Schwartz functions F we have

$$(V^* F)(x) = \int F(y,\eta) \mathcal{G}_{y,\eta}(x) \, dy \, d\eta,$$

it turns out that

$$(V a(x,D) V^* F)(y',\eta') = \int F(y,\eta) \, (a(x,D) \mathcal{G}_{y,\eta}) \, (x) \overline{\mathcal{G}_{y',\eta'}(x)} \, dx \, dy \, d\eta,$$

so that

$$K(y',\eta';y,\eta) = \int e^{ix\xi} a(x,\xi) \widehat{\mathcal{G}_{y,\eta}}(\xi) \overline{\mathcal{G}_{y',\eta'}(x)} \, dx \, d\xi.$$

An explicit computation shows that

$$\widehat{\mathcal{G}_{y,\eta}}(\xi) = e^{-iy(\xi-\eta)}\mathcal{G}_0(\xi - \eta).$$

Hence, by a linear change of variables we obtain

$$|K(y',\eta';y,\eta)| = \left| \int e^{ix\xi-iy\xi-ix\eta'} a(x,\xi)\mathcal{G}_0(\xi-\eta)\mathcal{G}_0(x-y')\,dx\,d\xi \right|$$

$$= \left| \int e^{ix(\eta-\eta')+i\xi(y'-y)} e^{ix\xi} a(x+y',\xi+\eta)\mathcal{G}_0(\xi)\mathcal{G}_0(x)\,dx\,d\xi \right|. \tag{1.7.27}$$

From this expression one easily deduces the estimate

$$|K(y',\eta';y,\eta)| \lesssim \|a\|_{4N,S(1;1,1)}\langle y-y'\rangle^{-2N}\langle \eta-\eta'\rangle^{-2N} \tag{1.7.28}$$

for every $N \in \mathbb{N}$. Indeed, we have

$$\langle y-y'\rangle^{2N}\langle \eta-\eta'\rangle^{2N} e^{ix(\eta-\eta')+i\xi(y'-y)} = (1-\Delta_x)^N(1-\Delta_\xi)^N e^{ix(\eta-\eta')+i\xi(y'-y)}.$$

Hence, (1.7.28) follows by repeated integrations by parts in (1.7.27), taking into account that, for every $N' \in \mathbb{N}$,

$$|(1-\Delta_x)^N(1-\Delta_\xi)^N\{e^{ix\xi}a(x+y',\xi+\eta)\mathcal{G}_0(\xi)\mathcal{G}_0(x)\}|$$

$$\lesssim \|a\|_{4N,S(1;1,1)}\langle x\rangle^{-N'}\langle \xi\rangle^{-N'},$$

due to Leibniz' formula, (1.7.24) and the fact that $\mathcal{G}_0 \in \mathcal{S}(\mathbb{R}^d)$.

Now, if $2N > d$ in (1.7.28), we see that the kernel $K(y',\eta';y,\eta)$ is dominated by a convolution kernel in $L^1(\mathbb{R}^{2d})$. Therefore (1.7.26) follows by Young's inequality. □

We now come to an application to the almost positivity of pseudo-differential operators.

Contrary to the result in Proposition 1.7.6 for the Anti-Wick quantization, pseudo-differential operators with positive Weyl symbol are not positive in general. Consider for example in dimension $d = 1$ the symbol $a(x,\xi) = x^2\xi^2 \geq 0$. The operator A with Weyl symbol a can be written as

$$A = \frac{1}{4}(xD_x + D_x x)^2 - \frac{1}{4},$$

and therefore

$$(Au,u)_{L^2} = \frac{1}{4}\|(xD_x + D_x x)u\|_{L^2}^2 - \frac{1}{4}\|u\|_{L^2}^2,$$

where the right-hand side is negative for suitable $u \in \mathcal{S}(\mathbb{R})$.

We have however the following lower bound, "with gain of one derivative", for pseudo-differential operators with positive Weyl symbol. It represents one of the basic applications of the Anti-Wick theory.

Theorem 1.7.15. (*Sharp Gårding inequality*) *Let* $a \in S(h^{-1}; \Psi, \Psi)$, $a(x, \xi) \geq 0$, $(x, \xi) \in \mathbb{R}^{2d}$. *There exists a constant* $C > 0$ *such that*

$$(a^w u, u)_{L^2(\mathbb{R}^d)} \geq -C\|u\|_{L^2(\mathbb{R}^d)}, \quad u \in S(\mathbb{R}^d). \tag{1.7.29}$$

Proof. By Proposition 1.7.11 we have $a^w = A_a + s^w$, where $s \in S(1; \Psi, \Psi)$. Since A_a satisfies (1.7.11) and s^w is bounded on $L^2(\mathbb{R}^d)$ by Theorem 1.7.14, the desired result follows. $\square$

Much more refined techniques allow one to prove (1.7.29) for symbols in $S(h^{-2}; \Phi, \Psi)$ (under additional assumptions on Φ and Ψ), which is the so-called Fefferman-Phong inequality.

1.7.4 Sobolev Spaces Revisited

As another application of Anti-Wick techniques we now show, as promised in Section 1.5, that the Sobolev spaces $H(M)$ introduced there are in fact independent of the choice of the weights Φ and Ψ.

Let M be a temperate weight, cf. (1.1.2). Let $L^2_M(\mathbb{R}^{2d})$ be the weighted Lebesgue space of measurable functions F in $\mathbb{R}^{2d}$ such that $MF \in L^2(\mathbb{R}^{2d})$, with the norm $\|F\|_{L^2_M(\mathbb{R}^{2d})} = \|MF\|_{L^2(\mathbb{R}^{2d})}$.

We define the space $\tilde{H}(M)$ as the completion of the Schwartz class $S(\mathbb{R}^d)$ with respect to the norm

$$\|u\|_{\tilde{H}(M)} = \|Vu\|_{L^2_M(\mathbb{R}^{2d})} = \left(\int M(y, \eta)^2 |Vu(y, \eta)|^2 \, dy \, d\eta \right)^{1/2}, \quad u \in S(\mathbb{R}^d),$$

where Vu is the short-time Fourier transform of u in (1.7.2).

Notice that the above norm is well defined for $u \in S(\mathbb{R}^d)$, because $Vu \in S(\mathbb{R}^{2d})$ and M has polynomial growth. In fact, we have a continuous inclusion $S(\mathbb{R}^d) \hookrightarrow \tilde{H}(M)$. Moreover, for $M(x, \xi) \equiv 1$ we get $\tilde{H}(1) = L^2(\mathbb{R}^d)$ by (1.7.3). Let us show that the space $H(M)$ in Definition 1.5.2 coincides with $\tilde{H}(M)$, and is therefore independent of Φ and Ψ. To this end, we need the following continuity result.

Theorem 1.7.16. *Let* M, M' *be temperate weights and* $a \in S(M'; 1, 1)$ $(\Phi(x, \xi) = \Psi(x, \xi) = 1)$. *Then* $a(x, D)$ *extends to a bounded operator* $\tilde{H}(M) \to \tilde{H}(M/M')$.

Proof. We argue as in the proof of Theorem 1.7.14. Namely, using (1.7.6) we write $a(x, D) = (2\pi)^{-4d} V^* V a(x, D) V^* V$ on $S(\mathbb{R}^d)$. Observe that the map $(2\pi)^{-d/2} V$ clearly extends to an isometry $\tilde{H}(M) \to \overline{V(S(\mathbb{R}^d))} \subset L^2_M(\mathbb{R}^{2d})$ and similarly $(2\pi)^{-d/2} V^*$ extends to an isometry $\overline{V(S(\mathbb{R}^d))} \subset L^2_{M/M'}(\mathbb{R}^{2d}) \to \tilde{H}(M/M')$. Hence we need only prove that $Va(x, D)V^*$ extends to a bounded map $L^2_M(\mathbb{R}^{2d}) \to L^2_{M/M'}(\mathbb{R}^{2d})$. We know that the integral kernel $K(y', \eta'; y, \eta)$ of this map has the expression in (1.7.27).

Repeated integrations by parts in (1.7.27) (as in the proof of Theorem 1.7.14), combined with the symbol estimates for $a(x, \xi)$ and the temperance $M'(x+y', \xi+\eta) \lesssim M'(y', \eta)(1 + |x| + |\xi|)^s$, show that for every N,

$$|K(y', \eta'; y, \eta)| \lesssim M'(y', \eta)\langle y - y'\rangle^{-N}\langle \eta - \eta'\rangle^{-N}.$$

As a consequence, for $F \in \mathcal{S}(\mathbb{R}^{2d})$, we get

$$\frac{M(y', \eta')}{M'(y', \eta')}|(Va(x, D)V^*F)(y', \eta')|$$

$$\lesssim \int \frac{M(y', \eta')}{M'(y', \eta')}\frac{M'(y', \eta)}{M(y, \eta)}\langle y - y'\rangle^{-N}\langle \eta - \eta'\rangle^{-N}M(y, \eta)|F(y, \eta)|\, dy\, d\eta.$$

Again by temperance we have

$$\frac{M(y', \eta')}{M(y, \eta)} \lesssim (1 + |y - y'| + |\eta - \eta'|)^s, \qquad \frac{M'(y', \eta)}{M'(y', \eta')} \lesssim (1 + |\eta - \eta'|)^s,$$

so that we get, for every N',

$$\frac{M(y', \eta')}{M'(y', \eta')}|(Va(x, D)V^*F)(y', \eta')|$$

$$\lesssim \int \langle y - y'\rangle^{-N'}\langle \eta - \eta'\rangle^{-N'}M(y, \eta)|F(y, \eta)|\, dy\, d\eta.$$

An application of Young's inequality then gives the estimate

$$\|Va(x, D)V^*F\|_{L^2_{M/M'}(\mathbb{R}^{2d})} \lesssim \|F\|_{L^2_M(\mathbb{R}^{2d})}, \quad F \in \mathcal{S}(\mathbb{R}^{2d}),$$

which concludes the proof. $\qquad\square$

Corollary 1.7.17. *Assume further that the temperate weight M is regular, in the sense of Definition 1.5.1. The Sobolev space $H(M)$ in Definition 1.5.2 coincides, for every choice of the weights Φ, Ψ, with the above space $\tilde{H}(M)$, with equivalent norms.*

Proof. Let $a \in S(M; \Phi, \Psi)$ be an elliptic symbol and set $A = a(x, D)$. Let $B = b(x, D)$, $b \in S(M^{-1}; \Phi, \Psi)$ be a (left) parametrix of A, so that $BA = I + R$, with R regularizing. By Theorem 1.7.16, A and R act $\tilde{H}(M) \to \tilde{H}(1) = L^2(\mathbb{R}^d)$ continuously. Hence we have the estimate

$$\|Au\|_{L^2(\mathbb{R}^d)} + \|Ru\|_{L^2(\mathbb{R}^d)} \lesssim \|u\|_{\tilde{H}(M)},$$

which is $\tilde{H}(M) \hookrightarrow H(M)$.

Similarly, by Theorem 1.7.16, B acts $L^2(\mathbb{R}^d) = \tilde{H}(1) \to \tilde{H}(M)$ continuously, so that using $u = BAu + Ru$ we get

$$\|u\|_{\tilde{H}(M)} \lesssim \|Au\|_{L^2(\mathbb{R}^d)} + \|Ru\|_{\tilde{H}(M)}.$$

Clearly $\|Au\|_{L^2(\mathbb{R}^d)} \lesssim \|u\|_{H(M)}$. On the other hand, since $\mathcal{S}(\mathbb{R}^d) \hookrightarrow \tilde{H}(M)$ and R is continuous $H(M) \to \mathcal{S}(\mathbb{R}^d)$, we get $\|Ru\|_{\tilde{H}(M)} \lesssim \|u\|_{H(M)}$, and the opposite inclusion $H(M) \hookrightarrow \tilde{H}(M)$ is therefore proved. $\square$

1.8 Quantizations of Polynomial Symbols

We conclude this chapter with a few remarks about relationships between several types of quantizations, especially for polynomial symbols.

First of all one could wonder about the reason for such a large number of quantization rules. Indeed, the main motivations come from Quantum Mechanics. In Classical Mechanics the state of a system is described by a point in $\mathbb{R}^d$ (or in a d-dimensional manifold); for example $d = 6N$ in the case of N particles in the space $\mathbb{R}^3$, the coordinates being the position and momentum of each particle. In Quantum Mechanics, the state of a system is instead described by a function ϕ in $L^2(\mathbb{R}^d)$ (up to a non-zero multiplicative constant). The (real) observables $a(x, \xi)$ are replaced by densely defined self-adjoint operators A, and $(\phi, A\phi)_{L^2}$ represents the mean value of the observable a. The main examples are given by the operators associated with the *position* observable $a(x, \xi) = x_j$ which corresponds to the operator $M_j\phi = x_j\phi$, and the *momentum* observable $a(x, \xi) = \xi_j$, which corresponds to the derivative $D_j = -i\partial_j$. Every reasonable quantization rule should verify these two axioms. The quantization of general observables is formally performed by replacing, in the expression of $a(x, \xi)$, the variables x_j and ξ_j by the above mentioned operators. This unfortunately leads to a non-trivial commutativity problem, as we have of course $x_j\xi_j = \xi_j x_j$, but $M_j D_j \neq D_j M_j$.

One preferred quantization rule is certainly the Weyl one since, as we know from Proposition 1.2.10, real-valued symbols a give rise to (formally) self-adjoint operators a^{w}. Another interesting quantization is the so-called *Feynmann quantization*, which is

$$a \mapsto \frac{1}{2}\left(a(x, D) + a(x, D)^*\right),$$

where $a(x, D)$ is the standard quantization. Observe that, if

$$a(x, D) = \sum_{|\alpha| \leq m} c_\alpha(x)D^\alpha$$

is a differential operator with smooth coefficients, then

$$a(x, D)^* f(x) = \sum_{|\alpha| \leq m} D^\alpha\left(\overline{c_\alpha(x)}f(x)\right).$$

Notice also that, if $a(x, \xi) = x_j\xi_j$,

$$a^{\mathrm{w}} = \frac{1}{2}(x_j D_j + D_j x_j) = \frac{1}{2}\left(a(x, D) + a(x, D)^*\right). \tag{1.8.1}$$

More generally (1.8.1) still holds for every polynomial $\sum_{\alpha,\beta} c_{\alpha,\beta} x^\beta \xi^\alpha$ with real coefficients, if each monomial has degree at most 1 either with respect to x or ξ. This follows by $\mathbb{R}$-linearity from the following result, which shows, among other things, that generally we have $a^w \neq \frac{1}{2}(a(x,D) + a(x,D)^*)$ even if a is real-valued.

Proposition 1.8.1. *Let a be a real-valued polynomial. Then*

$$a^w = \frac{1}{2}\left(a(x,D) + a(x,D)^*\right) + r(x,D),$$

with

$$r(x,\xi) = \sum_{|\alpha|\geq 2} (\alpha!)^{-1}\left(2^{-|\alpha|} - 2^{-1}\right) \partial_\xi^\alpha D_x^\alpha a(x,\xi),$$

where the above sum is finite because a is a polynomial.

Proof. The desired result follows by an application of the previous Remark 1.2.6 and Proposition 1.2.9 (or, better, from the proofs of those results). In fact, the implied asymptotic expansions become here finite and the corresponding asymptotic equivalences become equalities, because $a(x,\xi)$ is a polynomial. $\qquad\square$

For example, in dimension $d = 1$, for $a(x,\xi) = x^2\xi^2$, we have

$$a^w = \frac{1}{2}\left(a(x,D) + a(x,D)^* + I\right).$$

Another remark is that, if $a(x,\xi) = \sum_{|\beta|\leq m} c_\beta x^\beta$ is a polynomial with complex coefficients, then $a^w = a(x,D)$. That is, the Weyl quantization yields multiplication operators for polynomials independent of ξ. Notice that the same trivially happens on the frequency side, namely $a^w = a(x,D)$ if $a(x,\xi) = \sum_{|\alpha|\leq m} c_\alpha \xi^\alpha$, $c_\alpha \in \mathbb{C}$.

Let us now come to the relationship between Weyl and Anti-Wick quantization. We saw in Section 1.7.2 that any operator A_a with Anti-Wick symbol $a(x,\xi)$ can be written as a Weyl operator with symbol $b(x,\xi)$ given by (1.7.16), i.e.,

$$b(z) = (2\pi)^{-d}(a * \sigma_{0,0})(z), \quad z \in \mathbb{R}^{2d},$$

where $\sigma_{0,0}(z) = 2^d e^{-|z|^2}$. Vice-versa, as we already observed, it is not always possible to write an operator with a given Weyl symbol in the Anti-Wick form. In fact, using $\widehat{a * \sigma_{0,0}} = (2\pi)^d \hat{a}\,\widehat{\sigma_{0,0}}$, cf. (0.2.4), and $\widehat{\sigma_{0,0}}(\zeta) = e^{-|\zeta|^2/4}$, we have

$$b = e^{-|D|^2/4}a = e^{\Delta/4}a.$$

Hence one would be tempted to solve this equation as $a = e^{-\Delta/4}b$, but this does not make sense for general $b \in \mathcal{S}'(\mathbb{R}^{2d})$, because $e^{|\zeta|^2/4}$ is not a multiplier for $\mathcal{S}'(\mathbb{R}^{2d})$. However this formula holds when b is a polynomial, because its Fourier transform is then supported at the origin. Namely, we have the following result.

Proposition 1.8.2. *Let $b(z)$ be a polynomial, $z \in \mathbb{R}^{2d}$. Then the equation*

$$b = e^{\Delta/4} a$$

is satisfied by

$$a = \sum_{k=0}^{\infty} \frac{(-\Delta/4)^k}{k!} b,$$

where the sum is finite because b is a polynomial.

Proof. We have to verify that $e^{\Delta/4} \sum_{k=0}^{\infty} \frac{(-\Delta/4)^k}{k!} b = b$. But the Fourier transform of the left-hand side is

$$e^{-|\zeta|^2/4} \sum_{k=0}^{\infty} \frac{(|\zeta|^2/4)^k}{k!} \widehat{b}(\zeta) = \widehat{b}(\zeta). \qquad \square$$

Notes

Pseudo-differential operators were introduced explicitly by Kohn and Nirenberg [127] and Hörmander [115], although early related issues appeared in works by Calderón, Seeley, Zygmund and others. The first symbol classes considered, $S_{1,0}^m$, were suitable for the construction of a parametrix for elliptic differential operators, but in order to treat similarly non-elliptic differential operators, the more general classes $S_{\rho,\delta}^m$, $0 \leq \delta \leq \rho \leq 1$, defined by the estimates $|\partial_\xi^\alpha \partial_x^\beta a(x,\xi)| \lesssim (1 + |\xi|)^{m-\rho|\alpha|+\delta|\beta|}$, were soon revealed to be essential. The study of the local solvability problem for operators of principal type required Beals and Fefferman [11] to introduce much more general symbol classes; see also the subsequent works by Beals [9], [10], and Fefferman and Phong [78] for applications to sharp lower bounds for pseudo-differential operators. Meanwhile, in connection with problems in Quantum Mechanics and PDEs in unbounded domains, Berezin and Shubin [14], [15], Cordes [58], Grushin [103], Kumano-go [128], Parenti [156], and Shubin [182] studied pseudo-differential global calculi in $\mathbb{R}^d$, where x and ξ enter mostly in a symmetric way in the corresponding symbol estimates. Eventually, Hörmander [117] gave a systematic treatment of the whole theory, introducing general symbol classes which included all those considered until then. Applications to spectral theory of globally regular operators appeared in Hörmander [118] and to local solvability of pseudo-differential operators in Dencker [72].

The symbol classes considered in this chapter are inspired by [11], [117] and Hörmander [119, Vol III, Section 18.4]. More precisely, the class $S(M; \Phi, \Psi)$ corresponds to Hörmander's class $S(M, g)$, with the metric $g_{x,\xi} = |dx|^2/\Phi(x,\xi)^2 + |d\xi|^2/\Psi(x,\xi)^2$. Our calculus is in fact simpler, since we assume the conditions $\Phi \geq 1$ and $\Psi \geq 1$. On the other hand the above metric g, under our assumptions, does not need to be slowly varying or temperate in Hörmander's sense. The conditions of sub-linear growth $\Phi(x,\xi), \Psi(x,\xi) \lesssim 1 + |x| + |\xi|$ basically play a similar role to (but do not imply) the slow-variation condition for g (cf. Lemma 1.1.4 and Remark 1.1.7). The conditions in (1.3.8) are instead equivalent to saying that the metric g is slowly varying and the weight M is g-continuous, in Hörmander's terminology.

For the invariance of the Weyl calculus with respect to symplectic changes of variables and the connection with the theory of metaplectic operators, which is not discussed here, we refer the reader to Folland [82] and [119, Vol. III, Theorem 18.5.9].

The proof of the L^2-boundedness in Theorem 1.4.1 is a classical and nice application of the symbolic calculus, but it just works under the assumption of the strong uncertainty principle. Hence, that result does not contain the celebrated Calderón-Vaillancourt Theorem [34] for symbols in $S_{0,0}^0$, which is instead recaptured in Section 1.7.3.

The theory of the general Sobolev spaces tailored to a given symbol class in Section 1.5 is inspired by Shubin [183], Bony and Chemin [23], but is again much simpler here, because we assume the strong uncertainty principle and we can consequently rely on the existence of a parametrix. Similarly, Fredholm properties in the subsequent section are immediate once one has a full symbolic calculus and the continuity and compactness, on every Sobolev space, of operators "of order 0" and "of order lower than 0", respectively.

The presentation of the abstract theory of Fredholm operators in Hilbert spaces in Section 1.6.1 is inspired by Grubb [102] and [119, Vol. III] (where the case of operators on Banach spaces is considered), but the proof of the stability result for the index (Theorem 1.6.3) is extracted from Atiyah and Singer [6].

The short-time Fourier transform, or windowed transform, is a basic tool in Time-Frequency Analysis (see Gröchenig [97]), and plays a fundamental role in signal analysis. Indeed, contrary to the Fourier transform, its *magnitude* $|Vu(y,\eta)|$ contains the information of what frequencies of a signal are mostly present for a given time interval.

Anti-Wick operators were introduced as a quantization rule in Physics by Berezin [13]; see also Friedrichs [84]. They also appeared under the name of *localization operators* in Daubechies [67], which proposed them as a mathematical tool to localize a signal on the time-frequency plane. Since then they have been extensively studied in signal analysis and found many other applications (see Cordero and Gröchenig [52], [53], Cordero and Rodino [57], Ramanathan and Topiwala, [167], Wong [198] and the references therein). We also recall their employment as approximation of pseudo-differential operators (wave packets); see Córdoba and Fefferman [60], Folland [82], Lerner [133], Tataru [186]. Anti-Wick operators are also called *short-time Fourier transform multipliers*, see Feichtinger and Nowak [81] for a survey.

The proof of the L^2-boundedness in Theorem 1.7.14, for operators with symbols in $S_{0,0}^0$, is a continuous version of the almost diagonalization via Gabor frames in Rochberg and Tachizawa [171]. It is one of the most elementary proofs we know, since it avoids the use of the Cotlar-Stein lemma (and Schur's test). Moreover it generalizes to less regular symbols (see Gröchenig [98]) and even to Fourier integral operators (see Cordero, Nicola and Rodino [55] and the references therein). Different proofs can be found in the already quoted paper [34], Coifman, and Meyer [51], [97, Theorem 14.5.2], [119, Vol. III, Theorem 18.6.3], Hwang [121], Stein [185], Buzano and Toft [31].

Finally, the use of Anti-Wick techniques in the proof of the Sharp Gårding inequality (Theorem 1.7.15) appeared in [60] and [119, Vol. III, Theorem 18.1.4]. More general symbol classes were considered by similar techniques in [133]. The Fefferman-Phong inequality mentioned after Theorem 1.7.15 was proved in [78] for symbols in $S^2_{1,0}$ and in [119, Vol. III, Theorem 18.6.8] for the classes $S(h^{-2}, g)$; see also Fefferman and Phong [77], [79]. For recent work about lower bounds for pseudo-differential operators we refer the reader to the references in Nicola [153].

The spaces $\tilde{H}(M)$ in Section 1.7.4 are a type of *modulation spaces*, widely used in Time-Frequency Analysis. They were first introduced by H. Feichtinger in [80]; see [97, Chapter 11] and the references therein for a detailed study.

The results in Section 1.8 can also be proved without using the symbolic calculus, by direct computations, see Boggiatto and Rodino [21].

Chapter 2

Γ-Pseudo-Differential Operators and H-Polynomials

Summary

Let P be a linear operator, $P : S(\mathbb{R}^d) \to S(\mathbb{R}^d)$, with extension to a map from $S'(\mathbb{R}^d)$ to $S'(\mathbb{R}^d)$. According to Definition 1.3.8 we say that P is *globally regular* if, for any $f \in S(\mathbb{R}^d)$, all the solutions $u \in S'(\mathbb{R}^d)$ of the equation $Pu = f$ belong to $S(\mathbb{R}^d)$. In particular then, all the solutions $u \in S'(\mathbb{R}^d)$ of the equation $Pu = 0$ belong to $S(\mathbb{R}^d)$. An important tool for deducing global regularity, when P is a pseudo-differential operator, is given by Theorem 1.3.6, namely: the existence of a left parametrix $\tilde{P}$ of P, i.e., $\tilde{P}P = I + R$ where $R : S'(\mathbb{R}^d) \to S(\mathbb{R}^d)$, implies global regularity for P, as well as precise estimates in generalized Sobolev spaces, cf. Proposition 1.5.8. Besides, the simultaneous existence of a right parametrix gives Fredholmness, cf. Theorem 1.6.9.

The main subject of this chapter, and of Chapter 3, is the study of the above properties for relevant classes of operators, including as basic examples partial differential operators with polynomial coefficients in $\mathbb{R}^d$:

$$P = \sum_{|\alpha|+|\beta| \leq m} c_{\alpha\beta} x^\beta D^\alpha. \tag{2.0.1}$$

The key idea is to model on the structure of P, and define a suitable couple of weights Φ, Ψ satisfying the strong uncertainty principle (1.1.10), in order to obtain a pseudo-differential framework where parametrices of P can be constructed.

In this chapter attention is limited to Γ-pseudo-differential operators, corresponding to a case when Φ and Ψ coincide. A simple example is

$$\Phi(x,\xi) = \Psi(x,\xi) = (1 + |x|^2 + |\xi|^2)^{1/2}.$$

We review the corresponding pseudo-differential calculus in Section 2.1. In short, by writing $z = (x, \xi) \in \mathbb{R}^{2d}$, the symbols in $\Gamma^m(\mathbb{R}^d)$, $m \in \mathbb{R}$, are defined by the estimates, for $\gamma \in \mathbb{N}^{2d}$:

$$|\partial_z^\gamma a(z)| \lesssim \langle z \rangle^{m-|\gamma|}, \quad z \in \mathbb{R}^{2d}. \tag{2.0.2}$$

According to Definition 1.3.1, we call Γ-elliptic the symbols satisfying $|z|^m \lesssim |a(z)|$ for large z. A relevant subclass of $\Gamma^m(\mathbb{R}^d)$ is given by all the symbols admitting an asymptotic expansion $a \sim \sum_{k=0}^{\infty} a_{m-k}$, with $a_{m-k}(z)$ positively homogeneous of degree $m-k$ in $\mathbb{R}^{2d}$; for these symbols Γ-ellipticity amounts to assuming $a_m(z) \neq 0$ for every $z \neq 0$.

For $m \in \mathbb{N}$ the corresponding Sobolev spaces $H_\Gamma^m(\mathbb{R}^d)$ can be explicitly defined by setting

$$\|u\|_{H_\Gamma^m} = \sum_{|\alpha|+|\beta| \leq m} \|x^\beta D^\alpha u\|_{L^2} < \infty. \tag{2.0.3}$$

Section 2.2 and 2.3 are devoted to the study in this framework of the partial differential operators P in (2.0.1). With $z = (x, \xi)$ and $\gamma = (\beta, \alpha)$ the symbol of P, in the standard quantization, can be re-written in the form

$$p(z) = \sum_{|\gamma| \leq m} c_\gamma z^\gamma. \tag{2.0.4}$$

The Γ-ellipticity reads $p_m(z) = \sum_{|\gamma|=m} c_\gamma z^\gamma \neq 0$ for $z \neq 0$; this provides global regularity and Fredholm property of the map $P : H_\Gamma^m(\mathbb{R}^d) \to L^2(\mathbb{R}^d)$.

A motivating example is the harmonic oscillator of Quantum Mechanics:

$$H = -\Delta + |x|^2 \tag{2.0.5}$$

with elliptic symbol $|z|^2$. As appetizer to Chapter 4, devoted to Spectral Theory, we begin here to compute eigenvalues and eigenfunctions of H. We also study Γ-elliptic ordinary differential equations, by using the theory of Asymptotic Integration; we prove here some decay properties of the solutions, which we shall extend later in the book to linear and semi-linear partial differential equations, see Chapter 6.

In Section 2.4 we consider Γ-hypoelliptic symbols. Namely, we limit attention to polynomial symbols $p(z)$, as in (2.0.4), and assume

$$|\partial^\gamma p(z)| \lesssim |p(z)|\langle z \rangle^{-\rho|\gamma|} \tag{2.0.6}$$

for some ρ, $0 < \rho \leq 1$ and large z (the case $\rho = 1$ corresponds to Γ-ellipticity, defined as before). Polynomials $p(z)$ satisfying (2.0.6), called in the sequel *H-polynomials*, were first studied by Hörmander in connection with the Schwartz' problem of the interior regularity of the solutions of equations with constant coefficients. We recall in Section 2.4 some equivalent definitions and basic properties. Following Hörmander's presentation, we shall argue in arbitrary dimension

n, though in our applications $n = 2d$ is even. If (2.0.6) is satisfied, we can construct parametrices for the operator P in (2.0.1), hence in particular P is globally regular. However, with respect to the Γ-elliptic case, P is not Fredholm on the spaces $H_\Gamma^m(\mathbb{R}^d)$, cf. (2.0.3), and there is a loss of regularity in the related a priori estimates. It is then convenient to look for more appropriate weight functions, depending on the particular form of $p(z)$, so that $p(z)$ can be regarded as elliptic symbol in the new frame. This is performed in Sections 2.5, 2.6, 2.7 and 2.8 for some special subclasses of the H-polynomials.

Namely in Section 2.5 we consider the quasi-elliptic polynomials; representative examples for the corresponding class of operators are given by

$$D_x + r x^k, \quad x \in \mathbb{R}, \; k \geq 1, \; \operatorname{Im} r \neq 0, \tag{2.0.7}$$

and by the generalized harmonic oscillator

$$-\Delta + |x|^{2k}, \quad x \in \mathbb{R}^d, \; k \geq 1. \tag{2.0.8}$$

In Section 2.6 we study multi-quasi-elliptic polynomials, which contain quasi-elliptic polynomials as a particular case. Simple examples for the related class of operators are

$$D_x^6 + x^4 D_x^4 + x^6, \quad x \in \mathbb{R} \tag{2.0.9}$$

and

$$-\Delta + V(x_1, x_2), \quad (x_1, x_2) \in \mathbb{R}^2, \tag{2.0.10}$$

where the potential V is given by

$$V(x_1, x_2) = x_1^6 + x_1^4 x_2^4 + x_2^6. \tag{2.0.11}$$

In general, an H-polynomial is multi-quasi-elliptic if

$$|p(z)| \asymp \Lambda_{\mathcal{P}}(z) = \left(\sum_{\gamma \in V(\mathcal{P})} z^{2\gamma} \right)^{1/2} \tag{2.0.12}$$

for large z, the sum being extended to the set $V(\mathcal{P})$ of all the vertices γ of the Newton polyhedron $\mathcal{P}$ of $p(z)$. The corresponding pseudo-differential calculus is developed in Section 2.7. In fact $\Phi(z) = \Psi(z) = \Lambda_{\mathcal{P}}(z)^\rho$ give a couple of weights for $0 < \rho \leq 1/\mu$, where $\mu \geq 1$ is the so-called formal order of $\mathcal{P}$. The condition (2.0.12) can be read as ellipticity with respect to such a couple of weights; the Γ-ellipticity is recaptured when $\Lambda_{\mathcal{P}}(z) \asymp |z|^m$. The results of Chapter 1 apply therefore to the class $\Gamma_{\mathcal{P}}^m(\mathbb{R}^d)$, $m \in \mathbb{R}$, of all the symbols $a(z)$ satisfying

$$|\partial_z^\gamma a(z)| \lesssim \Lambda_{\mathcal{P}}(z)^{m - |\gamma|/\mu}, \quad z \in \mathbb{R}^{2d}. \tag{2.0.13}$$

Moreover, the weights $\Lambda_{\mathcal{P}}(z)$ in (2.0.12) enjoy peculiar properties, which suggest to consider also the subclass $M\Gamma_{\mathcal{P}}^m$ of all $a(z) \in \Gamma_{\mathcal{P}}^m(\mathbb{R}^d)$ satisfying

$$z^\gamma \partial_z^\gamma a(z) \in \Gamma_{\mathcal{P}}^m(\mathbb{R}^d) \tag{2.0.14}$$

for any $\gamma = (\gamma_1, \ldots, \gamma_{2d}) \in \mathbb{N}^{2d}$ with $\gamma_j \in \{0,1\}$. In particular, the symbol of the parametrix of a multi-quasi-elliptic operator belongs to this class.

This approach allows us to establish precise L^p estimates for multi-quasi-elliptic operators. Namely, in Section 2.8 we prove that every pseudo-differential operator with symbol in $M\Gamma_{\mathcal{P}}^0(\mathbb{R}^d)$ is bounded on L^p, $1 < p < \infty$. We may then consider, as generalization of (2.0.3), the Banach space $H_{\mathcal{P}}^p$ with norm

$$\|u\|_{p,\mathcal{P}} = \sum_{(\beta,\alpha)\in V(\mathcal{P})} \|x^\beta D^\alpha u\|_{L^p}, \tag{2.0.15}$$

cf. (2.0.12) with $z = (x,\xi)$, $\gamma = (\beta,\alpha)$. The operator P corresponding to the polynomial symbol $p(z)$ in (2.0.12) is then a Fredholm map $H_{\mathcal{P}}^p \to L^p$ for $1 < p < \infty$.

2.1   Γ-Pseudo-Differential Operators

The first basic examples of the general classes of symbols considered in the previous Chapter 1 are the Γ-classes, defined below. Write $z = (x,\xi) \in \mathbb{R}^{2d}$ and, according to the preceding notation,

$$\langle z \rangle = (1 + |z|^2)^{1/2} = \left(1 + |x|^2 + |\xi|^2\right)^{1/2}.$$

Definition 2.1.1. We define $\Gamma^m(\mathbb{R}^d)$, $m \in \mathbb{R}$, as the set of all functions $a(z) \in C^\infty(\mathbb{R}^{2d})$ satisfying, for all $\gamma \in \mathbb{N}^{2d}$,

$$|\partial_z^\gamma a(z)| \lesssim \langle z \rangle^{m-|\gamma|}, \quad z \in \mathbb{R}^{2d}. \tag{2.1.1}$$

As we observed before Definition 1.1.1, the weights $\Phi(z) = \Psi(z) = \langle z \rangle$ are sub-linear and temperate, cf. (1.1.1), (1.1.2). More generally we can define the following classes.

Definition 2.1.2. We define $\Gamma_\rho^m(\mathbb{R}^d)$, with $m \in \mathbb{R}$, $0 < \rho \leq 1$, as the set of all the functions $a(z) \in C^\infty(\mathbb{R}^{2d})$ satisfying, for all $\gamma \in \mathbb{N}^{2d}$,

$$|\partial_z^\gamma a(z)| \lesssim \langle z \rangle^{m-\rho|\gamma|}, \quad z \in \mathbb{R}^{2d}. \tag{2.1.2}$$

In the above definition we assumed $\rho > 0$, so the strong uncertainty principle (1.1.10) is satisfied. Note that $\Gamma_1^m(\mathbb{R}^d) = \Gamma^m(\mathbb{R}^d)$. Also, it is useful to introduce the subspace of $\Gamma^m(\mathbb{R}^d)$ of the polyhomogeneous symbols. Let us write $\mathcal{H}^m(\mathbb{R}^{2d} \setminus \{0\})$, $m \in \mathbb{R}$, for the class of functions $a(z) \in C^\infty(\mathbb{R}^{2d} \setminus \{0\})$ which are positively homogeneous of degree m in $\mathbb{R}^{2d}$, i.e., $a(tz) = t^m a(z)$ for $t > 0$, $z \in \mathbb{R}^{2d}$, $z \neq 0$.

Definition 2.1.3. We define $\Gamma_{cl}^m(\mathbb{R}^d)$ as the subset of $\Gamma^m(\mathbb{R}^d)$ of all the symbols $a(z)$ which admit asymptotic expansion

$$a(z) \sim \sum_{k=0}^\infty a_{m-k}(z) \tag{2.1.3}$$

for a sequence of functions $a_{m-k} \in \mathcal{H}^{m-k}(\mathbb{R}^{2d} \setminus \{0\})$, $k = 0, 1, \ldots$.

By (2.1.3) we mean that, considering $\chi(z) \in C^{\infty}(\mathbb{R}^{2d})$, with $\chi(z) = 1$ for $|z| \geq 2$, $\chi(z) = 0$ for $|z| \leq 1$, we have for every integer $N \geq 1$:

$$a(z) - \sum_{0 \leq k < N} \chi(z) a_{m-k}(z) \in \Gamma^{m-N}(\mathbb{R}^d). \tag{2.1.4}$$

In turn, given a sequence $\{a_{m-k}\}$ as in Definition 2.1.3, we may construct $a \in \Gamma_{cl}^m(\mathbb{R}^d)$ satisfying (2.1.4), uniquely determined modulo symbols in $\mathcal{S}(\mathbb{R}^{2d})$. The function $a_m \in \mathcal{H}^m(\mathbb{R}^{2d} \setminus \{0\})$ is called the principal symbol, or principal part of $a \in \Gamma_{cl}^m(\mathbb{R}^d)$.

Let us now pass to consideration of the corresponding classes of pseudo-differential operators, which we shall denote by $OP\Gamma^m(\mathbb{R}^d)$, $OP\Gamma_{\rho}^m(\mathbb{R}^d)$, $OP\Gamma_{cl}^m(\mathbb{R}^d)$ respectively. The rules of symbolic calculus in Section 1.2 apply obviously, in particular the definition of the classes does not depend on quantization, cf. Remark 1.2.6. Let us just recall the composition formula, limiting attention to the standard quantization (that we shall tacitly adopt in the following of the section):

$$a(x, D)u(x) = \int e^{ix\xi} a(x, \xi) \widehat{u}(\xi) \, d\xi, \tag{2.1.5}$$

with action $a(x, D) : \mathcal{S}(\mathbb{R}^d) \to \mathcal{S}(\mathbb{R}^d)$, $\mathcal{S}'(\mathbb{R}^d) \to \mathcal{S}'(\mathbb{R}^d)$. If $a \in \Gamma_{\rho}^m(\mathbb{R}^d)$, $b \in \Gamma_{\rho}^n(\mathbb{R}^d)$, $0 < \rho \leq 1$, then the operator $c(x, D) = a(x, D)b(x, D)$ belongs to $OP\Gamma_{\rho}^{m+n}(\mathbb{R}^d)$ with symbol

$$c(x, \xi) \sim \sum_{\alpha} (\alpha!)^{-1} \partial_{\xi}^{\alpha} a(x, \xi) D_x^{\alpha} b(x, \xi). \tag{2.1.6}$$

We leave to the reader to apply the other results of Section 1.2 for general quantizations, concerning symbols of compositions, transposed operators and formal adjoints. It is worth observing that the polyhomogeneous classes $\Gamma_{cl}^m(\mathbb{R}^d)$ are stable under such operations; so for example, taking $a \in \Gamma_{cl}^m(\mathbb{R}^d)$ with $a \sim \sum_{k=0}^{\infty} a_{m-k}$ and $b \in \Gamma_{cl}^n(\mathbb{R}^d)$ with $b \sim \sum_{j=0}^{\infty} b_{n-j}$, we get $c(x, D) = a(x, D)b(x, D)$ where $c \in \Gamma_{cl}^{m+n}(\mathbb{R}^d)$, $c \sim \sum_{s=0}^{\infty} c_{m+n-s}$, and

$$c_{m+n-s} = \sum_{k+j+2|\alpha|=s} (\alpha!)^{-1} \partial_{\xi}^{\alpha} a_{m-k} D_x^{\alpha} b_{n-j} \tag{2.1.7}$$

belonging to $\mathcal{H}^{m+n-s}(\mathbb{R}^{2d} \setminus \{0\})$. In particular:

$$c_{m+n} = a_m b_n \in \mathcal{H}^{m+n}(\mathbb{R}^{2d} \setminus \{0\}). \tag{2.1.8}$$

Let us now apply the results of Section 1.3.1, treating first the Γ-elliptic symbol. For simplicity, we shall refer only to the $\Gamma^m(\mathbb{R}^d)$ classes; the results can be obviously generalized to $\Gamma_{\rho}^m(\mathbb{R}^d)$.

Definition 2.1.4. Let $a \in \Gamma^m(\mathbb{R}^d)$, $m \in \mathbb{R}$. We say that a is Γ-elliptic if there exists $R > 0$ such that

$$|z|^m \lesssim |a(z)| \quad \text{for } |z| \geq R. \tag{2.1.9}$$

When applying this definition in $\Gamma^m_{cl}(\mathbb{R}^d) \subset \Gamma^m(\mathbb{R}^d)$, we shall rather rely on the following equivalent notion of Γ-ellipticity.

Proposition 2.1.5. *The symbol* $a \in \Gamma^m_{cl}(\mathbb{R}^d)$ *is Γ-elliptic if and only if the principal part a_m satisfies*

$$a_m(z) \neq 0 \quad \text{for every } z \neq 0. \tag{2.1.10}$$

Proof. Because of the homogeneity of a_m, the assumption (2.1.10) is equivalent to the estimate

$$|z|^m \lesssim |a_m(z)| \quad \text{for } z \neq 0. \tag{2.1.11}$$

On the other hand, from (2.1.4) with $N = 1$ we have, for any fixed $R > 0$,

$$|a(z) - a_m(z)| \lesssim |z|^{m-1} \quad \text{for } |z| \geq R.$$

Then, if (2.1.9) is satisfied, we deduce for suitable positive constants C and ϵ:

$$\epsilon |z|^m \leq |a(z)| \leq |a_m(z)| + C |z|^{m-1} \quad \text{for } |z| \geq R.$$

Since for sufficiently large $|z|$ we have

$$C |z|^{m-1} \leq \frac{\epsilon}{2} |z|^m ,$$

we conclude $\frac{\epsilon}{2} |z|^m \leq |a_m(z)|$, hence (2.1.11). Symmetrically we may prove that (2.1.11) implies (2.1.9). $\square$

From Theorem 1.3.6 we have the following result on the existence of parametrices. Let us observe that Γ-ellipticity in (2.1.9), (2.1.10) is invariant under different quantizations.

Theorem 2.1.6. *Let* $a \in \Gamma^m(\mathbb{R}^d)$ *be Γ-elliptic. Then there exists* $b \in \Gamma^{-m}(\mathbb{R}^d)$ *such that* $b(x, D)$ *is a parametrix of* $a(x, D)$, *i.e.,*

$$a(x, D)b(x, D) = I + S_1, \quad b(x, D)a(x, D) = I + S_2$$

where S_1 and S_2 are regularizing operators. Hence $a(x,D)$ is globally regular, i.e., $u \in \mathcal{S}'(\mathbb{R}^d)$ and $a(x, D)u \in \mathcal{S}(\mathbb{R}^d)$ imply $u \in \mathcal{S}(\mathbb{R}^d)$.

The result remains valid for symbols $a \in \Gamma^m_{cl}(\mathbb{R}^d)$ satisfying (2.1.10), the symbol of the parametrix belonging then to $\Gamma^{-m}_{cl}(\mathbb{R}^d)$, with principal part $1/a_m$.

We may now apply the results of Section 1.7, concerning the Anti-Wick quantization, to symbols in the Γ-classes. In particular we shall use the following version of Proposition 1.7.12.

Theorem 2.1.7. *Let $a \in \Gamma^s(\mathbb{R}^d)$ be Γ-elliptic, $s \in \mathbb{R}$, with $a(z) > 0$ for every $z \in \mathbb{R}^{2d}$. Denote by A_a the operator with Anti-Wick symbol a. The map $A_a : \mathcal{S}(\mathbb{R}^d) \to \mathcal{S}(\mathbb{R}^d)$ is an isomorphism, extending to an isomorphism on $\mathcal{S}'(\mathbb{R}^d)$. Moreover, the inverse A_a^{-1} belongs to $\mathrm{OP}\Gamma^{-m}(\mathbb{R}^d)$.*

Take for example, for $s \in \mathbb{R}$,

$$a_s(z) = \langle z \rangle^s = (1 + |x|^2 + |\xi|^2)^{s/2}, \tag{2.1.12}$$

satisfying the assumptions of Theorem 2.1.7. We may refer to such symbols in the following definition of Γ-Sobolev spaces.

Definition 2.1.8. Let $a_s(z)$, $s \in \mathbb{R}$, be the symbol in (2.1.12) and write W_s for the operator with Anti-Wick symbol a_s. Then we set

$$H_\Gamma^s(\mathbb{R}^d) = W_s^{-1}\left(L^2(\mathbb{R}^d)\right) = \{u \in \mathcal{S}'(\mathbb{R}^d) : W_s u \in L^2(\mathbb{R}^d)\}. \tag{2.1.13}$$

All the properties in Sections 1.5 and 1.6 apply. In particular H_Γ^s are Hilbert spaces with respect to the scalar product

$$(u, v)_{H_\Gamma^s} = (W_s u, W_s v)_{L^2} \tag{2.1.14}$$

and the corresponding norm

$$\|u\|_{H_\Gamma^s} = \|W_s u\|_{L^2}. \tag{2.1.15}$$

In Definition 2.1.8 we may of course replace $a_s(z)$ with any other strictly positive Γ-elliptic symbol, with equivalence of norms. More generally we can define the $H_\Gamma^s(\mathbb{R}^d)$ spaces using arbitrary Γ-elliptic symbols, as stated in the following proposition, cf. Definition 1.5.2.

Proposition 2.1.9. *Let $T \in \mathrm{OP}\Gamma^s(\mathbb{R}^d)$ have Γ-elliptic symbol, cf. Definition 2.1.4. Then*

$$H_\Gamma^s(\mathbb{R}^d) = \{u \in \mathcal{S}'(\mathbb{R}^d) : Tu \in L^2(\mathbb{R}^d)\}$$

and on $H_\Gamma^s(\mathbb{R}^d)$ an equivalent Hilbert space structure is given by the scalar product

$$(u, v)_{H_\Gamma^s} = (Tu, Tv)_{L^2} + (Ru, Rv)_{L^2} \tag{2.1.16}$$

where R is a regularizing operator associated to a parametrix $\tilde{T} \in \mathrm{OP}\Gamma^{-s}(\mathbb{R}^d)$ of T, namely $\tilde{T}T = I + R$.

From Proposition 1.5.5, we have:

Theorem 2.1.10. *Every $A \in \mathrm{OP}\Gamma^m(\mathbb{R}^d)$ defines, for all $s \in \mathbb{R}$, a continuous operator*

$$A : H_\Gamma^s(\mathbb{R}^d) \to H_\Gamma^{s-m}(\mathbb{R}^d). \tag{2.1.17}$$

Later on in the book we shall also use the following result of boundedness in the standard Sobolev spaces $H^s(\mathbb{R}^d)$.

Theorem 2.1.11. *Every $A \in \mathrm{OP\Gamma}^0(\mathbb{R}^d)$ defines, for all $s \in \mathbb{R}$, a continuous operator*

$$A : H^s(\mathbb{R}^d) \to H^s(\mathbb{R}^d). \tag{2.1.18}$$

Proof. Consider the pseudo-differential operator $\langle D \rangle^t$ with standard symbol $\langle \xi \rangle^t$, $t \in \mathbb{R}$. We know that $u \in H^s(\mathbb{R}^d)$ if and only if $\langle D \rangle^s u \in L^2(\mathbb{R}^d)$, cf. Section 0.2. Hence, by writing

$$\langle D \rangle^s A u = \langle D \rangle^s A \langle D \rangle^{-s} \langle D \rangle^s u,$$

we have to prove the boundedness of the map

$$\langle D \rangle^s A \langle D \rangle^{-s} : L^2(\mathbb{R}^d) \to L^2(\mathbb{R}^d). \tag{2.1.19}$$

The operator in (2.1.19) does not belong to $\mathrm{OP\Gamma}^0(\mathbb{R}^d)$, however we may regard $\langle D \rangle^s$, A and $\langle D \rangle^{-s}$ as operators with symbols in the classes $S(\langle \xi \rangle^s; 1, 1)$, $S(1; 1, 1)$ and $S(\langle \xi \rangle^{-s}; 1, 1)$ respectively. Hence $\langle D \rangle^s A \langle D \rangle^{-s} \in \mathrm{OP}S(1; 1, 1)$ by Theorem 1.2.16 and (2.1.19) is granted, cf. Theorem 1.7.14.   □

Other properties of the $H_\Gamma^s(\mathbb{R}^d)$ spaces, following from Section 1.5, are compact immersions $j : H_\Gamma^s(\mathbb{R}^d) \to H_\Gamma^t(\mathbb{R}^d)$ for $s > t$, and compactness of the maps $A : H_\Gamma^s(\mathbb{R}^d) \to H_\Gamma^t(\mathbb{R}^d)$ for $A \in \mathrm{OP\Gamma}^m(\mathbb{R}^d)$ whenever $s - t > m$. In particular if A is regularizing, then A belongs to $\mathrm{OP\Gamma}^{-\infty}(\mathbb{R}^d) := \bigcap_m \mathrm{OP\Gamma}^m(\mathbb{R}^d)$, hence it is continuous and compact from $H_\Gamma^s(\mathbb{R}^d)$ to $H_\Gamma^t(\mathbb{R}^d)$ for any $s, t \in \mathbb{R}$. Moreover, for every $s \in \mathbb{R}$ we have continuous immersions

$$j : \mathcal{S}(\mathbb{R}^d) \to H_\Gamma^s(\mathbb{R}^d), \qquad j : H_\Gamma^s(\mathbb{R}^d) \to \mathcal{S}'(\mathbb{R}^d)$$

and moreover

$$\bigcap_{s \in \mathbb{R}} H_\Gamma^s(\mathbb{R}^d) = \mathcal{S}(\mathbb{R}^d), \qquad \bigcup_{s \in \mathbb{R}} H_\Gamma^s(\mathbb{R}^d) = \mathcal{S}'(\mathbb{R}^d).$$

Finally, the spaces $H_\Gamma^s(\mathbb{R}^d)$, $H_\Gamma^{-s}(\mathbb{R}^d)$ are dual to each other with respect to the bilinear pairing $\langle u, v \rangle = \int u(x) v(x) \, dx$.

The following equivalent definition when $m \in \mathbb{N}$ is peculiar for the spaces $H_\Gamma^s(\mathbb{R}^d)$.

Theorem 2.1.12. *Let $m \in \mathbb{N}$. An equivalent definition of the space $H_\Gamma^m(\mathbb{R}^d)$ is given by*

$$H_\Gamma^m(\mathbb{R}^d) = \{ u \in \mathcal{S}'(\mathbb{R}^d) : x^\beta D^\alpha u \in L^2(\mathbb{R}^d) \text{ for } |\alpha| + |\beta| \le m \} \tag{2.1.20}$$

with equivalent norm

$$\|u\|_{H_\Gamma^m}^* = \sum_{|\alpha| + |\beta| \le m} \|x^\beta D^\alpha u\|_{L^2}. \tag{2.1.21}$$

Proof. Assume first $u \in H^m_\Gamma(\mathbb{R}^d)$, according to the Definition 2.1.8. Since $x^\beta D^\alpha \in$ $\mathrm{OP\Gamma}^m(\mathbb{R}^d)$ if $|\alpha| + |\beta| \leq m$, then $x^\beta D^\alpha u \in H^0_\Gamma(\mathbb{R}^d) = L^2(\mathbb{R}^d)$ in view of Theorem 2.1.10, and $\|u\|^*_{H^m_\Gamma}$ can be estimated in terms of $\|u\|_{H^m_\Gamma}$.

In the opposite direction, it will be convenient to estimate the norm coming from (2.1.16) in terms of $\|u\|^*_{H^m_\Gamma}$. Namely, we shall fix an operator $T \in \mathrm{OP\Gamma}^m(\mathbb{R}^d)$ with Γ-elliptic symbol, see below, and consider

$$\|u\|^2_{H^m_\Gamma} = \|Tu\|^2_{L^2} + \|Ru\|^2_{L^2}, \tag{2.1.22}$$

where R is a regularizing operator associated to a left parametrix $\tilde{T} \in \mathrm{OP\Gamma}^{-m}(\mathbb{R}^d)$ of T. We may then estimate, for some $C > 0$,

$$\|u\|_{H^m_\Gamma} \leq \|Tu\|_{L^2} + C\|u\|_{L^2}. \tag{2.1.23}$$

Choose T as follows:

$$T = A_{-m}\, P \tag{2.1.24}$$

where A_{-m} is any operator in $\mathrm{OP\Gamma}^{-m}_{\mathrm{cl}}(\mathbb{R}^d)$ with elliptic symbol and

$$P = \sum_{|\alpha|+|\beta|\leq m} x^\beta D^\alpha x^\beta D^\alpha. \tag{2.1.25}$$

In fact, using Leibniz' rule we have that the principal symbol of $P \in \mathrm{OP\Gamma}^{2m}_{\mathrm{cl}}(\mathbb{R}^d)$ is given by

$$p_{2m}(x,\xi) = \sum_{|\alpha|+|\beta|=2m} x^{2\beta}\xi^{2\alpha},$$

which satisfies the Γ-ellipticity condition in Proposition 2.1.5. Hence T in (2.1.24) belongs to $\mathrm{OP\Gamma}^m_{\mathrm{cl}}(\mathbb{R}^d)$ with elliptic symbol, cf. (2.1.10). Combining (2.1.23) with (2.1.24), (2.1.25), and observing that $A_{-m}x^\beta D^\alpha \in \mathrm{OP\Gamma}^0(\mathbb{R}^d)$, we obtain

$$\begin{aligned}
\|u\|_{H^m_\Gamma} &\leq \sum_{|\alpha|+|\beta|\leq m} \|A_{-m}x^\beta D^\alpha(x^\beta D^\alpha u)\|_{L^2} + C\|u\|_{L^2} \\
&\leq C'\|u\|^*_{H^m_\Gamma}
\end{aligned}$$

for a new constant C'. The theorem is therefore proved. $\qquad\square$

Referring to the Sobolev-type spaces $H^s_\Gamma(\mathbb{R}^d)$, we may now give a more precise frame to Theorem 2.1.6. Namely, particularizing the results of Sections 1.5, 1.6, we have the following two theorems.

Theorem 2.1.13. *Consider $A \in \mathrm{OP\Gamma}^m(\mathbb{R}^d)$ with Γ-elliptic symbol and assume $u \in \mathcal{S}'(\mathbb{R}^d)$, $Au \in H^s_\Gamma(\mathbb{R}^d)$. Then $u \in H^{s+m}_\Gamma(\mathbb{R}^d)$ and, for every $t < s + m$,*

$$\|u\|_{H^{s+m}_\Gamma} \leq C \left(\|Au\|_{H^s_\Gamma} + \|u\|_{H^t_\Gamma} \right) \tag{2.1.26}$$

for a positive constant C depending on s and t.

In particular if m is a positive integer, we may refer to the equivalent norm (2.1.21) and re-write (2.1.26) for $s = 0$, $t = 0$:

$$\sum_{|\alpha|+|\beta|\leq m} \|x^\beta D^\alpha u\|_{L^2} \leq C\left(\|Au\|_{L^2} + \|u\|_{L^2}\right). \tag{2.1.27}$$

To be definite, in the following statement we shall denote by A_s the restriction of $A : \mathcal{S}'(\mathbb{R}^d) \to \mathcal{S}'(\mathbb{R}^d)$ to $H^s_\Gamma(\mathbb{R}^d)$, $s \in \mathbb{R}$, or equivalently the extension of $A : \mathcal{S}(\mathbb{R}^d) \to \mathcal{S}(\mathbb{R}^d)$ to $H^s_\Gamma(\mathbb{R}^d)$.

Theorem 2.1.14. *Consider $A \in \mathrm{OP\Gamma}^m(\mathbb{R}^d)$ with Γ-elliptic symbol. Then:*

(i) $A_s \in \mathrm{Fred}(H^s_\Gamma(\mathbb{R}^d), H^{s-m}_\Gamma(\mathbb{R}^d))$.

(ii) *$\mathrm{ind}\, A_s = \dim \mathrm{Ker}\, A - \dim \mathrm{Ker}\, A^*$, $\mathrm{ind}\, A_s = \dim \mathrm{Ker}\, A - \dim \mathrm{Ker}\, {}^t A$, where A^* is the formal adjoint and ${}^t A$ is the transposed operator. Observe that the index is then independent of s.*

(iii) *If $T \in \mathrm{OP\Gamma}^{m'}(\mathbb{R}^d)$ with $m' < m$, then $A_s + T_s \in \mathrm{Fred}\,(H^s_\Gamma(\mathbb{R}^d), H^{s-m}_\Gamma(\mathbb{R}^d))$ and $\mathrm{ind}\,(A_s + T_s) = \mathrm{ind}\, A_s$.*

(iv) *Suppose $A_s : H^s_\Gamma(\mathbb{R}^d) \to H^s_\Gamma(\mathbb{R}^d)$ is invertible for some $s \in \mathbb{R}$; then it is invertible for all $s \in \mathbb{R}$, and the inverse is an operator in $\mathrm{OP\Gamma}^{-m}(\mathbb{R}^d)$.*

In conclusion, we pass to consider the hypoelliptic case, the natural frame being now the classes $\Gamma^m_\rho(\mathbb{R}^d)$, $0 < \rho \leq 1$.

Definition 2.1.15. We say that $a \in \Gamma^m_\rho(\mathbb{R}^d)$, $0 < \rho \leq 1$, $m \in \mathbb{R}$, is Γ_ρ-hypoelliptic if there exist $m' \in \mathbb{R}$, $m' \leq m$, and $R > 0$ such that

$$|z|^{m'} \lesssim |a(z)|, \quad \text{for } |z| \geq R \tag{2.1.28}$$

and, for every $\gamma \in \mathbb{N}^{2d}$,

$$|\partial^\gamma_z a(z)| \lesssim |a(z)| \langle z \rangle^{-\rho|\gamma|}, \quad \text{for } |z| \geq R. \tag{2.1.29}$$

Note that the assumption of Γ-ellipticity (2.1.9) for a symbol $a \in \Gamma^m_\rho(\mathbb{R}^d)$ corresponds to taking $m = m'$ in (2.1.28) and gives automatically (2.1.29). So the Γ-ellipticity implies Γ_ρ-hypoellipticity. The new interesting case is then $m' < m$, for which we have the following counterpart of Theorems 2.1.6 and 2.1.13.

Theorem 2.1.16. *Let $a \in \Gamma^m_\rho(\mathbb{R}^d)$, $0 < \rho \leq 1$, be Γ_ρ-hypoelliptic, for some $m' < m$ in (2.1.28). Then there exists $b \in \Gamma^{-m'}_\rho(\mathbb{R}^d)$ such that, denoted $A = a(x, D)$, $B = b(x, D)$, we have*

$$AB = I + S_1, \qquad BA = I + S_2,$$

with S_1 and S_2 regularizing operators. Hence A is globally regular. Moreover, if we assume $u \in \mathcal{S}'(\mathbb{R}^d)$, $Au \in H^s_\Gamma(\mathbb{R}^d)$, then we have $u \in H^{s+m'}_\Gamma(\mathbb{R}^d)$ and, for every $t < s + m'$,

$$\|u\|_{H^{s+m'}_\Gamma} \leq C\left(\|Au\|_{H^s_\Gamma} + \|u\|_{H^t_\Gamma}\right) \tag{2.1.30}$$

for a positive constant C depending on s and t.

The estimate (2.1.30) presents a loss of $m - m'$ in the order of the Γ-Sobolev spaces, with respect to the Γ-elliptic case. So, if $m' < m$ we cannot regard A as a Fredholm map from $H_\Gamma^s(\mathbb{R}^d)$ to $H_\Gamma^{s-m}(\mathbb{R}^d)$.

2.2 Γ-Elliptic Differential Operators; the Harmonic Oscillator

All *differential* operators in $\mathrm{OP}\Gamma_\rho^m(\mathbb{R}^d)$, $0 < \rho \leq 1$, have polynomial coefficients. Namely:

Proposition 2.2.1. *Assume* $p(x, \xi) \in \Gamma_\rho^m(\mathbb{R}^d)$ *is of the form*

$$p(x, \xi) = \sum_{|\alpha| \leq m} a_\alpha(x) \xi^\alpha$$

for some $a_\alpha(x) \in C^\infty(\mathbb{R}^d)$. *Then* $a_\alpha(x)$ *is a polynomial.*

Proof. From (2.1.2) we have, for all $\beta \in \mathbb{N}^d$,

$$|\partial_x^\beta p(x, \xi)| \lesssim \langle z \rangle^{m - \rho|\beta|}, \quad z = (x, \xi) \in \mathbb{R}^{2d}.$$

Hence, taking $|\beta| > m/\rho$, we have, for $\epsilon = \rho|\beta| - m$,

$$\left| \sum_{|\alpha| \leq m} \partial_x^\beta a_\alpha(x) \xi^\alpha \right| \lesssim \langle z \rangle^{-\epsilon},$$

which can be satisfied only if $\partial_x^\beta a_\alpha(x) = 0$ for all $x \in \mathbb{R}^d$, i.e., a_α is a polynomial. $\square$

Let us then consider

$$P = \sum_{|\alpha| + |\beta| \leq m} c_{\alpha\beta} x^\beta D^\alpha, \qquad c_{\alpha\beta} \in \mathbb{C}, \tag{2.2.1}$$

with symbol, according to the standard quantization:

$$p(x, \xi) = \sum_{|\alpha| + |\beta| \leq m} c_{\alpha\beta} x^\beta \xi^\alpha. \tag{2.2.2}$$

Setting $z = (x, \xi)$, $\gamma = (\beta, \alpha)$ we shall also write for short

$$p(z) = \sum_{|\gamma| \leq m} c_\gamma z^\gamma. \tag{2.2.3}$$

We have $p \in \Gamma_{\mathrm{cl}}^m(\mathbb{R}^d)$ and the Γ-ellipticity condition (2.1.9) reads

$$|z|^m \lesssim \left| \sum_{|\gamma| \leq m} c_\gamma z^\gamma \right|, \quad \text{for } |z| \geq R \tag{2.2.4}$$

with $R > 0$ sufficiently large. The principal part is given by

$$p_m(z) = \sum_{|\gamma|=m} c_\gamma z^\gamma, \tag{2.2.5}$$

and the equivalent Γ-ellipticity condition (2.1.10) is

$$\sum_{|\gamma|=m} c_\gamma z^\gamma \neq 0 \quad \text{for } z \neq 0. \tag{2.2.6}$$

Under the assumption (2.2.6) we may apply Theorems 2.1.6, 2.1.13 and 2.1.14.

In the following, we discuss some relevant examples of Γ-elliptic differential operators. Consider first the one-dimensional case. Define for $x \in \mathbb{R}$ and $D = -id/dx$ as standard:

$$L = D_x - rx, \quad r \in \mathbb{C}. \tag{2.2.7}$$

The principal symbol is given by $\xi + rx$ and (2.2.6) is satisfied if and only if $\operatorname{Im} r \neq 0$. The transposed operator is given by

$${}^t L = -D_x - rx. \tag{2.2.8}$$

The classical solutions of $Lu = 0$ are the functions

$$u(t) = C \exp[irx^2/2], \quad C \in \mathbb{C}, \tag{2.2.9}$$

which for $C \neq 0$ belong to $\mathcal{S}(\mathbb{R})$, or $\mathcal{S}'(\mathbb{R})$, if and only if $\operatorname{Im} r > 0$. We may regard L as a Fredholm operator, setting for example

$$L : H_\Gamma^1(\mathbb{R}) \to L^2(\mathbb{R}), \tag{2.2.10}$$

with $\operatorname{ind} L = \dim \operatorname{Ker} L - \dim \operatorname{Ker} {}^t L$ given by

$$\operatorname{ind} L = \begin{cases} 1 & \text{for } \operatorname{Im} r > 0, \\ -1 & \text{for } \operatorname{Im} r < 0. \end{cases} \tag{2.2.11}$$

We may now consider the case of a generic ordinary differential operator with polynomial coefficients. The Γ-ellipticity condition (2.2.6) on the principal symbol reads

$$p_m(x, \xi) = \sum_{\alpha+\beta=m} c_{\alpha\beta} x^\beta \xi^\alpha \neq 0 \quad \text{for } (x, \xi) \in \mathbb{R}^2, \ (x, \xi) \neq (0, 0). \tag{2.2.12}$$

Factorizing we then obtain

$$p_m(x, \xi) = c(\xi - r_1 x) \cdots (\xi - r_m x) \tag{2.2.13}$$

with $\operatorname{Im} r_j \neq 0$, $j = 1, \ldots, m$, $c \neq 0$. Hence we may write after multiplication by c^{-1}:

$$P = (D_x - r_1 x) \cdots (D_x - r_m x) + \sum_{\alpha+\beta<m} a_{\alpha\beta} x^\beta D^\alpha \tag{2.2.14}$$

for some constants $a_{\alpha\beta} \in \mathbb{C}$. We may regard P as a Fredholm operator

$$P : H_\Gamma^m(\mathbb{R}) \to L^2(\mathbb{R}). \tag{2.2.15}$$

The index does not depend on the lower order terms in the sum in (2.2.14), in view of Theorem 2.1.14, (iii), and it is then given by the sum of the indices of the factors $D_x - r_j x$, $j = 1, \ldots, m$, cf. Corollary 1.6.5, which we may compute by (2.2.11). We therefore obtain the following result.

Theorem 2.2.2. *Consider P in (2.2.14), (2.2.15) and assume* $\operatorname{Im} r_j > 0$ *for* $j = 1, \ldots, m^+$, $\operatorname{Im} r_j < 0$ *for* $j = m^+ + 1, \ldots, m$, $m = m^+ + m^-$ *(we do not exclude the case $m^+ = 0$ or $m^- = 0$). Then P is a Fredholm operator with*

$$\operatorname{ind} P = m^+ - m^-. \tag{2.2.16}$$

Concerning the existence of a non-trivial solution $u \in \mathcal{S}'(\mathbb{R})$, hence $u \in \mathcal{S}(\mathbb{R})$, of $Pu = 0$, we may obtain the following conclusions. If $m^+ - m^- > 0$, then $\operatorname{ind} P > 0$, therefore $\dim \operatorname{Ker} P > 0$ and a non-trivial solution exists. Moreover, since $\dim \operatorname{Ker} P$ and $\dim \operatorname{Ker}{}^t P$ cannot exceed m, if $m^+ = 0$, $m^- = m$, then $\operatorname{ind} P = -m$ and a non-trivial solution does not exist. See the next section for further investigation.

As an example in dimension $d \geq 1$, we then fix attention on the harmonic oscillator of Quantum Mechanics,

$$H = -\Delta + |x|^2, \quad x \in \mathbb{R}^d, \tag{2.2.17}$$

which will play an important role in the sequel of the book. Consider for $\lambda \in \mathbb{C}$:

$$P = H - \lambda = -\Delta + |x|^2 - \lambda, \quad P : H_\Gamma^2(\mathbb{R}^d) \to L^2(\mathbb{R}^d). \tag{2.2.18}$$

The principal symbol $|z|^2 = |x|^2 + |\xi|^2$ satisfies obviously the Γ-ellipticity condition, and P is a Fredholm operator with $\operatorname{ind} P = \operatorname{ind} H = 0$ because ${}^t H = H$. Non-trivial solutions $u \in \mathcal{S}'(\mathbb{R}^d)$ of $Pu = Hu - \lambda u = 0$ belong to $\mathcal{S}(\mathbb{R}^d)$ and exist or not, according to the values of λ. Namely, let us return to the one-dimensional case and consider first the equation

$$Hu - \lambda u = -u'' + x^2 u - \lambda u = 0, \quad x \in \mathbb{R}. \tag{2.2.19}$$

It is convenient to introduce

$$\Psi_+ = x - \frac{d}{dx}, \quad \Psi_- = x + \frac{d}{dx} \tag{2.2.20}$$

with respective Γ-elliptic symbols

$$x - i\xi, \quad x + i\xi.$$

The operators $\Psi_\pm$ are the celebrated creation (Ψ_+) and annihilation (Ψ_-) operators of Quantum Mechanics. They are of type (2.2.7), and the preceding arguments apply obviously. Since

$$\Psi_+\Psi_- = H - 1,$$

then for $\lambda = 1$ the equation (2.2.19) admits the solution

$$u_0(x) = e^{-x^2/2}. \tag{2.2.21}$$

We may then prove that for

$$\lambda = 2n + 1, \quad n = 0, 1, \ldots, \tag{2.2.22}$$

a solution is given by

$$u_n(x) = \Psi_+^n u_0(x) \in S(\mathbb{R}). \tag{2.2.23}$$

In fact, arguing by induction on n, we set $u_n = \Psi_+ u_{n-1}$ and compute

$$Hu_n = (\Psi_+\Psi_- + 1)u_n = (\Psi_+\Psi_- + 1)\Psi_+ u_{n-1}$$
$$= \Psi_+(\Psi_-\Psi_+ - 1)u_{n-1} + 2\Psi_+ u_{n-1}.$$

Since $\Psi_-\Psi_+ = H + 1$, and $Hu_{n-1} = (2n-1)u_{n-1}$ by the inductive hypothesis, we obtain

$$\begin{aligned} Hu_n &= \Psi_+ Hu_{n-1} + 2\Psi_+ u_{n-1} \\ &= \Psi_+(2n-1)u_{n-1} + 2\Psi_+ u_{n-1} = (2n+1)\Psi_+ u_{n-1}. \end{aligned}$$

Also, let us compute

$$\|u_n\|_{L^2}^2 = (u_n, u_n)_{L^2} = (\Psi_+^n u_0, \Psi_+^n u_0)_{L^2}.$$

Since the formal adjoint of Ψ_+ is Ψ_-, we have

$$\|u_n\|_{L^2}^2 = (\Psi_-\Psi_+^n u_0, \Psi_+^{n-1} u_0)_{L^2} = ((H+1)\Psi_+^{n-1} u_0, \Psi_+^{n-1} u_0)_{L^2}$$
$$= 2n(\Psi_+^{n-1} u_0, \Psi_+^{n-1} u_0)_{L^2} = 2n\|u_{n-1}\|_{L^2}^2.$$

Hence

$$\|u_n\|_{L^2}^2 = 2^n n! \|u_0\|_{L^2}^2 = 2^n n! \sqrt{\pi}. \tag{2.2.24}$$

Since $\Psi_- u_0 = 0$, a similar computation also shows that

$$(u_n, u_m)_{L^2} = 0 \quad \text{for } n \neq m, \tag{2.2.25}$$

that yields the orthogonality of the system $\{u_n\}$, $n \in \mathbb{N}$.

Let us now define the n-th order *Hermite polynomial* by

$$P_n(x) = c_n e^{x^2/2}\left(x - \frac{d}{dx}\right)^n e^{-x^2/2}, \tag{2.2.26}$$

$$c_n = 2^{-n/2}(n!)^{-1/2}\pi^{-1/4}.$$

We have therefore proved that the *Hermite functions*, still written by u_n by abuse,

$$u_n(x) = P_n(x)e^{-x^2/2}, \quad n = 0, 1, \ldots, \tag{2.2.27}$$

give an orthonormal system of $L^2(\mathbb{R})$. To prove completeness, it will be sufficient to argue on $g \in \mathcal{S}(\mathbb{R})$ and assume

$$\int_{-\infty}^{+\infty} g(x)u_n(x)dx = 0 \quad \text{for all } n \in \mathbb{N}. \tag{2.2.28}$$

Since we may write x^n as a linear combination of $P_j(x)$, $0 \le j \le n$, then (2.2.28) implies

$$\int_{-\infty}^{+\infty} g(x)x^n e^{-x^2/2}dx = 0 \quad \text{for all } n \in \mathbb{N}.$$

Compute now the Fourier transform

$$\mathcal{F}_{x\to\xi}\left(g(x)e^{-x^2/2}\right) = \int_{-\infty}^{+\infty} \sum_{n=0}^{\infty} \frac{(-ix\xi)^n}{n!} g(x)e^{-x^2/2}\,dx$$

$$= \lim_{N\to\infty} \sum_{n\le N} \frac{(-i\xi)^n}{n!} \int_{-\infty}^{+\infty} g(x)x^n e^{-x^2/2}\,dx = 0.$$

A function in $\mathcal{S}(\mathbb{R})$ which has zero Fourier transform must be the zero function, hence $g(x)e^{-x^2/2} = 0$ for all $x \in \mathbb{R}$, and therefore $g(x) \equiv 0$.

Let us pass to consider the higher dimensional case. It will be sufficient to search for solutions of $Hu - \lambda u = 0$ of the form

$$u(x) = u_k(x) = \Pi_{j=1}^{d} P_{k_j}(x_j)e^{-\frac{|x|^2}{2}},$$

where $k = (k_1, \ldots, k_d) \in \mathbb{N}^d$ and P_n stands for the n-th Hermite polynomial. Writing $H_j = D_{x_j}^2 + x_j^2$, since $H_j u_k = (2k_j + 1)u_k$, we obtain

$$Hu_k = \sum_{j=1}^{d} H_j u_k = \sum_{j=1}^{d}(2k_j + 1)u_k.$$

We may summarize the preceding results as follows.

Theorem 2.2.3. *The equation*

$$Pu = Hu - \lambda u = -\triangle u + |x|^2 u - \lambda u = 0, \quad u \in \mathcal{S}'(\mathbb{R}^d),$$

admits for

$$\lambda = \lambda_k = \sum_{j=1}^{d}(2k_j + 1), \quad k = (k_1, \ldots, k_d) \in \mathbb{N}^d, \tag{2.2.29}$$

the solutions in $S(\mathbb{R}^d)$,

$$u_k(x) = \prod_{j=1}^{d} P_{k_j}(x_j)e^{-\frac{|x|^2}{2}}, \tag{2.2.30}$$

which form an orthonormal system in $L^2(\mathbb{R}^d)$.

Because of the completeness of the Hermite functions u_k, $k \in \mathbb{N}^d$, we know from Spectral Theory, see Theorem 4.2.9 in the next Chapter 4, that for $\lambda \neq \lambda_k$ the map

$$P = H - \lambda : H_\Gamma^2(\mathbb{R}^d) \to L^2(\mathbb{R}^d)$$

is an isomorphism, with inverse

$$P^{-1} = (H - \lambda)^{-1} : L^2(\mathbb{R}^d) \to H_\Gamma^2(\mathbb{R}^d)$$

belonging to $\mathrm{OP}\Gamma^{-2}(\mathbb{R}^d)$, cf. Theorem 2.1.14, (iv). Returning to (2.2.29), (2.2.30) we observe that the eigenvalues $\lambda = 2K + d$, $K \in \mathbb{N}$, appear with multiplicity if $d > 1$. Precisely, corresponding to $\lambda = 2K + d$ we have in (2.2.30)

$$\sharp\left\{k \in \mathbb{N}^d, \ \sum_{j=1}^{d} k_j = K\right\} = \binom{K+d-1}{d-1}$$

different eigenfunctions, cf. (0.3.16). Let us also observe that, taking into account multiplicity, the number of the eigenvalues which do not exceed $\lambda \in \mathbb{R}_+$ is given by

$$N(\lambda) = \sharp\left\{k \in \mathbb{N}^d, \ \sum_{j=1}^{d} k_j \leq K\right\}$$

$$= \sum_{h=0}^{K} \binom{h+d-1}{d-1} = \binom{K+d}{d}, \tag{2.2.31}$$

cf. (0.3.15). Here we have set $K = [(\lambda - d)/2]$, the integer part of $(\lambda - d)/2$. Hence for $\lambda \to +\infty$,

$$N(\lambda) = \frac{(K+d)(K+d-1)\cdots(K+1)}{d!} \sim \frac{K^d}{d!} \sim \frac{\lambda^d}{2^d d!}. \tag{2.2.32}$$

2.3 Asymptotic Integration and Solutions of Exponential Type

Let us return to consider the generic Γ-elliptic ordinary differential operator, $P \in \mathrm{OP}\Gamma^m(\mathbb{R})$. In view of the discussion in the preceding section, we may write, for

$x \in \mathbb{R}$,

$$P = (D_x - r_1 x) \dots (D_x - r_m x) + \sum_{\alpha+\beta<m} a_{\alpha\beta} x^\beta D^\alpha, \qquad (2.3.1)$$

with $\operatorname{Im} r_j \neq 0$, $j = 1, \dots, m$, $a_{\alpha\beta} \in \mathbb{C}$. The classical solutions of $Pu = 0$ extend to entire functions of $x \in \mathbb{C}$. A precise analysis in the complex domain is given by the theory of Asymptotic Integration. Assuming initially that all the r_j, $j = 1, \dots, m$, are distinct, we formally solve $Pu = 0$ in $\mathbb{C}$ by

$$u_j(x) = x^{s_j} \exp\left[i r_j \frac{x^2}{2} + \eta_j x\right] \sum_{p=0}^{\infty} \beta_{-p,j} x^{-p}, \qquad j = 1, \dots, m. \qquad (2.3.2)$$

We may first determine η_j and s_j by computing

$$\exp\left[-i r_j \frac{x^2}{2} - \eta x\right] P \exp\left[i r_j \frac{x^2}{2} + \eta x\right] x^s$$
$$= g(\eta) x^{s+m-1} + f(s) x^{s+m-2} + O(x^{s+m-3}).$$

The function $g(\eta)$ is linear in η. Namely, setting $g(\eta) = 0$ we obtain

$$\eta = \eta_j = -i \sum_{\alpha+\beta=m-1} \frac{a_{\alpha\beta} r_j^\alpha}{\prod_{h \neq j}(r_j - r_h)}.$$

The expression of $f(s)$, involving $a_{\alpha\beta}$ with $\alpha + \beta \geq m - 2$, is linear in s and imposing $f(s) = 0$ we determine $s = s_j$. Similarly we can find $\beta_{-p,j}$, assuming $\beta_{0,j} = 1$. Note however that the series in (2.3.2) is not in general convergent, and the preceding expressions must be understood as asymptotic expansions of the actual solutions in suitable sectors of the complex plane. Namely, we recall the following basic result, see for example Wasow [194].

Proposition 2.3.1. *Given any sector* $\Lambda = \{x \in \mathbb{C} \setminus \{0\} : \varphi_1 < \arg x < \varphi_2\}$ *with* $\varphi_2 - \varphi_1 < \frac{\pi}{2}$, *there exist* $u_{\Lambda,j}(x)$, $j = 1, \dots, m$, $x \in \mathbb{C}$, *linearly independent solutions of* $Pu = 0$, *such that for every* $n \geq 0$, $x \in \Lambda$, $x \to \infty$:

$$u_{\Lambda,j}(x) = x^{s_j} \exp\left[i r_j \frac{x^2}{2} + \eta_j x\right] \left(\sum_{p=0}^{n} \beta_{-p,j} x^{-p} + o(x^{-n})\right). \qquad (2.3.3)$$

In the case some r_j coincide, we may replace (2.3.3) with the weaker information, valid for any linear combination of the corresponding independent solutions:

$$u_{\Lambda,j}(x) = \exp\left[i r_j \frac{x^2}{2}\right] \tilde{u}_j(x), \qquad |\tilde{u}_j(x)| \lesssim e^{\nu|x|^{2-\epsilon}}, \qquad x \in \Lambda, \qquad (2.3.4)$$

for some $\nu > 0$, $\epsilon > 0$. Observe that the solutions $u_{\Lambda,j}$ in (2.3.3), (2.3.4) depend on the sector Λ. Consider now two sectors Λ_+, Λ_- satisfying the hypotheses of

Proposition 2.3.1 and containing $\mathbb{R}_+$, $\mathbb{R}_-$, respectively. We may conclude the existence of two systems of solutions, u_j^+, u_j^-, $j = 1, \ldots, m$, satisfying (2.3.3) or (2.3.4) in Λ_+, Λ_-. Any classical solution u of $Pu = 0$ in $\mathbb{R}$ can be written

$$u = \sum_{j=1}^{m} \mu_j^+ u_j^+ = \sum_{j=1}^{m} \mu_j^- u_j^-, \quad \mu_j^+, \mu_j^- \in \mathbb{C}.$$

Assume as in Theorem 2.2.2 that $\operatorname{Im} r_j > 0$ for $j = 1, \ldots, m^+$, $\operatorname{Im} r_j < 0$ for $j = m^+ + 1, \ldots, m$, $m = m^+ + m^-$. Then from (2.3.3), (2.3.4) we have that u belongs to $\mathcal{S}(\mathbb{R})$, or equivalently to $\mathcal{S}'(\mathbb{R})$, if and only if $\mu_j^+ = \mu_j^- = 0$ for $m^+ < j \le m$. Summing up:

$$\text{If } u \in \mathcal{S}(\mathbb{R}), \text{ then } u = \sum_{j=1}^{m^+} \mu_j^+ u_j^+ = \sum_{j=1}^{m^+} \mu_j^- u_j^-. \tag{2.3.5}$$

From (2.3.5), (2.3.3) and (2.3.4) we obtain the following results, which complete the information from Theorem 2.2.2.

Theorem 2.3.2. *Let $u \in \mathcal{S}'(\mathbb{R})$, hence $u \in \mathcal{S}(\mathbb{R})$, be a solution of $Pu = 0$. Then*

$$|u(x)| \lesssim e^{-\delta x^2}, \quad \text{for } x \in \mathbb{R} \tag{2.3.6}$$

for some positive constant $\delta > 0$. Moreover

$$m^+ - m^- \le \dim(\operatorname{Ker} P \cap \mathcal{S}(\mathbb{R})) \le m^+. \tag{2.3.7}$$

In view of (2.3.3), (2.3.4) we may take in (2.3.6) any $\delta > 0$ with

$$\delta < \min_{j=1,\ldots,m^+} \operatorname{Im} r_j / 2.$$

With respect to Theorem 2.2.2, the new information in (2.3.7) is that $\dim(\operatorname{Ker} P \cap \mathcal{S}(\mathbb{R})) \le m^+$. When $m^+ > 0$, but $\operatorname{ind} P = m^+ - m^- \le 0$, the existence of non-trivial solutions $u \in \mathcal{S}(\mathbb{R})$ of $Pu = 0$ depends on the coefficients $a_{\alpha\beta}$ in (2.3.1). We shall now prove that in the case $m^+ = 1$ all the possible solutions in $\mathcal{S}(\mathbb{R})$ are of the form

$$u(x) = Q(x)e^{ir_1 \frac{x^2}{2} + \eta_1 x}, \tag{2.3.8}$$

for some polynomials $Q(x)$. This reduces the computation of the eigenvalues to a purely algebraic matter, and generalizes what we proved for the one-dimensional harmonic oscillator (2.2.19).

Theorem 2.3.3. *Consider*

$$P = (D_x - r_1 x) \ldots (D_x - r_m x) + \sum_{\alpha + \beta < m} a_{\alpha\beta} x^\beta D^\alpha \tag{2.3.9}$$

with $\text{Im} \, r_1 > 0$, $\text{Im} \, r_j < 0$ *for* $j = 2, \ldots, m$, $a_{\alpha\beta} \in \mathbb{C}$. *If* $u \in \mathcal{S}(\mathbb{R})$ *is a solution of* $Pu = 0$, *then* u *is of exponential type, i.e.,* u *has the form* (2.3.8) *where*

$$\eta_1 = -i \sum_{\alpha+\beta=m-1} \frac{a_{\alpha\beta} r_1^{\alpha}}{\prod_{j=2}^{m}(r_1 - r_j)}$$

and $Q(x)$ *is a polynomial.*

In the proof we shall use the following classical Phragmen-Lindelöf result; see e.g. Titchmarsh [188, page 177].

Lemma 2.3.4. *Let* $U(x)$, $x \in \mathbb{C}$, *be analytic for*

$$R \le |x| < \infty, \quad \varphi_1 \le \arg x \le \varphi_2, \tag{2.3.10}$$

where R, φ_1, φ_2 *are real costants,* $R > 0$. *Let*

$$|U(x)| \le C \exp[\nu \, |x|^{\eta}] \tag{2.3.11}$$

in the same region, for some positive constants C, ν *and* η *such that* $\varphi_2 - \varphi_1 < \pi/\eta$. *If* U *is bounded as* $x \to \infty$ *on the lines* $\arg x = \varphi_1$ *and* $\arg x = \varphi_2$, *then* U *is bounded uniformly in the region* (2.3.10).

Proof of Theorem 2.3.3. Let $\Lambda_+^1, \Lambda_+^2, \Lambda_-^1, \Lambda_-^2, \Lambda_{+i}, \Lambda_{-i}$ be 6 sectors in $\mathbb{C}$, with vertex at the origin and a positive central angle not exceeding $\frac{\pi}{2}$, such that

$$\Lambda_+^1 \cup \Lambda_+^2 \cup \Lambda_-^1 \cup \Lambda_-^2 \cup \Lambda_{+i} \cup \Lambda_{-i} = \mathbb{C} \setminus \{0\}. \tag{2.3.12}$$

Moreover: Λ_+^1 and Λ_+^2 contain $\mathbb{R}_+$; Λ_-^1 and Λ_-^2 contain $\mathbb{R}_-$; Λ_{+i} contains $i\mathbb{R}_+$; Λ_{-i} contains $i\mathbb{R}_-$. To be definite, let us define

$$\Lambda_+^1 = \left\{ x \in \mathbb{C} \setminus \{0\} : \; -\epsilon < \arg x < \frac{\pi}{2} - 2\epsilon \right\},$$

$$\Lambda_+^2 = \left\{ x \in \mathbb{C} \setminus \{0\} : \; -\frac{\pi}{2} + 2\epsilon < \arg x < \epsilon \right\},$$

$$\Lambda_{+i} = \left\{ x \in \mathbb{C} \setminus \{0\} : \; \frac{\pi}{2} - 3\epsilon < \arg x < \frac{\pi}{2} + 3\epsilon \right\},$$

and symmetrically Λ_-^1, Λ_-^2, Λ_{-i}. Fix in particular $0 < \epsilon < \pi/12$.

Let $u \in \mathcal{S}(\mathbb{R})$ be a solution of $Pu = 0$. Then we may apply (2.3.5) with $m^+ = 1$:

$$u(x) = \mu_1^+ u_1^+(x) = \mu_1^- u_1^-(x).$$

More precisely, in the four sectors Λ_+^1, Λ_+^2, Λ_-^1, Λ_-^2, we have for u, up to multiplicative constants, the asymptotic expansion given by (2.3.3):

$$x^{s_1} \exp\left[ir_1 \frac{x^2}{2} + \eta_1 x \right] \left(\sum_{p=0}^{n} \beta_{-p,1} x^{-p} + o(x^{-n}) \right), \tag{2.3.13}$$

with η_1 as in Theorem 2.3.3. Take now an integer N such that $N \geq |s_1|$ and consider, for $x \in \mathbb{C}$,

$$U(x) = x^{-N} \exp\left[-ir_1\frac{x^2}{2} - \eta_1 x\right] u(x) , \qquad (2.3.14)$$

which is an analytic function for $x \neq 0$. It follows from (2.3.13) that $U(x)$ is bounded in $\Lambda^1_+ \cup \Lambda^2_+ \cup \Lambda^1_- \cup \Lambda^2_-$, for $|x| \geq 1$ say. We may actually prove that U is bounded in the whole $\mathbb{C}$ for $|x| \geq 1$, by applying Lemma 2.3.4 to the remaining sectors Λ_{+i}, Λ_{-i}. Let us fix attention on Λ_{+i}, $|x| \geq 1$. We know already that $U(x)$ is bounded on the lines $\arg x = \frac{\pi}{2} - 3\epsilon$, $\arg x = \frac{\pi}{2} + 3\epsilon$, since they belong to Λ^1_+, Λ^1_-, respectively. On the other hand, we may apply Proposition 2.3.1 and (2.3.4) to the sector Λ_{+i} as well. Since $u(x)$ is a linear combination of the corresponding functions in (2.3.3), (2.3.4) we have in Λ_{+i} an estimate

$$|U(x)| \leq C \exp[\nu \, |x|^2],$$

for suitable positive constants C and ν. Hence (2.3.11) is satisfied with $\eta = 2$. Since we assume $6\epsilon < \pi/2$, we may apply Lemma 2.3.4 and conclude that $U(x)$ is bounded in Λ_{+i}, and similarly in Λ_{-i}. In conclusion: $U(x)$ in (2.3.14) is bounded in $\mathbb{C}$ for $|x| \geq 1$, therefore the function

$$Q(x) = \exp\left[-ir_1\frac{x^2}{2} - \eta_1 x\right] u(x) = U(x)x^N$$

is analytic in $\mathbb{C}$ with a pole at ∞; thus it is indeed a polynomial. For the sake of completeness, we report the classical proof. It is sufficient to show that $(d^n Q/dx^n)(0) = 0$ for $n \geq n_0$, with n_0 sufficiently large. Now, by the Cauchy formula,

$$\frac{d^n Q}{dx^n}(0) = \frac{(-1)^n n!}{2\pi i} \int_\gamma Q(z) z^{-n-1} dz$$

where γ is a closed path around the origin. Take n_0 such that $\left|Q(z)z^{-n_0-1}\right| \leq C |z|^{-2}$ for large $|z|$. Then for any fixed $n \geq n_0$ and every $\delta > 0$, there exists $R > 0$ such that, by choosing $\gamma = \{x \in \mathbb{C} : |x| = R\}$, we have $|(d^n Q/dx^n)(0)| < \delta$. This ends the proof of Theorem 2.3.3. $\qquad\square$

The preceding Theorem 2.3.3 can be extended to the case when in (2.2.13), (2.3.1) we have one root with positive imaginary part, appearing with multiplicity $m^+ \geq 2$. Let us write $r_1 = \cdots = r_{m^+} = r_0$, $\mathrm{Im}\, r_0 > 0$. We need however some conditions on the lower order terms, which are expressed by imposing on the operator P the following particular form:

$$P = \sum_{0 \leq j < m^+} \left(\sum_{\alpha+\beta=m^--j} c_{\alpha\beta} x^\beta D_x^\alpha\right)(D_x - r_0 x)^{m^+-j}$$

$$+ \sum_{\alpha+\beta \leq m^- - m^+} c_{\alpha\beta} x^\beta D_x^\alpha , \qquad (2.3.15)$$

where we understand $1 \leq m^+ \leq m^-$ and

$$\sum_{\alpha+\beta=m^-} c_{\alpha\beta} x^\beta \xi^\alpha = \prod_{m^+<j\leq m} (\xi - r_j x) \tag{2.3.16}$$

with $\operatorname{Im} r_j < 0$ for $j = m^+ + 1, \ldots, m$. In the case $m^+ = 1$ we recapture the operator P in Theorem 2.3.3, with $a_{\alpha\beta} = 0$ for $\alpha + \beta = m - 1$, hence $\eta_1 = 0$.

Theorem 2.3.5. *Consider P as in* (2.3.15), (2.3.16) *with* $1 \leq m^+ \leq m^-$, $\operatorname{Im} r_0 > 0$, $\operatorname{Im} r_j < 0$ *for* $j = m^+ + 1, \ldots, m$, $c_{\alpha\beta} \in \mathbb{C}$. *If* $u \in \mathcal{S}(\mathbb{R})$ *is a solution of* $Pu = 0$, *then*

$$u(x) = Q(x) \exp\left[i r_0 \frac{x^2}{2}\right] \tag{2.3.17}$$

for some polynomial $Q(x)$.

The rough estimate (2.3.4), valid in the case of multiple roots, is not sufficient for the proof of Theorem 2.3.5. Taking advantage of the particular form of P in (2.3.15), however, we have from the asymptotic integration the following more precise result.

Lemma 2.3.6. *Let P be as in Theorem 2.3.5. Given any sector $\Lambda \subset \mathbb{C}$ with vertex at the origin and a positive central angle not exceeding $\pi/2$, there exist $u_{\Lambda,j}(x)$, $j = 1, \ldots, m^+$, linearly independent solutions of $Pu = 0$, of the form*

$$u_{\Lambda,j}(x) = \exp\left[i r_0 \frac{x^2}{2}\right] \tilde{u}_j(x), \quad j = 1, \ldots, m^+ \tag{2.3.18}$$

where $\tilde{u}_j(x)$ are entire functions, satisfying for some integer N:

$$|\tilde{u}_j(x)| \lesssim |x|^N, \quad x \in \Lambda, \ |x| \geq 1. \tag{2.3.19}$$

In fact, when integrating asymptotically $Pu = 0$, we have

$$\exp\left[-i r_0 \frac{x^2}{2}\right] P \exp\left[i r_0 \frac{x^2}{2}\right] x^s = f(s) x^{s-m^++m^-} + O(x^{s-m^++m^--1}),$$

where

$$f(s) = \sum_{0 \leq j < m^+} \left(\sum_{\alpha+\beta=m^--j} c_{\alpha\beta} r_0^\alpha\right) s(s-1)\ldots(s-m^++j+1).$$

This is a polynomial of degree m^+, corresponding to the multiplicity of the root r_0. That grants the lack of lower order exponential terms. Moreover, if the equation $f(s) = 0$ has m^+ distinct roots $s_1, \ldots, s_{m^+}$, we have

$$u_{\Lambda,j}(x) = x^{s_j} \exp\left[i r_0 \frac{x^2}{2}\right](1 + o(1)), \quad j = 1, \ldots, m^+.$$

When some s_j coincide, logarithmic terms appear; the estimate (2.3.19) keeps valid anyhow for a sufficiently large N.

Proof of Theorem 2.3.5. We cover $\mathbb{C}\setminus\{0\}$ by six sectors as in (2.3.12). Let $u \in \mathcal{S}(\mathbb{R})$ be a solution of $Pu = 0$. Then we may apply (2.3.5):

$$u = \sum_{j=1}^{m^+} \mu_j^+ u_j^+ = \sum_{j=1}^{m^+} \mu_j^- u_j^- .$$

More precisely, using Lemma 2.3.6 in the four sectors Λ_+^1, Λ_+^2, Λ_-^1, Λ_-^2, we have

$$u(x) = \exp\left[ir_0 \frac{x^2}{2} \right] \tilde{u}(x)$$

where the entire function $\tilde{u}(x)$ satisfies

$$|\tilde{u}(x)| \lesssim |x|^N, \quad x \in \Lambda_+^1 \cup \Lambda_+^2 \cup \Lambda_-^1 \cup \Lambda_-^2, \ |x| \geq 1.$$

We may then argue exactly as in the proof of Theorem 2.3.3 and obtain (2.3.17).   □

2.4   H-Polynomials

We pass now to consider Γ_ρ-hypoelliptic partial differential operators, $0 < \rho \leq 1$, cf. Definition 2.1.15. In view of Proposition 2.2.1, we may again limit our attention to operators with polynomial coefficients:

$$P = \sum_{|\alpha|+|\beta|\leq m} c_{\alpha\beta} x^\beta D_x^\alpha . \tag{2.4.1}$$

Setting $z = (x, \xi)$, we write their symbols in the standard quantization:

$$p(z) = \sum_{|\gamma|\leq m} c_\gamma z^\gamma . \tag{2.4.2}$$

In the following we shall assume $m \geq 1$ and $c_\gamma \neq 0$ for some γ with $|\gamma| = m$.
    The Γ_ρ-hypoellipticity is expressed by

$$|\partial_z^\gamma p(z)| \lesssim |p(z)| \langle z \rangle^{-\rho|\gamma|}, \quad |z| \geq R \tag{2.4.3}$$

for some $R > 0$. In fact, condition (2.1.29) implies for a polynomial $p(z)$ condition (2.1.28), since for some γ with $|\gamma| = m$ we have $\partial^\gamma p(z) =$ constant$\neq 0$, hence

$$0 < c = |\partial^\gamma p(z)| \lesssim |p(z)| \langle z \rangle^{-\rho m}, \quad |z| \geq R,$$

and therefore

$$|z|^{\rho m} \lesssim |p(z)|, \quad |z| \geq R. \tag{2.4.4}$$

So, under the assumption (2.4.3) we may apply Theorem 2.1.16 and deduce existence of parametrices for P in (2.4.1). In particular, P is globally regular and the following a priori estimate is valid for $u \in \mathcal{S}(\mathbb{R}^d)$:

$$\|u\|_{H^{\rho m}_\Gamma} \leq C \left(\|Pu\|_{L^2} + \|u\|_{L^2} \right). \tag{2.4.5}$$

In the sequel of this section we want to clarify the algebraic meaning of (2.4.3). We shall argue on polynomials in arbitrary dimension n though in our applications $n = 2d$ is even. Consider the subset of $\mathbb{C}^n$,

$$V = \{\zeta \in \mathbb{C}^n : p(\zeta) = 0\}, \tag{2.4.6}$$

and define, for $z \in \mathbb{R}^n$,

$$d(z) = \text{distance}_{\mathbb{C}^n}(z, V) = \inf_{\zeta \in V} |z - \zeta|. \tag{2.4.7}$$

Proposition 2.4.1. *The following properties are equivalent for a polynomial $p(z)$:*

(i) $\zeta \in V$, $\zeta \to \infty$ *implies* $|\text{Im}\,\zeta| \to +\infty$.

(ii) $z \in \mathbb{R}^n$, $z \to \infty$ *implies* $d(z) \to +\infty$.

(iii) $|\partial_z^\gamma p(z)/p(z)| \to 0$ *when* $z \to \infty$ *in* $\mathbb{R}^n$, *if* $\gamma \neq 0$.

If (i), (ii), (iii) *are satisfied, we say that $p(z)$ is H-type.*

Let ρ be fixed, $\rho > 0$; we say that $p(z)$ is ρ-H-type if the following equivalent properties are satisfied:

(i)$_\rho$ *We have:*

$$|\zeta|^\rho \lesssim 1 + |\text{Im}\,\zeta| \quad \text{for } \zeta \in V.$$

(ii)$_\rho$ *There is a constant $R > 0$ such that*

$$|z|^\rho \lesssim d(z) \quad \text{for } |z| \geq R.$$

(iii)$_\rho$ *There is a constant $R > 0$ such that*

$$|\partial^\gamma p(z)| \lesssim |p(z)| \langle z \rangle^{-\rho|\gamma|} \quad \text{for } |z| \geq R,$$

i.e., $p(z)$ is Γ_ρ-hypoelliptic.

Obviously, if $p(z)$ is ρ-H-type for some ρ, $0 < \rho \leq 1$, then it is H-type. In the opposite direction, if $p(z)$ is H-type, then it is ρ-H-type for some $\rho > 0$; moreover, the numbers ρ for which (i)$_\rho$, (ii)$_\rho$, (iii)$_\rho$ *are valid form an interval* $]0, \rho_0]$, *with ρ_0 a rational number ≤ 1.*

To prove Proposition 2.4.1 we need two auxiliary results.

Lemma 2.4.2. *Let $m \geq 1$ be an integer. There is a constant $C > 0$, depending only on m and the dimension n, such that for all polynomials $p(z)$ of degree m we have*

$$C^{-1} \leq d(z) \sum_{|\gamma| \neq 0} |\partial^\gamma p(z)/p(z)|^{1/|\gamma|} \leq C, \qquad (2.4.8)$$

for $z \in \mathbb{R}^n$, $p(z) \neq 0$.

Proof. Setting, for $p(z) \neq 0$,

$$A(z) = \sum_{|\gamma| \neq 0} |\partial^\gamma p(z)/p(z)|^{1/|\gamma|},$$

we have

$$|\partial^\gamma p(z)| \leq A(z)^{|\gamma|} |p(z)|.$$

From the Taylor expansion we obtain, for $\zeta \in \mathbb{C}^n$,

$$|p(z + \zeta) - p(z)| \leq |p(z)| \sum_{1 \leq |\gamma| \leq m} (A(z) |\zeta|)^{|\gamma|}/\gamma!.$$

Choose $c > 0$ such that

$$\sum_{1 \leq |\gamma| \leq m} c^{|\gamma|}/\gamma! \leq 1.$$

If $A(z) |\zeta| < c$ we must have $p(z + \zeta) \neq 0$ and therefore $A(z)d(z) \geq c$. We obtain thus the left-hand side inequality in (2.4.8) for any $C \geq 1/c$.

To establish the right-hand side inequality, choose $\zeta \in \mathbb{C}^n$ such that $|\zeta| \leq d(z)/2$ and regard $p(z + t\zeta)$ as a polynomial $Q(t)$ in t. The roots t_j of $Q(t)$ satisfy the inequality $|t_j \zeta| \geq d(z)$, and therefore $|t_j| \geq 2$. It follows that

$$|p(z + \zeta)/p(z)| = |Q(1)/Q(0)| = \left| \prod_{j=1}^{m} (t_j - 1)/t_j \right| \leq (3/2)^m. \qquad (2.4.9)$$

Regarding $p(z + \zeta)$ as a holomorphic function of $\zeta \in \mathbb{C}^n$ and applying Cauchy's inequality on a polycylinder in the ball $|\zeta| \leq d(z)/2$, we obtain from (2.4.9)

$$|\partial^\gamma p(z)| \leq C d(z)^{-|\gamma|} |p(z)|,$$

for a suitable C depending only on m and n, which implies the right-hand side inequality of (2.4.8). $\qquad \square$

The second auxiliary result is the following version of the classical Seidenberg-Tarski Theorem. For the proof see for example Hörmander [119, Vol. II].

Theorem 2.4.3. *If A, subset of $\mathbb{R}^{n+m} = \mathbb{R}^n \oplus \mathbb{R}^m$, is semi-algebraic (i.e., finite union of finite intersections of sets defined by a polynomial equation or inequality), then the projection of A in $\mathbb{R}^m$ is also semi-algebraic.*

Proof of Proposition 2.4.1. The equivalence between (i) and (ii) is easily estab-
lished, while the equivalence between (ii) and (iii) is a consequence of Lemma
2.4.2.

Let us detail the proof of $(i)_\rho \Leftrightarrow (ii)_\rho$. Observe first that $(ii)_\rho$ cannot be valid
for $\rho > 1$. Indeed, the function $d(z)$ is Lipschitz, therefore

$$d(z) \leq C(1 + |z|),$$

for some $C > 0$. This forces $\rho \leq 1$ in $(ii)_\rho$. It is now clear that $(ii)_\rho$ implies $(i)_\rho$;
in fact we have for $\zeta \in V$:

$$|\operatorname{Re} \zeta|^\rho \leq C(1 + d(\operatorname{Re} \zeta)) \leq C(1 + |\operatorname{Im} \zeta|).$$

On the other hand, if $(i)_\rho$ holds we can take $\zeta \in V$ with $|\zeta - z| \leq 2d(z)$, say, and
write

$$|z| \leq |\zeta| + |z - \zeta| \leq C^{1/\rho}(1 + |\operatorname{Im} \zeta|)^{1/\rho} + 2d(z);$$

since $|\operatorname{Im} \zeta| \leq 2d(z)$, we obtain $(ii)_\rho$.

The equivalence between $(ii)_\rho$ and $(iii)_\rho$ follows easily from Lemma 2.4.2.

Finally, let us prove that (ii) implies $(ii)_\rho$ for some $\rho > 0$.

Let us begin by considering the set A of points $(z, \eta, \theta, \tau, \delta) \in \mathbb{R}^n \times \mathbb{R}^n \times \mathbb{R}^n \times \mathbb{R} \times \mathbb{R}$ defined by the following equations and inequalities:

$$p(\eta + i\theta) = 0, \quad \tau > 0, \quad |z - \eta|^2 + |\theta|^2 \leq \tau^{-2}, \quad \delta > 0, \quad |z|\,\delta = 1.$$

By Theorem 2.4.3 the image B of A by the projection $(z, \eta, \theta, \tau, \delta) \mapsto (z, \tau, \delta)$ is
also semi-algebraic. It is easily seen that B is defined by the inequalities

$$\tau > 0, \quad d(z) \leq \tau^{-1}, \quad \delta > 0, \quad |z|\,\delta = 1.$$

If (ii) holds, there is $\delta_0 > 0$ such that if $0 < \delta < \delta_0$ and $|z| = \delta^{-1}$ we have $d(z) > 0$.
For all such δ the function

$$\tau(\delta) = \sup_{|z|\delta=1} d(z)^{-1}$$

is well defined and continuous. The image E of B by the projection $(z, \tau, \delta) \mapsto (\tau, \delta)$ is a semi-algebraic set. Moreover, we can show that $(\tau(\delta), \delta)$ varies on the
boundary of E when $0 < \delta < \delta_0$. Shrinking this interval, if necessary, it follows
that $Q(\tau(\delta), \delta) = 0$ for a suitable polynomial Q in two variables. Then, $\tau(\delta)$
admits a Puiseux expansion in some neighbourhood of the origin in the complex
δ-plane:

$$\tau(\delta) = a_k(\delta^{1/q})^k + a_{k+1}(\delta^{1/q})^{k+1} + \cdots,$$

where $q > 0$ and k are integers. We can assume $a_k \neq 0$ and, by choosing for $\delta^{1/q}$
the branch which is real and positive for $\delta > 0$, also $a_k > 0$. In view of (ii), we

have $\tau(\delta) \to 0$ as $\delta \to 0$, hence $k > 0$. It follows easily that, for some $c > 0$ and sufficiently large $|z|$,

$$\inf_{|z|\delta=1} d(z) = \tau(\delta)^{-1} \geq c|z|^{k/q} .$$

This implies (ii)$_\rho$ with $\rho = k/q$ and also the fact that the set of numbers for which (ii)$_\rho$ holds is a closed interval with a rational number as upper limit. The proof of Proposition 2.4.1 is therefore complete. □

As we show now, to prove that $p(z)$ is H-type, it is sufficient to check (iii) in Proposition 2.4.1 for first order derivatives. This simplifies a lot the reasoning in the applications.

Proposition 2.4.4. *The following properties of the polynomial $p(z)$ are equivalent (and provide Γ_ρ-hypoellipticity of $p(z)$ for some ρ, $0 < \rho \leq 1$):*

(a) $\partial_z^\gamma p(z)/p(z) \to 0$ *when $z \to \infty$ in $\mathbb{R}^n$, if $\gamma \neq 0$.*

(b) *For all $\theta \in \mathbb{R}^n$, $p(z + \theta)/p(z) \to 1$ when $z \to \infty$ in $\mathbb{R}^n$.*

(c) $\partial_{z_j} p(z)/p(z) \to 0$ *when $z \to \infty$ in $\mathbb{R}^n$, for $j = 1, \dots, n$.*

We need the following lemma, which allows us to write each γ-derivative of a polynomial $p(z)$ as a linear combination of translations of $p(z)$.

Lemma 2.4.5. *Let $p(z) = \sum_{|\gamma| \leq m} c_\gamma z^\gamma$, $z \in \mathbb{R}^n$, be a polynomial of degree $m \geq 0$, and $N = \binom{m+n}{n}$. Then for all $(\theta_1, \theta_2, \dots, \theta_N)$ in an open dense subset of $(\mathbb{R}^n)^N$ and all $\gamma \in \mathbb{N}^n$ there exist real numbers $t_1, t_2, \dots, t_N$, such that*

$$\partial^\gamma p(z) = \sum_{k=1}^{N} t_k p(z + \theta_k), \tag{2.4.10}$$

and satisfying further $\sum_{k=1}^{N} t_k = 0$ if $\gamma \neq 0$.

Proof. For all $k = 1, \dots, N$, let $\theta_k \in \mathbb{R}^n$. Consider Taylor's formula for $p(z + \theta_k)$:

$$p(z + \theta_k) = p(z) + \sum_{\beta \neq 0} \frac{\partial^\beta p(z)}{\beta!} \theta_k^\beta,$$

and multiply the left-hand and right-hand sides by a real variable t_k. Summing up on k from 1 to N we obtain

$$\sum_{k=1}^{N} t_k p(z + \theta_k) = p(z) \sum_{k=1}^{N} t_k + \sum_{\beta \neq 0} \frac{\partial^\beta p(z)}{\beta!} \sum_{k=1}^{N} t_k \theta_k^\beta. \tag{2.4.11}$$

In order to get (2.4.10) when $\gamma \neq 0$, it suffices that the following inhomogeneous linear system is solvable:

$$\begin{cases} \sum_{k=1}^{N} t_k = 0, \\ \sum_{k=1}^{N} t_k \theta_k^\gamma = \gamma!, \\ \sum_{k=1}^{N} t_k \theta_k^\beta = 0 \quad \text{for } \beta \neq \gamma, \ \beta \neq 0. \end{cases} \tag{2.4.12}$$

This is an $N \times N$ system, cf. (0.3.15), whose coefficients matrix is $A = (\theta_k^\beta)$, $k = 1, \ldots, N$, $|\beta| \leq m$. Here the column index is k, whereas β plays the role of row index and it is understood that the set of these multi-indices is ordered in some way. Now, the determinant of A is a polynomial in the components of each θ_k, $k = 1, \ldots, N$. It also clear that it does not vanish identically because, up to the sign, it is given by $\sum_{(\sigma_1,\ldots,\sigma_N)} \text{sign}(\sigma_1, \ldots, \sigma_N)\theta_1^{\sigma_1} \cdots \theta_N^{\sigma_N}$, where $(\sigma_1, \ldots, \sigma_N)$ is a permutation of $\{\beta \in \mathbb{N}^n : |\beta| \leq m\}$, and there are no similar terms in this sum. Hence the above system is solvable for every choice of $\theta_1, \ldots, \theta_N$ for which such a determinant does not vanish, which gives an open dense subset of $(\mathbb{R}^n)^N$.

Similarly, for $\gamma = 0$ one considers, in place of (2.4.12), the system $\sum_{k=1}^N t_k = 1$ and $\sum_{k=1}^N t_k \theta_k^\beta = 0$ for all $\beta \neq 0$, which concludes the proof. □

Proof of Proposition 2.4.4. Of course (a) $\Rightarrow$ (c). We just observe that by Proposition 2.4.1 and (2.4.4) it follows from (a) that $p(z) \to \infty$ as $z \to \infty$.

(c) $\Rightarrow$ (b) Let $n \geq 3$. Since $p(z) \neq 0$ on the simply connected open set $|z| > R$, for R large enough, we can consider a branch of $\log p(z)$, for $|z| > R$. For any fixed $\theta = (b_1, b_2, \ldots, b_n) \in \mathbb{R}^n$ we can write

$$\log p(z + \theta) - \log p(z) = \int_0^1 \frac{d}{dt} \log p(z + t\theta)\, dt,$$

for $|z| > R + |\theta|$. We have

$$\left| \frac{d}{dt} \log p(z + t\theta) \right| = \left| \sum_{j=1}^n \frac{\partial_{z_j} p(z + t\theta)}{p(z + t\theta)} b_j \right| \leq \sum_{j=1}^n \left| \frac{\partial_{z_j} p(z + t\theta)}{p(z + t\theta)} \right| |b_j|,$$

and it follows from (c) that

$$\frac{p(z + \theta)}{p(z)} \to 1 \quad \text{as } z \to \infty.$$

In the cases $n = 1, 2$ one can repeat the above argument separately on two simply connected open subsets which cover $\mathbb{R}$ and $\mathbb{R}^2$, respectively.

(b) $\Rightarrow$ (a) From Lemma 2.4.5 it follows that for each $\gamma \in \mathbb{N}^n$, $\gamma \neq 0$, there exist real numbers $t_1, \ldots, t_N$ and vectors $\theta_1, \ldots, \theta_N$ in $\mathbb{R}^n$, $N = \binom{m+n}{n}$, such that

$$\partial^\gamma p(z) = \sum_{k=1}^N t_k p(z + \theta_k), \quad \text{with } \sum_{k=1}^N t_k = 0.$$

For large $|z|$ and $\gamma \neq 0$ we obtain

$$\frac{\partial^\gamma p(z)}{p(z)} = \sum_{k=1}^N t_k \frac{p(z + \theta_k)}{p(z)} \to \sum_{k=1}^N t_k = 0, \quad \text{as } z \to \infty. \quad □$$

2.5   Quasi-Elliptic Polynomials

We may now discuss some relevant subclasses of the H-polynomials, namely the quasi-elliptic ones and, as a generalization in the next section, the multi-quasi-elliptic polynomials. Such polynomials, regarded as symbols of linear partial differential operators, are obviously Γ_ρ-hypoelliptic for some $\rho > 0$, cf. (2.4.3). Their importance, however, resides in the fact that we may easily restore for them a notion of ellipticity, cf. Definition 1.3.1, by introducing adapted weight functions. In the present and next sections we fix attention on the algebraic aspects, and address to the next Section 2.7 for the related pseudo-differential calculus.

We first define the quasi-elliptic polynomials. Let $M = (M_1, \ldots, M_n)$ be an n-tuple of rational numbers $M_j \geq 1$, $j = 1, \ldots, n$. Assume also $\min_j M_j = 1$. Every polynomial $p(z)$ in $\mathbb{R}^n$ can be written

$$p(z) = \sum_{\langle \gamma, M \rangle \leq m} c_\gamma z^\gamma, \tag{2.5.1}$$

for a sufficiently large integer m, the M-order of $p(z)$. In the sum we mean $\langle \gamma, M \rangle = \gamma_1 M_1 + \cdots + \gamma_n M_n$.

The *quasi-principal part of $p(z)$, with respect to M*, will be then defined as

$$p_{M,m}(z) = \sum_{\langle \gamma, M \rangle = m} c_\gamma z^\gamma. \tag{2.5.2}$$

Definition 2.5.1. We shall say that the polynomial $p(z)$ in (2.5.1) is quasi-elliptic with respect to M if

$$p_{M,m}(z) \neq 0 \quad \text{for all } z \neq 0. \tag{2.5.3}$$

If $M = (1, \ldots, 1)$, then $p_{M,m}(z)$ in (2.5.2) is $p_m(z)$, the standard principal part of $p(z)$, and (2.5.3) means that $p(z)$ is Γ-elliptic, cf. (2.2.5), (2.2.6).

Let us recall that a function $f(z)$ in $\mathbb{R}^n \setminus \{0\}$ is called *quasi-homogeneous of degree $k \in \mathbb{R}$, with respect to M*, if

$$f(t^{M_1} z_1, \ldots, t^{M_n} z_n) = t^k f(z_1, \ldots, z_n) \quad \text{for all } t > 0, \ z \neq 0.$$

Observe that if f is quasi-homogeneous of degree k, then $\partial^\gamma f$ is quasi-homogeneous of degree $k - \langle \gamma, M \rangle$. According to the previous definition, the quasi-principal part $p_{M,m}(z)$ in (2.5.2) is quasi-homogeneous of degree m with respect to M.

Let us define, for $z \in \mathbb{R}^n$,

$$|z|_M = \sum_{j=1}^n |z_j|^{1/M_j}. \tag{2.5.4}$$

The function $|z|_M$ is quasi-homogeneous of degree 1. Any quasi-homogeneous function $f(z)$ of degree k is identified by its values on the compact manifold $\{|z|_M = 1\}$; namely we may write

$$f(z) = |z|_M^k f(\tilde{z}) \tag{2.5.5}$$

where

$$\tilde{z} = (z_1/|z|_M^{M_1}, \ldots, z_n/|z|_M^{M_n}),$$

hence $|\tilde{z}|_M = 1$. If $f(z)$ is continuous in $\mathbb{R}^n \setminus \{0\}$, it follows that

$$|f(z)| \le C |z|_M^k \tag{2.5.6}$$

with $C = \max_{|z|_M=1} |f(z)|$. If in addition $f(z) \ne 0$, then

$$|z|_M^k \le \epsilon^{-1} |f(z)| \tag{2.5.7}$$

with $\epsilon = \min_{|z|_M=1} |f(z)|$.

Since $p(z)$ in (2.5.1) can be regarded as a sum of quasi-homogeneous terms of degree $\le m$, we deduce from (2.5.6) that, for every $R > 0$,

$$|p(z)| \lesssim |z|_M^m, \quad |z| \ge R. \tag{2.5.8}$$

On the other hand, using (2.5.7) and arguing as in the proof of Proposition 2.1.5 we have:

Proposition 2.5.2. *The polynomial $p(z)$ in (2.5.1) is quasi-elliptic with respect to M if and only if there exists $R > 0$ such that*

$$|z|_M^m \lesssim |p(z)|, \quad |z| \ge R, \tag{2.5.9}$$

where m is the M-order of $p(z)$.

Summing up, we have that quasi-elliptic polynomials $p(z)$, of order m with respect to M, are characterized by the asymptotic equivalence $|p(z)| \asymp |z|_M^m$.

We can now prove that quasi-elliptic polynomials are H-type.

Proposition 2.5.3. *Let $p(z)$ in (2.5.1) be quasi-elliptic with respect to M. Then*

$$|\partial_z^\gamma p(z)| \lesssim |p(z)| |z|_M^{-|\gamma|}, \quad |z| \ge R, \tag{2.5.10}$$

for some $R > 0$, and $p(z)$ is ρ-H-type for

$$\rho = \min_j 1/M_j. \tag{2.5.11}$$

Proof. Since $p(z)$ in (2.5.1) can be written as a sum of quasi-homogeneous terms of degree $\le m$, then $\partial^\gamma p(z)$ is a sum of quasi-homogeneous terms of degree $\le m - \langle \gamma, M \rangle$. It follows from the estimates (2.5.6) that

$$|\partial_z^\gamma p(z)| \lesssim |z|_M^{m-\langle \gamma, M \rangle}, \quad |z| \ge R.$$

Hence by applying (2.5.9),

$$|\partial_z^\gamma p(z)| \lesssim |p(z)| |z|_M^{-\langle \gamma, M \rangle}, \quad |z| \ge R \tag{2.5.12}$$

which implies (2.5.10) since $\langle \gamma, M \rangle \geq |\gamma|$. On the other hand, we have $\langle z \rangle^\rho \lesssim |z|_M$ for ρ as in (2.5.11) and large $|z|$; therefore

$$|\partial_z^\gamma p(z)| \lesssim |p(z)| \langle z \rangle^{-\rho|\gamma|}, \quad |z| \geq R,$$

i.e., $p(z)$ is ρ-H-type.   □

Example 2.5.4. Let us return to split $z = (x, \xi)$ and consider, for $x \in \mathbb{R}$, $\xi \in \mathbb{R}$,

$$p(x, \xi) = \xi^h + rx^k, \tag{2.5.13}$$

where h, k are positive integers and $r \in \mathbb{C}$. The associated ordinary differential operator is

$$P = D_x^h + rx^k. \tag{2.5.14}$$

If $k \geq h$, we fix the weight 1 for the x variable and the weight k/h for the ξ variable, i.e., $M = (1, k/h)$. If $h \geq k$, then $M = (h/k, 1)$. The M-order of $p(x, \xi)$ in (2.5.13) is $\max\{h, k\}$ and, in view of (2.5.3), quasi-ellipticity amounts to assuming $\xi^h + rx^k \neq 0$ for $(x, \xi) \neq (0, 0)$. In particular, if h or k is odd, $p(x, \xi)$ is quasi-elliptic if and only if $\operatorname{Im} r \neq 0$.

Example 2.5.5. For $z = (x, \xi) \in \mathbb{R}^{2d}$ consider

$$p(x, \xi) = |\xi|^2 + V(x) \tag{2.5.15}$$

where

$$V(x) = \sum_{|\beta| \leq 2k} a_\beta x^\beta, \tag{2.5.16}$$

for some integer $k \geq 1$. Assume

$$V_{2k}(x) = \sum_{|\beta| = 2k} a_\beta x^\beta \neq 0 \quad \text{for } x \neq 0. \tag{2.5.17}$$

The associated operator is

$$P = -\Delta + V(x). \tag{2.5.18}$$

We give now the weight 1 to the x variables and the weight k to the ξ variables, i.e., $M = (1, \dots, 1, k, \dots, k)$. Then $p(x, \xi)$ in (2.5.15), (2.5.16), (2.5.17) is quasi-elliptic if and only if $|\xi|^2 + V_{2k}(x) \neq 0$ for $(x, \xi) \neq (0, 0)$, that is $V_{2k}(x)$ does not take values in $\mathbb{R}_- \cup \{0\}$ for $x \neq 0$.

Remark 2.5.6. In the estimates (2.5.8), (2.5.9), (2.5.10), the function $|z|_M$ can be replaced by any asymptotically equivalent function. Observe in particular that

$$\left(1 + \sum_{j=1}^n z_j^{2m/M_j} \right)^{1/2} \asymp |z|_M^m \asymp |p(z)|, \quad |z| \geq R. \tag{2.5.19}$$

Note that m/M_j, $j = 1, \dots, n$, are integers; in fact, the quasi-ellipticity assumption (2.5.3) forces the presence of the monomials z_j^{m/M_j}, $j = 1, \dots, n$, in the expression of $p(z)$.

2.6 Multi-Quasi-Elliptic Polynomials

We now extend the arguments of the preceding section. Namely, we try to repro-
duce the estimates (2.5.8), (2.5.9), (2.5.10) by replacing the function $|z|_M^m \asymp |p(z)|$
with weight functions suited to more general H-polynomials. In particular, looking
to the weight in the left-hand side of (2.5.19), asymptotically equivalent to $|z|_M^m$,
one is led in a natural way to consider the function $(\sum_\gamma z^{2\gamma})^{1/2}$, where γ runs over
indices appearing in the expression of $p(z)$. To be precise, we begin by defining
the Newton polyhedron of a polynomial $p(z)$, and then recalling terminology and
notation from the general theory of the convex polyhedra.

Definition 2.6.1. The Newton polyhedron $\mathcal{P}$ of a polynomial

$$p(z) = \sum_{|\gamma| \le m} c_\gamma z^\gamma, \quad z \in \mathbb{R}^n,$$

is the convex hull of $\mathcal{A} \cup \{0\}$ with

$$\mathcal{A} = \{\gamma \in \mathbb{N}^n : c_\gamma \ne 0\}. \tag{2.6.1}$$

We recall that, in general, a convex polyhedron $\mathcal{P} \subset \mathbb{R}^n$ is defined as the
convex hull of a finite set of points in $\mathbb{R}^n$. One can show that $\mathcal{P}$ can be obtained
as the convex hull of a finite subset $V(\mathcal{P}) \subset \mathbb{R}^n$ of convex-linearly independent
points, called the vertices of $\mathcal{P}$ and univocally determinated by $\mathcal{P}$. Moreover there
exists a finite set of normal vectors $N(\mathcal{P}) = N_0(\mathcal{P}) \cup N_1(\mathcal{P}) \subset \mathbb{R}^n$ such that

$$|\nu| = 1, \quad \text{for all } \nu \in N_0(\mathcal{P}),$$

$$\mathcal{P} = \{z \in \mathbb{R}^n : \nu \cdot z \ge 0, \, \forall \nu \in N_0(\mathcal{P})\} \cap \{z \in \mathbb{R}^n : \nu \cdot z \le 1, \, \forall \nu \in N_1(\mathcal{P})\}.$$

$N_0(\mathcal{P})$ and $N_1(\mathcal{P})$ are univocally determined by $\mathcal{P}$, if $\mathcal{P}$ has non-empty interior.
The boundary of $\mathcal{P}$ is made of faces F_ν which are the convex hull of the vertices
of $\mathcal{P}$ lying on the hyperplane H_ν orthogonal to $\nu \in N(\mathcal{P})$ and of equation

$$\nu \cdot z = 0, \quad \text{if } \nu \in N_0(\mathcal{P}),$$

$$\nu \cdot z = 1, \quad \text{if } \nu \in N_1(\mathcal{P}).$$

If $\mathcal{P} \subset (\mathbb{R}_+ \cup \{0\})^n$ and $V(\mathcal{P}) \subset \mathbb{N}^n$, as we have for the Newton polyhedron of a
polynomial, we define

$$\Lambda_{\mathcal{P}}(z) = \left(\sum_{\gamma \in V(\mathcal{P})} z^{2\gamma} \right)^{1/2}, \quad z \in \mathbb{R}^n. \tag{2.6.2}$$

Definition 2.6.2. A complete polyhedron is a convex polyhedron $\mathcal{P} \subset (\mathbb{R}_+ \cup \{0\})^n$
such that

(1) $V(\mathcal{P}) \subset \mathbb{N}^n$;

(2) $(0, \ldots, 0) \in V(\mathcal{P})$;

(3) $V(\mathcal{P}) \neq \{(0, \ldots, 0)\}$;

(4) $N_0(\mathcal{P}) = \{e_1, \ldots, e_n\}$, with $e_j = (0, \ldots, 0, \underset{j\text{-}th\ entry}{1}, 0, \ldots, 0) \in \mathbb{R}^n$ for $j = 1, \ldots, n$;

(5) $N_1(\mathcal{P}) \subset \mathbb{R}_+^n$, i.e., every $\nu \in N_1(\mathcal{P})$ has strictly positive components ν_j, $j = 1, \ldots, n$.

One easily proves:

$$\langle z \rangle^{\mu_0} \lesssim \Lambda_{\mathcal{P}}(z) \lesssim \langle z \rangle^{\mu_1} \quad \text{for all } z \in \mathbb{R}^n,$$

with

$$\mu_0 = \min_{\gamma \in V(\mathcal{P}) \setminus \{0\}} |\gamma|, \quad \mu_1 = \max_{\gamma \in V(\mathcal{P})} |\gamma| = \max_{\gamma \in \mathcal{P}} |\gamma|. \tag{2.6.3}$$

μ_0 and μ_1 are called the minimum and the maximum order of $\mathcal{P}$.

We also introduce the *formal order* of $\mathcal{P}$:

$$\mu = \max \left\{ \frac{1}{\nu_j} : j = 1, \ldots, n, \ \nu \in N_1(\mathcal{P}) \right\}. \tag{2.6.4}$$

We have $0 < \mu_0 \leq \mu_1 \leq \mu$.

Theorem 2.6.3. *The Newton polyhedron of an H-type polynomial is complete.*

Proof. Let $p(z)$ satisfy the equivalent properties (i), (ii), (iii) in Proposition 2.4.1. Let us prove by induction on $n \geq 2$ that its Newton polyhedron $\mathcal{P}$ is complete. If $n = 2$ let

$$p(z) = \sum_{j=1}^{l} b_j(z_1) z_2^{s_j}$$

with $s_1 < s_2 < \ldots < s_l$. By (iii) in Proposition 2.4.1 we have that

$$\frac{\partial_{z_1} p(z)}{p(z)} \to 0, \quad \text{as } z \to \infty.$$

In particular we have that

$$\frac{\partial_{z_1} b_l(z_1)}{b_l(z_1)} = \lim_{z_2 \to \infty} \frac{\partial_{z_1} p(z)}{p(z)} = 0.$$

It follows that $s_l > 0$ and that $b_l(z_1)$ does not depend on z_1. Therefore $\mathcal{P}$ contains the point $(0, s_l)$ and all other points $(q_1, q_2) \in \mathcal{P}$ are such that $q_2 < s_l$. In a similar way we have that there exists $(r, 0) \in \mathcal{P}$, with $r > 0$, such that $q_1 < r$ for all other points $(q_1, q_2) \in \mathcal{P}$. The completeness of $\mathcal{P}$ follows now easily from the convexity of $\mathcal{P}$.

Let us now assume that the theorem is true in dimension $n \geq 2$ and let us prove it in dimension $n + 1$. Let us first show that there are no faces passing through the origin and not lying on the coordinate hyperplanes.

In fact if F_ν is such a face, ν should have at least two components, say, ν_q and ν_{n+1} different from zero and at least one, say ν_q, negative. The reason is that F_ν intersects $(\mathbb{R}_+)^{n+1}$. The intersection of F_ν with the coordinate hyperplane $z_{n+1} = 0$ is the face F_σ with $\sigma = |\nu'|^{-1}\nu'$ and $\nu = (\nu', \nu_{n+1})$, of the Newton polyhedron of $p(z', 0)$ with $z = (z', z_{n+1})$. It is obvious that $p(z', 0)$ is H–type, so by induction σ should not have negative components, in contrast to the fact that $\sigma_q = |\nu'|^{-1}\nu_q < 0$. This shows that there are no faces through the origin that do not lie on the coordinate hyperplanes.

Now we prove that $\nu \in \mathbb{R}_+^{n+1}$ for all $\nu \in N_1(\mathcal{P})$ such that F_ν intersects a coordinate hyperplane, say $z_{n+1} = 0$. Set $z = (z', z_{n+1})$ and $\nu = (\nu', \nu_{n+1})$ and observe that $p(z', 0)$ is H-type and the intersection of F_ν with the hyperplane $z_{n+1} = 0$ is the face $F_{\nu'}$ of the Newton polyhedron of $p(z', 0)$ with normal ν'. By induction we have that $\nu' \in \mathbb{R}_+^n$. So we have to prove that $\nu_{n+1} > 0$. Let

$$p(z) = \sum_{j=1}^{m} b_j(z') z_{n+1}^{r_j}$$

with $r_1 < r_2 < \ldots < r_m$. It is clear that the exponents of the monomials $z'^{\alpha'}$ in $b_1(z')$ are such that $\alpha' \cdot \nu' \leq 1$ and that if α' belongs to $F_{\nu'}$, then $\alpha' \cdot \nu' = 1$. Hence $r_1 = 0$. Moreover, if γ is in $F_\nu \cap \mathbb{N}^{n+1}$ with $\gamma_{n+1} > 0$, then there exists $r_j = \gamma_{n+1}$ and γ' is the exponent of a monomial in $b_j(z')$. Now $r_j! b_j(z') = \partial_{z_{n+1}}^{r_j} p(z', 0)$, thus by (iii) in Proposition 2.4.1 we have that $b_j(z')/b_1(z') \to 0$ as $z' \to \infty$. In particular, if we set $z' = (a_1 t^{\nu_1}, \ldots, a_n t^{\nu_n})$ and let $t \to \infty$ for different choices of $u_1, \ldots, u_n$, we obtain that $\gamma' \cdot \nu' < 1$. But then $1 = \gamma \cdot \nu = \gamma' \cdot \nu' + \gamma_{n+1}\nu_{n+1}$ implies that $\nu_{n+1} > 0$.

Finally we prove that $\nu \in \mathbb{R}_+^{n+1}$ for all $\nu \in N_1(\mathcal{P})$. In fact if this were not the case, we would have that there exists $\nu \in N_1(\mathcal{P})$ with a non-positive component, say $\nu_{n+1} \leq 0$. Let $w = (w', w_{n+1})$ be an internal point to F_ν. Let F_σ, $\sigma = (\sigma', \sigma_{n+1})$, be any face intersecting the coordinate hyperplane $z_{n+1} = 0$. We must have $w \cdot \sigma \leq 1$ because $w \in \mathcal{P}$ and, because $\sigma_{n+1} > 0$ by what we have just proved, we have $w' \cdot \sigma' < 1$. So $(w', 0)$ is internal to the intersection of $\mathcal{P}$ with the hyperplane $z_{n+1} = 0$ and hence, in particular, we must have $w' \cdot \nu' < 1$, but this is impossible because $w' \cdot \nu' = 1 - w_{n+1}\nu_{n+1} \geq 1$, since $\nu_{n+1} \leq 0$. $\qquad\square$

Remark 2.6.4. The Newton polyhedron $\mathcal{P}$ may be complete without $p(z)$ being H-type. For example $p(x, \xi) = (x - \xi)^2$, $z = (x, \xi) \in \mathbb{R}^2$, is not H-type but its Newton polyhedron is the triangle with vertices $(2, 0)$, $(0, 2)$, $(0, 0)$, which is complete.

Example 2.6.5. As in the preceding Section 2.5 consider

$$p(z) = \sum_{\langle \gamma, M \rangle \leq m} c_\gamma z^\gamma.$$

If we assume quasi-ellipticity, the terms z_j^{m/M_j} appear in the expression, with non-zero coefficients, cf. Remark 2.5.6. Hence the Newton polyhedron is given by

$$\mathcal{P} = \Big\{ z \in \mathbb{R}^n : \ z \geq 0, \ \sum_{j=1}^n M_j z_j/m \leq 1 \Big\},$$

which is complete, according to the previous Theorem 2.6.3, with vertices

$$V(\mathcal{P}) = \{0, me_1/M_1, \ldots, me_n/M_n\}.$$

We have $\mu_0 = \min\{m/M_j\}$, $\mu_1 = \mu = m$. The function $\Lambda_{\mathcal{P}}(z)$ is given by

$$\Big(1 + \sum_{j=1}^n z_j^{2m/M_j}\Big)^{1/2} \asymp |z|_M^m$$

for large $|z|$, cf. Remark 2.5.6. The Γ-elliptic case, considered in Section 2.2, corresponds to taking $M_1 = \ldots = M_n = 1$ in the preceding formulas.   □

Theorem 2.6.6. *Let $p(z)$ be a polynomial and let $\mathcal{P}$ be its Newton polyhedron. Define as in (2.6.2)*

$$\Lambda_{\mathcal{P}}(z) = \Big(\sum_{\gamma \in V(\mathcal{P})} z^{2\gamma} \Big)^{1/2}.$$

Then we have

$$|p(z)| \lesssim \Lambda_{\mathcal{P}}(z), \quad z \in \mathbb{R}^n. \tag{2.6.5}$$

The proof is not so obvious as for (2.5.8), because of the lack of the quasi-homogeneous structure. We need the following lemma.

Lemma 2.6.7. *Given $z \in (\mathbb{R}_+ \cup \{0\})^n$, a finite subset $A \subset (\mathbb{R}_+ \cup \{0\})^n$ and a convex linear combination $\beta = \sum_{\alpha \in A} c_\alpha \alpha$, we have*

$$z^\beta \leq \sum_{\alpha \in A} c_\alpha z^\alpha. \tag{2.6.6}$$

Proof. We argue by induction on the number N of the elements of A. If $N = 1$, the estimate (2.6.6) is obvious. If $N > 1$, choosing any $\gamma \in A$, let $A' = A \setminus \{\gamma\}$. If $c_\gamma = 1$, then $c_\alpha = 0$ for all $\alpha \in A'$ and there is nothing to prove.

If $c_\gamma < 1$, then $c' = \sum_{\alpha \in A'} c_\alpha > 0$. If we set $\gamma' = \frac{1}{c'} \sum_{\alpha \in A'} c_\alpha \alpha$, we have $\beta = c'\gamma' + (1 - c')\gamma$. Let us consider the function

$$f(t) = z^{t\gamma' + (1-t)\gamma} = \big(z^{\gamma'}\big)^t \big(z^\gamma\big)^{1-t}.$$

We have that $f'' \geq 0$, so that f is convex and

$$z^\beta \leq c' z^{\gamma'} + (1 - c')z^\gamma = c' z^{\gamma'} + c_\gamma z^\gamma.$$

On the other hand, by the inductive hypothesis, we have

$$z^{\gamma'} \leq \sum_{\alpha \in A'} c'_\alpha z^\alpha$$

with $c'_\alpha = c_\alpha/c'$. Therefore

$$z^\beta \leq c' \sum_{\alpha \in A'} c'_\alpha z^\alpha + c_\gamma z^\gamma = \sum_{\alpha \in A} c_\alpha z^\alpha. \qquad \square$$

Proof of Theorem 2.6.6. Observe first that

$$\Lambda_{\mathcal{P}}(z) \asymp \sum_{\gamma \in V(\mathcal{P})} |z^\gamma|. \qquad (2.6.7)$$

To prove (2.6.5), it will be sufficient to estimate every term of the type $|z^\beta|$ with $\beta \in \mathbb{N}^n$ belonging to the Newton polyhedron $\mathcal{P}$. Such β can be regarded as a convex linear combination $\beta = \sum_{\alpha \in A} c_\alpha \alpha$ with $A = V(\mathcal{P})$. Hence applying Lemma 2.6.7 we obtain

$$|z^\beta| \lesssim \sum_{\gamma \in V(\mathcal{P})} |z^\gamma|.$$

In view of (2.6.7), this concludes the proof of Theorem 2.6.6 $\qquad \square$

Having as model (2.5.9) in Proposition 2.5.2, we now define the multi-quasi-elliptic polynomials.

Definition 2.6.8. We say that the polynomial $p(z)$ is multi-quasi-elliptic if the corresponding Newton polyhedron is complete and for some $R > 0$:

$$\Lambda_{\mathcal{P}}(z) \lesssim |p(z)|, \quad |z| \geq R. \qquad (2.6.8)$$

Theorem 2.6.9. *Let $p(z)$ be multi-quasi-elliptic. Then*

$$|\partial_z^\gamma p(z)| \lesssim |p(z)| \Lambda_{\mathcal{P}}(z)^{-|\gamma|/\mu}, \quad |z| \geq R, \qquad (2.6.9)$$

for some $R > 0$, and $p(z)$ is ρ-H–type for

$$\rho = \mu_0/\mu \qquad (2.6.10)$$

where μ_0 and μ are the minimum and the formal order of $\mathcal{P}$, cf. (2.6.3), (2.6.4).

Proof. Let us show that, for each $\beta \in \mathbb{N}^n$,

$$|\partial^\beta p(z)| \lesssim \Lambda_{\mathcal{P}}(z)^{1-|\beta|/\mu}, \quad \text{for all } z \in \mathbb{R}^n. \qquad (2.6.11)$$

Let $\gamma \in A$ with A given by (2.6.1), and consider $\beta \in \mathbb{N}^n$. We have to prove that

$$|\partial^\beta z^\gamma| \lesssim \Lambda_{\mathcal{P}}(z)^{1-|\beta|/\mu}.$$

If we do not have $\beta \leq \gamma$, then $\partial^\beta z^\gamma = 0$ and there is nothing to prove. Since $\mu \geq \mu_1 \geq |\gamma|$, cf. (2.6.3), the conclusion is obvious also for $\beta = \gamma$. Assume that $\beta < \gamma$. Then for each $\nu \in N_1(\mathcal{P})$ we have

$$0 < (\gamma - \beta) \cdot \nu \leq 1 - \beta \cdot \nu \leq 1 - \frac{|\beta|}{\mu}.$$

Hence

$$\frac{\mu}{\mu - |\beta|}(\gamma - \beta) \in \mathcal{P}$$

and, by Lemma 2.6.7,

$$\left| z^{\gamma - \beta} \right|^{\frac{\mu}{\mu - |\beta|}} \lesssim \Lambda_{\mathcal{P}}(z),$$

which gives

$$\left| \partial^\beta z^\gamma \right| \lesssim \Lambda_{\mathcal{P}}(z)^{1 - |\beta|/\mu}.$$

This completes the proof of (2.6.11).

Using the multi-quasi-ellipticity assumption, from (2.6.11) we obtain (2.6.9). Finally, since $\langle z \rangle^{\mu_0} \lesssim \Lambda_{\mathcal{P}}(z)$, we have from (2.6.9),

$$\left| \partial^\gamma p(z) \right| \lesssim |p(z)| \langle z \rangle^{-|\gamma|\mu_0/\mu}, \tag{2.6.12}$$

i.e., $p(z)$ is ρ-H-type with ρ as in (2.6.10). □

Example 2.6.10. Quasi-elliptic polynomials are multi-quasi-elliptic, since the estimate (2.5.9) is equivalent to (2.6.8), cf. (2.5.19) and Example 2.6.5. A simple example of multi-quasi-elliptic polynomial, which is not quasi-elliptic, is given for $z = (x, \xi) \in \mathbb{R}^2$ by

$$\xi^{h_1} + x^{k_2}\xi^{h_2} + x^{k_1}, \tag{2.6.13}$$

where h_j, k_j, $j = 1, 2$, are positive integers, $h_1 > h_2$, $k_1 > k_2$, and $\frac{k_2}{k_1} + \frac{h_2}{h_1} > 1$. The associated ordinary differential operator is

$$P = D_x^{h_1} + x^{k_2} D_x^{h_2} + x^{k_1}. \tag{2.6.14}$$

The Newton polyhedron is given by

$$\mathcal{P} = \{(x, \xi) \in \mathbb{R}^2 : x \geq 0, \; \xi \geq 0, \; \frac{x}{k_1} + \frac{(k_1 - k_2)}{h_2 k_1}\xi \leq 1, \; \frac{(h_1 - h_2)}{k_2 h_1}x + \frac{\xi}{h_1} \leq 1\}, \tag{2.6.15}$$

which is complete, with vertices

$$V(\mathcal{P}) = \{(0, 0), \; (k_1, 0), \; (0, h_1), \; (h_2, k_2)\}. \tag{2.6.16}$$

We have $\mu_0 = \min\{h_1, k_1\}$, $\mu_1 = h_2 + k_2$,

$$\mu = \max\left\{ \frac{h_2 k_1}{k_1 - k_2}, \; \frac{k_2 h_1}{h_1 - h_2} \right\}. \tag{2.6.17}$$

If we assume that h_j, k_j, $j = 1, 2$, are even, we have

$$\Lambda_{\mathcal{P}}(x, \xi) \asymp \xi^{h_1} + x^{k_2}\xi^{h_2} + x^{k_1}, \quad |x| + |\xi| \geq R, \tag{2.6.18}$$

for every $R > 0$, and the polynomial in (2.6.13) is obviously multi-quasi-elliptic.

An alternative definition of multi-quasi-ellipticity is given by the following proposition.

Proposition 2.6.11. *The polynomial $p(z)$ is multi-quasi-elliptic if and only if the corresponding Newton polyhedron $\mathcal{P}$ is complete and there exists $R > 0$ such that*

$$\Lambda_{\mathcal{P}}(z) \lesssim |p_{\mathcal{P},\mu}(z)|, \quad |z| \geq R, \tag{2.6.19}$$

where $p_{\mathcal{P},\mu}(z)$, for short $p_\mu(z)$ in the following, is the $\mathcal{P}$-principal part of $p(z)$:

$$p_\mu(z) = \sum_{\gamma \in F(\mathcal{P})} c_\gamma z^\gamma \tag{2.6.20}$$

with

$$F(\mathcal{P}) = \bigcup_{\nu \in N_1(\mathcal{P})} F_\nu(\mathcal{P}). \tag{2.6.21}$$

We recall that μ denotes the formal order of $\mathcal{P}$ and for $\nu \in N_1(\mathcal{P})$ we have $F_\nu(\mathcal{P}) = \{z \in \mathcal{P} : \nu \cdot z = 1\}$.

We need the following lemma, whose proof is an obvious application of Lemma 2.6.7.

Lemma 2.6.12. *For every $\gamma \in (\mathbb{R}_+ \cup \{0\})^n$, in particular for all $\gamma \in \mathbb{N}^n$, we have*

$$|z^\gamma| \lesssim \Lambda_{\mathcal{P}}(z)^{k(\mathcal{P},\gamma)} \tag{2.6.22}$$

where

$$k(\mathcal{P}, \gamma) = \inf\{t > 0, \ t^{-1}\gamma \in \mathcal{P}\} = \max_{\nu \in N_1(\mathcal{P})} \nu \cdot \gamma. \tag{2.6.23}$$

Proof of Proposition 2.6.11. We split $p(z) = p_\mu(z) + \tilde{p}(z)$ with

$$\tilde{p}(z) = \sum_{\gamma \in \mathcal{P} \setminus F(\mathcal{P})} c_\gamma z^\gamma.$$

From Lemma 2.6.12 we have for some $C > 0$

$$|\tilde{p}(z)| \leq C\Lambda_{\mathcal{P}}(z)^\delta, \quad z \in \mathbb{R}^n,$$

with

$$\delta = \max_{\gamma \in \mathcal{P} \setminus F(\mathcal{P})} k(\mathcal{P}, \gamma) < 1,$$

in view of (2.6.20), (2.6.21) and (2.6.22), (2.6.23). For any fixed $\epsilon > 0$, we then obtain in the region $|z| \geq R$:

$$|p(z) - p_\mu(z)| = |\tilde{p}(z)| \leq \epsilon \Lambda_{\mathcal{P}}(z)$$

provided R is so large that $\Lambda_{\mathcal{P}}(z)^{\delta-1} \leq \epsilon C^{-1}$ in the same region. Taking sufficiently small ϵ, we get the equivalence of (2.6.19) and (2.6.8). Hence Proposition 2.6.11 is proved. □

We may now explain the name "multi-quasi-elliptic polynomials" attributed to polynomials satisfying (2.6.19), i.e., (2.6.8).

Proposition 2.6.13. *Let $p(z)$ be a polynomial of the variable $z = (x, \xi) \in \mathbb{R}^2$. Assume $p(z)$ is multi-quasi-elliptic, with complete Newton polyhedron $\mathcal{P}$. Then we can always find a product of quasi-elliptic polynomials of type (2.5.13), whose $\mathcal{P}$-principal part coincides with that of $p(x, \xi)$; precisely:*

$$p(x, \xi) = c_0(\xi^{h_1} + r_1 x^{k_1}) \cdots (\xi^{h_M} + r_M x^{k_M}) + \sum_{(\beta,\alpha)\in\mathcal{P}\backslash F(\mathcal{P})} c_{\alpha\beta} x^\beta \xi^\alpha, \quad (2.6.24)$$

for suitable positive integers h_j, k_j prime to each other and complex constants c_0, r_j, $c_{\alpha\beta}$, with $c_0 \neq 0$, $\operatorname{Im} r_j \neq 0$ for $j = 1,\ldots, M$. Correspondently, $\mathcal{P}$ can be decomposed into an algebraic sum of triangles:

$$\mathcal{P} = \mathcal{P}_{h_1,k_1} + \ldots + \mathcal{P}_{h_M,k_M}, \quad (2.6.25)$$

where $\mathcal{P}_{h_j,k_j}$ has vertices $\{(0,0), (0,h_j), (k_j,0), \ j = 1,\ldots, M\}$.

Proof. To prove (2.6.24), (2.6.25) we consider the principal part p_μ of p, by ordering $(\beta, \alpha) \in F(\mathcal{P})$:

$$p_\mu(x, \xi) = \sum_{i=0}^{M} a_i x^{\beta_i} \xi^{\alpha_i}$$

so that $\alpha_M > \alpha_{M-1} > \ldots > \alpha_0 = 0$, $0 = \beta_M < \beta_{M-1} < \ldots < \beta_0$. Note that $a_0 \neq 0$, $a_M \neq 0$. We begin to determine the integers h, k, with no common factor, and $t \geq 1$ by means of the conditions

$$k = |\beta_M - \beta_{M-1}| = \ldots = |\beta_{M-t+1} - \beta_{M-t}|,$$
$$h = \alpha_M - \alpha_{M-1} = \ldots = \alpha_{M-t+1} - \alpha_{M-t},$$
$$k \neq |\beta_{M-t} - \beta_{M-t-1}| \text{ or else } h \neq \alpha_{M-t} - \alpha_{M-t-1}.$$

We can then write

$$p_\mu(x, \xi) = q(x, \xi)(\xi^h + rx^k) + \sum_{(\beta,\alpha)\in\mathcal{P}\backslash F(\mathcal{P})} \tilde{c}_\beta x^\beta \xi^\alpha,$$

with

$$q(x,\xi) = \sum_{i=0}^{M-1} b_i x^{\beta_i} \xi^{\alpha_i - h},$$

where $r, b_i \in \mathbb{C}$ are obtained by solving the system

$$\begin{cases} a_M = r b_{M-1} \\ a_{M-1} = b_{M-1} + r b_{M-2} \\ \cdots \\ a_{M-t+1} = b_{M-t+1} + r b_{M-t} \\ a_{M-t} = b_{M-t} \\ \cdots \\ a_0 = b_0 \end{cases}$$

and $\tilde{c}_{\alpha\beta}$ are suitable complex constants. Writing respectively $\mathcal{Q}$ and $\mathcal{P}_{h,k}$ for the Newton polyhedron of $q(x,\xi)$ and $\xi^h + r x^k$, we have

$$\mathcal{P} = \mathcal{Q} + \mathcal{P}_{h,k}.$$

It is easy to see that the multi-quasi-ellipticity of p implies the multi-quasi ellipticity of q and $\xi^h + r x^k$, which in turn forces $\operatorname{Im} r \neq 0$, cf. Example 2.5.4.

Iterating the procedure, we get (2.6.24), (2.6.25). □

Example 2.6.14. In $\mathbb{R}^n$, $n > 2$, there exist multi-quasi-elliptic polynomials which cannot be factorized in terms of quasi-elliptic polynomials. Consider for example in $\mathbb{R}^4$, with $z = (x_1, x_2, \xi_1, \xi_2)$,

$$p(x,\xi) = \xi_1^2 + \xi_2^2 + V(x_1, x_2), \tag{2.6.26}$$

corresponding to the operator in $\mathbb{R}^2$,

$$-\Delta + V(x_1, x_2). \tag{2.6.27}$$

Let us assume that the potential V is of the form (2.6.13) with respect to the variables x_1, x_2, say to be definite

$$V(x_1, x_2) = x_1^6 + x_1^4 x_2^4 + x_2^6. \tag{2.6.28}$$

Then $p(x,\xi)$ in (2.6.26) is multi-quasi-elliptic, but it cannot be factorized as before. □

Finally, we observe that multi-quasi-elliptic polynomials do not exhaust the class of the H-polynomials, as the following example shows.

Example 2.6.15. Let, for $(x,\xi) \in \mathbb{R}^2$,

$$p(x,\xi) = \xi^2 + i x \xi - x^3, \tag{2.6.29}$$

corresponding in the standard quantization to the ordinary differential operator

$$P = D_x^2 + ixD_x - x^3.$$ (2.6.30)

We have that p is not multi-quasi-elliptic. In fact, let $\mathcal{P}$ be the Newton polyhedron of p. Then $\mathcal{P}$ is complete, with vertices

$$(0,0),\ (3,0),\ (0,2).$$

Assuming p multi-quasi-elliptic, we would have

$$|p(x,\xi)|^2 = (\xi^2 - x^3)^2 + x^2\xi^2 \gtrsim 1 + \xi^4 + x^6$$

for $x^2 + \xi^2 \geq R > 0$. But this is contradicted on the curve $\xi^2 = x^3$.

Let us show that $p(x,\xi)$ is H-type. Using Proposition 2.4.4 (c), we may limit ourselves to checking that

$$\partial_x p(x,\xi)/p(x,\xi) \to 0 \quad \text{and} \quad \partial_\xi p(x,\xi)/p(x,\xi) \to 0$$ (2.6.31)

for $(x,\xi) \to \infty$. In turn, (2.6.31) reduces readily to prove that

$$\xi^2/|p(x,\xi)|^2 \to 0, \quad x^4/|p(x,\xi)|^2 \to 0 \quad \text{as } (x,\xi) \to \infty.$$ (2.6.32)

If $x < 0$ the proof of (2.6.32) is easy. Assume $x \geq 0$ and, without loss of generality, also $\xi \geq 0$. We discuss the validity of (2.6.32) separately in three regions:

$$D_1 = \left\{\xi^2 \leq \frac{1}{2}x^3\right\}, \quad D_2 = \left\{\frac{1}{2}x^3 \leq \xi^2 \leq 2x^3\right\}, \quad D_3 = \left\{x^3 \leq \frac{1}{2}\xi^2\right\}.$$

In D_1 we have

$$|p(x,\xi)|^2 \geq (\xi^2 - x^3)^2 \gtrsim x^6.$$

In D_2 we estimate

$$|p(x,\xi)|^2 \geq x^2\xi^2 \asymp x^5 \asymp \xi^{10/3}.$$

In D_3 we have

$$|p(x,\xi)|^2 \geq (\xi^2 - x^3)^2 \gtrsim \xi^4.$$

Taking into account that $\xi^2 \lesssim x^3$ in D_1 and $x^4 \lesssim \xi^{8/3}$ in D_3, we obtain (2.6.32).

2.7 Γ_𝒫-Pseudo-Differential Operators

Quasi-elliptic and multi-quasi-elliptic polynomials in $\mathbb{R}^{2d}$ are Γ_ρ-hypoelliptic symbols in $\Gamma_\rho^m(\mathbb{R}^d)$ for some ρ with $0 < \rho \leq 1$, see Definition 2.1.15, Proposition 2.5.3 and Theorem 2.6.9. We may then apply Theorem 2.1.16 to the corresponding operators, obtaining in particular global regularity. However, the algebraic estimates (2.5.10) and (2.6.9) suggest that we may organize a more precise symbolic calculus, provided $|z|_M$ and $\Lambda_\mathcal{P}(z)$ are admissible weight functions. Namely, we have to check that $\Phi(z) = \Psi(z) = |z|_M$, or $\Phi(z) = \Psi(z) = \Lambda_{\mathcal{P}}(z)^\rho$, for some range of exponents ρ, are sub-linear and temperate weights, cf. (1.1.1), (1.1.2), satisfying the strong uncertainty principle (1.1.10).

To this end, re-starting from scratch, we consider a generic complete polyhedron $\mathcal{P} \subset (\mathbb{R}_+ \cup \{0\})^n$, cf. Definition 2.6.2, and we define as in (2.6.2):

$$\Lambda_{\mathcal{P}}(z) = \left(\sum_{\gamma \in V(\mathcal{P})} z^{2\gamma} \right)^{1/2}, \quad z \in \mathbb{R}^n, \tag{2.7.1}$$

where $V(\mathcal{P})$ is the set of the vertices of $\mathcal{P}$. As we observed after Definition 2.6.2 we have

$$\langle z \rangle^{\mu_0} \lesssim \Lambda_{\mathcal{P}}(z) \lesssim \langle z \rangle^{\mu_1}, \quad z \in \mathbb{R}^n, \tag{2.7.2}$$

where μ_0 and μ_1 are the minimum and the maximum order of $\mathcal{P}$, $0 < \mu_0 \le \mu_1$. Another obvious remark is that, for every $t \in \mathbb{R}$,

$$\Lambda_{\mathcal{P}}(tz) \le C\Lambda_{\mathcal{P}}(z), \quad z \in \mathbb{R}^n, \tag{2.7.3}$$

where the constant C depends on t. The next lemma summarizes the main properties of the function $\Lambda_{\mathcal{P}}(z)$. We use below the notation of Section 2.6, in particular the formal order μ of $\mathcal{P}$ is defined by

$$\mu = \max \left\{ \frac{1}{\nu_j} : j = 1, \ldots, n, \ \nu \in N_1(\mathcal{P}) \right\}. \tag{2.7.4}$$

Lemma 2.7.1. *Let $\mathcal{P}$ be a complete polyhedron in $\mathbb{R}^n$ and let $\Lambda_{\mathcal{P}}(z)$ be defined as in (2.7.1). Then for every $\alpha \in \mathbb{N}^n$, $\beta \in \mathbb{N}^n$ the following estimate holds:*

$$|\xi^\alpha \partial^{\alpha+\beta} \Lambda_{\mathcal{P}}(z)| \lesssim \Lambda_{\mathcal{P}}(z)^{1-|\beta|/\mu}, \quad z \in \mathbb{R}^n, \tag{2.7.5}$$

where μ is the formal order of $\mathcal{P}$.

Proof. We argue by induction on $k = |\alpha + \beta|$. For $k = 0$ the estimate (2.7.5) is trivially satisfied with $C_{0,0} = 1$. For a fixed $k \in \mathbb{N}$, let us assume that (2.7.5) holds for any $\alpha, \beta \in \mathbb{N}^n$ with $|\alpha + \beta| \le k$. Consider now $\alpha, \beta \in \mathbb{N}^n$ such that $|\alpha + \beta| = k + 1$. From (2.7.1) we obtain:

$$\partial^{\alpha+\beta} (\Lambda_{\mathcal{P}}(z)^2) = \sum_{\substack{\chi \in V(\mathcal{P}) \\ 2\chi \ge \alpha+\beta}} (\alpha + \beta)! \binom{2\chi}{\alpha + \beta} z^{2\chi-\alpha-\beta}. \tag{2.7.6}$$

So, by Leibniz' formula, we get

$$\partial^{\alpha+\beta} \Lambda_{\mathcal{P}}(z) = \frac{1}{2\Lambda_{\mathcal{P}}(z)} \left\{ \sum_{\substack{\chi \in V(\mathcal{P}) \\ 2\chi \ge \alpha+\beta}} (\alpha + \beta)! \binom{2\chi}{\alpha + \beta} z^{2\chi-\alpha-\beta} \right.$$

$$\left. - \sum_{\substack{\delta \le \beta, \eta \le \alpha \\ (\eta,\delta) \ne (0,0),\ (\eta,\delta) \ne (\alpha,\beta)}} \binom{\alpha}{\eta} \binom{\beta}{\delta} \partial^{\eta+\delta} \Lambda_{\mathcal{P}}(z) \partial^{\alpha-\eta+\beta-\delta} \Lambda_{\mathcal{P}}(z) \right\}, \tag{2.7.7}$$

whence

$$|z^\beta \partial^{\alpha+\beta} \Lambda_{\mathcal{P}}(z)| \leq \frac{1}{2\Lambda_{\mathcal{P}}(z)} \left\{ \sum_{\substack{\chi \in V(\mathcal{P}) \\ 2\chi \geq \alpha+\beta}} (\alpha+\beta)! \binom{2\chi}{\alpha+\beta} |z^{2\chi-\alpha}| \right.$$

$$+ \sum_{\substack{\delta \leq \beta, \eta \leq \alpha \\ (\eta,\delta) \neq (0,0), \ (\eta,\delta) \neq (\alpha,\beta)}} \binom{\alpha}{\eta}\binom{\beta}{\delta} \left. |z^\delta \partial^{\eta+\delta} \Lambda_{\mathcal{P}}(z)| \, |z^{\beta-\delta} \partial^{\alpha-\eta+\beta-\delta} \Lambda_{\mathcal{P}}(z)| \right\}. \quad (2.7.8)$$

From inductive assumption, we have

$$|z^\delta \partial^{\eta+\delta} \Lambda_{\mathcal{P}}(z)| \lesssim \Lambda_{\mathcal{P}}(z)^{1-\frac{|\eta|}{\mu}}, \quad z \in \mathbb{R}^n \qquad (2.7.9)$$

and

$$|z^{\beta-\delta} \partial^{\alpha-\eta+\beta-\delta} \Lambda_{\mathcal{P}}(z)| \lesssim \Lambda_{\mathcal{P}}(z)^{1-\frac{|\alpha-\eta|}{\mu}}, \quad z \in \mathbb{R}^n. \qquad (2.7.10)$$

We can prove now that

$$|z^{2\chi-\alpha}| \lesssim \Lambda_{\mathcal{P}}(z)^{2-\frac{|\alpha|}{\mu}}, \quad z \in \mathbb{R}^n. \qquad (2.7.11)$$

The argument follows closely the proof of (2.6.11). In fact, if $2\chi = \alpha$ ($\beta = 0$), $z^{2\chi-\alpha} \equiv 1$ and $|\alpha| = 2|\chi| \leq 2\mu_1 \leq 2\mu$ ($\mu_1 := \max_{\chi \in V(\mathcal{P})} |\chi|$), so that $\Lambda_{\mathcal{P}}(z)^{2-\frac{|\alpha|}{\mu}} \geq 1$ and the inequality (2.7.11) is trivially verified.

Let us consider now the case when $2\chi > \alpha$. As $\chi \in V(\mathcal{P}) \subset \mathcal{P}$, we have $\chi \cdot \nu \leq 1$ and, from the definition of μ, $\alpha \cdot \nu \geq \frac{1}{\mu}|\alpha|$, when $\nu \in N_1(\mathcal{P})$. Since $2\mu - |\alpha| > 0$, the previous inequalities yield $\frac{\mu}{2\mu-|\alpha|}(2\chi - \alpha) \cdot \nu \leq 1$, as $\nu \in N_1(\mathcal{P})$, and then $\frac{\mu}{2\mu-|\alpha|}(2\chi - \alpha) \in \mathcal{P}$. So $|z^{2\chi-\alpha}|^{\frac{\mu}{2\mu-|\alpha|}} \leq \Lambda_{\mathcal{P}}(z)$, whence the estimate (2.7.11) follows.

So estimates (2.7.9), (2.7.10) and (2.7.11), jointly with (2.7.8), give (2.7.5) for $|\alpha+\beta| = k+1$ and conclude the proof. □

Let us detail some consequences of Lemma 2.7.1.

Through the following we will set for brevity $\mathbb{K} := \{\gamma \in \mathbb{N}^n : \gamma_j \in \{0,1\}, j = 1, \ldots, n\}$.

Proposition 2.7.2. *Let $\Lambda_{\mathcal{P}}(z)$ be defined as in (2.7.1). Then for any $m \in \mathbb{R}$, $\alpha \in \mathbb{N}^n$, $\gamma \in \mathbb{K}$ we have*

$$|z^\gamma \partial^{\alpha+\gamma} \Lambda_{\mathcal{P}}(z)^m| \lesssim \Lambda_{\mathcal{P}}(z)^{m-|\alpha|/\mu}, \quad z \in \mathbb{R}^n. \qquad (2.7.12)$$

Proof. By induction one can easily prove the estimates

$$|z^\gamma \partial^{\alpha+\gamma} (\Lambda_{\mathcal{P}}(z))^m|$$

$$\leq \sum_{k=1}^{|\alpha+\gamma|} C_{m,k} \Lambda_{\mathcal{P}}(z)^{m-k} \sum C_{\alpha,\gamma,k} |z^{\gamma_1} \partial^{\alpha_1+\gamma_1} \Lambda_{\mathcal{P}}(z)| \cdots |z^{\gamma_k} \partial^{\alpha_k+\gamma_k} \Lambda_{\mathcal{P}}(z)|,$$

for any $\alpha \in \mathbb{N}^n$ and $\gamma \in \mathbb{K}$ with $|\alpha + \gamma| > 0$. For any $1 \leq k \leq |\alpha + \gamma|$, the multi-indices in the second sum of the right-hand side span over all the $2k$-tuples $(\gamma_1, \ldots, \gamma_k, \alpha_1, \ldots \alpha_k) \in \mathbb{K}^k \times (\mathbb{N}^n)^k$ such that $\gamma_1 + \ldots + \gamma_k = \gamma$ and $\alpha_1 + \ldots + \alpha_k = \alpha$, cf. the proof of Lemma 1.3.5.

In order to complete the proof, it suffices now to apply the estimates (2.7.5) to each term $|z^{\gamma_J} \partial^{\alpha_J + \gamma_J} \Lambda_\mathcal{P}(z)|$, $J = 1, \ldots, k$. $\qquad\square$

Proposition 2.7.3. *For $\Lambda_\mathcal{P}(z)$ defined as before, there exist positive constants C and ϵ such that for all $z, \zeta \in \mathbb{R}^n$:*

(i) $\left| \Lambda_\mathcal{P}(z)^{\frac{1}{\mu}} - \Lambda_\mathcal{P}(\zeta)^{\frac{1}{\mu}} \right| \leq C |z - \zeta|$;

(ii) $C^{-1} \leq \frac{\Lambda_\mathcal{P}(z)}{\Lambda_\mathcal{P}(\zeta)} \leq C$ *if* $|z - \zeta| \leq \epsilon \Lambda_\mathcal{P}(z)^{\frac{1}{\mu}}$.

Proof. By means of Taylor's expansion we have

$$\Lambda_\mathcal{P}(\zeta)^{\frac{1}{\mu}} - \Lambda_\mathcal{P}(z)^{\frac{1}{\mu}} = \sum_{|\alpha|=1} \frac{1}{\alpha!} (\zeta - z)^\alpha \partial^\alpha \Lambda_\mathcal{P}(\theta)^{\frac{1}{\mu}}, \qquad (2.7.13)$$

where $\theta = t\zeta + (1-t)z$, for some $0 < t < 1$. Then:

$$\left| \Lambda_\mathcal{P}(\zeta)^{\frac{1}{\mu}} - \Lambda_\mathcal{P}(z)^{\frac{1}{\mu}} \right| \leq \sum_{|\alpha|=1} \frac{1}{\alpha!} |\zeta - z| \left| \partial^\alpha \Lambda_\mathcal{P}(\theta)^{\frac{1}{\mu}} \right|, \qquad (2.7.14)$$

proving assertion (i), since for $|\alpha| = 1$, $\left| \partial^\alpha \Lambda_\mathcal{P}(\theta)^{\frac{1}{\mu}} \right| \leq c_\alpha$ in view of (2.7.12).

Let us assume now that $|\zeta - z| \leq \epsilon \Lambda_\mathcal{P}(z)^{\frac{1}{\mu}}$. From (2.7.14) again, we have $\left| \Lambda_\mathcal{P}(\zeta)^{\frac{1}{\mu}} - \Lambda_\mathcal{P}(z)^{\frac{1}{\mu}} \right| \leq C_1 \epsilon \Lambda_\mathcal{P}(z)^{\frac{1}{\mu}}$ that is $(1 - C_1\epsilon)^\mu \Lambda_\mathcal{P}(z) \leq \Lambda_\mathcal{P}(\zeta) \leq (1 + C_1\epsilon)^\mu \Lambda_\mathcal{P}(z)$, which for a suitable $\epsilon > 0$ shows (ii). $\qquad\square$

Corollary 2.7.4. *Let $0 < \rho \leq \frac{1}{\mu}$. Then $\Phi(z) = \Psi(z) = \Lambda_\mathcal{P}(z)^\rho$, $z = (x, \xi) \in \mathbb{R}^{2d}$, are sub-linear and temperate weights, cf. (1.1.1), (1.1.2), satisfying the strong uncertainty principle (1.1.10).*

Proof. Since in (2.7.1) $0 \in V(\mathcal{P})$ (cf. (2) in Definition 2.6.2), we have $\Lambda_\mathcal{P}(z) \geq 1$. More precisely, $\Lambda_\mathcal{P}(z) \gtrsim \langle z \rangle^{\mu_0}$ with $\mu_0 > 0$, cf. (2.7.2). Sub-linear growth in (1.1.1) follows from (i) in Proposition 2.7.3, by fixing there $\zeta = 0$.

It is clear that it suffices to verify the temperance estimate (1.1.2) for a fixed power of $\Lambda_\mathcal{P}(z)$. Now, by (i) in Proposition 2.7.3,

$$\Lambda_\mathcal{P}(\zeta)^{\frac{1}{\mu}} \lesssim \Lambda_\mathcal{P}(z)^{\frac{1}{\mu}} + |z - \zeta|, \quad z \in \mathbb{R}^{2d}, \zeta \in \mathbb{R}^{2d}, \qquad (2.7.15)$$

which gives the temperance estimate (1.1.2) for $\Lambda_\mathcal{P}(z)^{\frac{1}{\mu}}$ when $|z - \zeta| \lesssim \Lambda_\mathcal{P}(z)^{\frac{1}{\mu}}$. On the other hand, when $\Lambda_\mathcal{P}(z)^{\frac{1}{\mu}} \lesssim |z - \zeta|$, (2.7.15) yields

$$\Lambda_\mathcal{P}(\zeta)^{\frac{1}{\mu}} \lesssim |z - \zeta| \leq \Lambda_\mathcal{P}(z)^{\frac{1}{\mu}}(1 + |z - \zeta|),$$

since $\Lambda_{\mathcal{P}}(z)^{\frac{1}{\mu}} \geq 1$. Hence (1.1.2) is proved for $\Lambda_{\mathcal{P}}(z)^{\frac{1}{\mu}}$.

Finally, since $\rho > 0$ the strong uncertainty principle (1.1.10) is satisfied as well, by the lower bound in (2.7.2). □

Remark 2.7.5. Note that the slow variation conditions (1.3.8) are also satisfied by $\Phi(z) = \Psi(z) = \Lambda_{\mathcal{P}}(z)^{\rho}$, and $M(z) = \Lambda_{\mathcal{P}}(z)^m$, $m \in \mathbb{R}$. Indeed they reduce to prove that there exists $\epsilon > 0$ such that

$$\Lambda_{\mathcal{P}}(z) \lesssim \Lambda_{\mathcal{P}}(\zeta) \lesssim \Lambda_{\mathcal{P}}(z) \quad \text{for} \quad |\zeta - z| \leq \epsilon \Lambda_{\mathcal{P}}(z)^{\frac{1}{\mu}}, \qquad (2.7.16)$$

which coincides with (ii) in the previous Proposition 2.7.3.

Example 2.7.6. As in Section 2.5, let $M = (M_1, \ldots, M_n)$ be an n-tuple of rational numbers $M_j \geq 1$ with $\min_j M_j = 1$, and define

$$|z|_M = \sum_{j=1}^{n} |z_j|^{1/M_j} . \qquad (2.7.17)$$

We may choose $m \in \mathbb{N}$ in such a way that all the m/M_j, $M_j = 1, \ldots, n$, are integers. Then we have from Remark 2.5.6 and Example 2.6.5:

$$|z|_M^m \asymp \Lambda_{\mathcal{P}}(\zeta) = \left(1 + \sum_{j=1}^{n} z_j^{2m/M_j}\right)^{1/2}, \quad |z| \geq R > 0, \qquad (2.7.18)$$

with

$$\mathcal{P} = \left\{z \in \mathbb{R}^n : z \geq 0, \ \sum_{j=1}^{n} M_j z_j/m \leq 1\right\}. \qquad (2.7.19)$$

For such $\mathcal{P}$ the formal order μ is given by m.

In view of Corollary 2.7.4, taking $\Phi(z) = \Psi(z) = \Lambda_{\mathcal{P}}(z)^{\frac{1}{m}} \asymp |z|_M$ we obtain a couple of sub-linear and temperate weights. Note that slow variation and temperance can be deduced for $|z|_M$ directly from (2.7.17); however, in view of Proposition 2.7.2, the equivalent function $\Lambda_{\mathcal{P}}(z)^{\frac{1}{m}}$ satisfies the estimates (2.7.12), providing more precise information. □

In the sequel, $\Lambda_{\mathcal{P}}(z)$ is defined as in (2.7.1), corresponding to a complete polyhedron $\mathcal{P}$ in $\mathbb{R}^n$, with $n = 2d$. We understand $z = (x, \xi)$, $x \in \mathbb{R}^d$, $\xi \in \mathbb{R}^d$. As before μ denotes the formal order of $\mathcal{P}$.

Definition 2.7.7. Let $m \in \mathbb{R}$ and $0 < \rho \leq \frac{1}{\mu}$. We denote by $\Gamma^m_{\rho,\mathcal{P}}(\mathbb{R}^d)$, or for short $\Gamma^m_{\rho,\mathcal{P}}$, the class of functions $a \in C^{\infty}(\mathbb{R}^{2d})$ such that, for all $\alpha \in \mathbb{N}^{2d}$,

$$|\partial^{\alpha} a(z)| \lesssim \Lambda_{\mathcal{P}}(z)^{m - \rho|\alpha|}, \quad z \in \mathbb{R}^{2d}. \qquad (2.7.20)$$

We also write $\Gamma^m_{\mathcal{P}}$ for the class $\Gamma^m_{\frac{1}{\mu},\mathcal{P}}$.

In view of Corollary 2.7.4, we may develop a symbolic calculus for the corresponding pseudo-differential operators, by applying the results of Chapter 1. Note that we assume $\rho > 0$, and the strong uncertainty principle (1.1.10) is satisfied, cf. (2.7.2). We may also define Sobolev spaces corresponding to $\Lambda_{\mathcal{P}}(z)$ and obtain Fredholmness on them for the $\Lambda_{\mathcal{P}}$-elliptic operators. The results can be obviously particularized to the case $\Lambda_{\mathcal{P}}(z) \asymp |z|_M^m$. We leave all this to the reader.

In the following we prefer to limit attention to $M\Gamma_{\rho,\mathcal{P}}^m$, subclass of $\Gamma_{\rho,\mathcal{P}}^m$, whose definition below is suggested by the sharp estimates (2.7.12). On the one hand, parametrices of operators with multi-quasi-elliptic polynomials as symbols belong to the corresponding pseudo-differential subclass $OPM\Gamma_{\rho,\mathcal{P}}^m$, so nothing is lost in the application. On the other hand, operators in the subclass $OPM\Gamma_{\rho,\mathcal{P}}^0$ turn out to be bounded on $L^p(\mathbb{R}^d)$, $1 < p < \infty$, as we shall prove in Section 2.8.

Definition 2.7.8. For $m \in \mathbb{R}$ and $0 < \rho \le \frac{1}{\mu}$, we denote by $M\Gamma_{\rho,\mathcal{P}}^m(\mathbb{R}^d)$, for short $M\Gamma_{\rho,\mathcal{P}}^m$, the class of the functions $a(z) \in C^\infty(\mathbb{R}^{2d})$ such that

$$z^\gamma \partial^\gamma a(z) \in \Gamma_{\rho,\mathcal{P}}^m \qquad (2.7.21)$$

for any $\gamma \in \mathbb{K} = \{\gamma \in \mathbb{N}^{2d} : \gamma_j \in \{0,1\}, \ j = 1,\dots,2d\}$. We also write $M\Gamma_{\mathcal{P}}^m$ for the class $M\Gamma_{\frac{1}{\mu},\mathcal{P}}^m$.

Example 2.7.9. Given a complete polyhedron $\mathcal{P}$ in $\mathbb{R}^{2d}$, consider the polynomial

$$p(z) = \sum_{\beta \in \mathcal{P}} c_\beta z^\beta, \quad c_\beta \in \mathbb{C}, \ z \in \mathbb{R}^{2d}. \qquad (2.7.22)$$

We have $p \in M\Gamma_{\mathcal{P}}^1$. In fact, denoting by $\mathcal{Q}$ the Newton polyhedron of $p(z)$, we have $\mathcal{Q} \subset \mathcal{P}$, hence $\Lambda_{\mathcal{Q}}(z) \le \Lambda_{\mathcal{P}}(z)$. From Theorem 2.6.6 we then obtain

$$|p(z)| \lesssim \Lambda_{\mathcal{Q}}(z) \le \Lambda_{\mathcal{P}}(z), \quad z \in \mathbb{R}^{2d}.$$

From the first part of the proof of Theorem 2.6.9, cf. in particular (2.6.11), we have the more precise information

$$|\partial^\alpha p(z)| \lesssim \Lambda_{\mathcal{P}}(z)^{1-|\alpha|/\mu}, \quad z \in \mathbb{R}^{2d}.$$

This shows that $p \in \Gamma_{\mathcal{P}}^1$. On the other hand, for any $\gamma \in \mathbb{N}^{2d}$ we have

$$z^\gamma \partial^\gamma p(z) = \sum_{\beta \in \mathcal{P}} c_{\gamma\beta} z^\beta$$

for new coefficients $c_{\gamma\beta} \in \mathbb{C}$. The preceding arguments give then $z^\gamma \partial^\gamma p(z) \in \Gamma_{\mathcal{P}}^1$. Hence we obtain $p \in M\Gamma_{\mathcal{P}}^1$. Note that this conclusion does not require the multi-quasi-ellipticity of $p(z)$ in (2.7.22).

We list some propositions; they are obvious variants of the standard symbolic calculus and proofs are omitted.

Proposition 2.7.10. *The following statements are equivalent:*

(1) $a(z) \in M\Gamma^m_{\rho,\mathcal{P}}$;

(2) $z^\gamma \partial^{\alpha+\gamma} a(z) \in \Gamma^{m-\rho|\alpha|}_{\rho,\mathcal{P}}$ *for all* $\alpha \in \mathbb{N}^{2d}$ *and* $\gamma \in \mathbb{K}$;

(3) *for every* $\alpha \in \mathbb{N}^{2d}$ *and* $\gamma \in \mathbb{K}$,

$$|z^\gamma \partial^{\alpha+\gamma} a(z)| \lesssim \Lambda_\mathcal{P}(z)^{m-\rho|\alpha|}, \quad z \in \mathbb{R}^{2d}. \tag{2.7.23}$$

Proposition 2.7.11. *For* $m, m' \in \mathbb{R}$, $0 < \rho' \le \rho \le \frac{1}{\mu}$ *the following properties hold:*

(1) *if* $m \le m'$, *then* $M\Gamma^m_{\rho,\mathcal{P}} \subset M\Gamma^{m'}_{\rho',\mathcal{P}}$;

(2) *if* $a(z) \in M\Gamma^m_{\rho,\mathcal{P}}$ *and* $b(z) \in M\Gamma^{m'}_{\rho,\mathcal{P}}$, *then* $(ab)(z) \in M\Gamma^{m+m'}_{\rho,\mathcal{P}}$;

(3) *if* $a(z) \in M\Gamma^m_{\rho,\mathcal{P}}$, *then* $\partial^\alpha a(z) \in M\Gamma^{m-\rho|\alpha|}_{\rho,\mathcal{P}}$, *for any* $\alpha \in \mathbb{N}^{2d}$.

Proposition 2.7.12. *Let* $m \in \mathbb{R}$ *and* $0 < \rho \le \frac{1}{\mu}$. *Then the following inclusions hold:*

$$\Gamma^{m-N_0}_{\rho,\mathcal{P}} \subset M\Gamma^m_{\rho,\mathcal{P}} \subset \Gamma^m_{\rho,\mathcal{P}}, \tag{2.7.24}$$

where $N_0 := 2d(\frac{1}{\mu_0} - \rho)$, *cf.* (2.7.2).

Definition 2.7.13. We say that a sequence $\{a_j\}_{j\in\mathbb{N}}$ of symbols $a_j(z) \in M\Gamma^{m_j}_{\rho,\mathcal{P}}$, such that $m_j > m_{j+1}$ and $\lim_{j\to\infty} m_j = -\infty$, is an asymptotic expansion for $a(z) \in M\Gamma^{m_0}_{\rho,\mathcal{P}}$ and write

$$a(z) \sim \sum_j a_j(z),$$

if, for any integer $N \ge 1$, $a(z) - \sum_{j<N} a_j(z) \in M\Gamma^{m_N}_{\rho,\mathcal{P}}$.

Proposition 2.7.14. *If* $\{a_j\}_{j\in\mathbb{N}}$ *is a sequence of symbols as in Definition 2.7.13, there exists* $a(z) \in M\Gamma^{m_0}_{\rho,\mathcal{P}}$ *for which* $\{a_j\}_{j\in\mathbb{N}}$ *is an asymptotic expansion. Moreover the symbol* $a(z)$ *is unique modulo* $\Gamma^{-\infty}_{\rho,\mathcal{P}} := \bigcap_{m\in\mathbb{R}} \Gamma^m_{\rho,\mathcal{P}} = \mathcal{S}(\mathbb{R}^{2d})$.

The following notion of *ellipticity* is the same as that in Definition 1.3.1.

Definition 2.7.15. The symbol $a(z) \in \Gamma^m_{\rho,\mathcal{P}}$ is $\mathcal{P}$-elliptic of order m if, for some constant $R > 0$,

$$|a(z)| \gtrsim \Lambda_\mathcal{P}(z)^m, \quad |z| \ge R. \tag{2.7.25}$$

We will write $E\Gamma^m_{\rho,\mathcal{P}}$ for the class of the $\mathcal{P}$-elliptic symbols of order m and $EM\Gamma^m_{\rho,\mathcal{P}} = E\Gamma^m_{\rho,\mathcal{P}} \cap M\Gamma^m_{\rho,\mathcal{P}}$. We also set $E\Gamma^m_\mathcal{P} = E\Gamma^m_{1/\mu,\mathcal{P}}$, $EM\Gamma^m_\mathcal{P} = EM\Gamma^m_{1/\mu,\mathcal{P}}$.

Example 2.7.16. We know from Example 2.7.9 that the polynomial

$$p(z) = \sum_{\beta\in\mathcal{P}} c_\beta z^\beta, \quad c_\beta \in \mathbb{C}, \ z \in \mathbb{R}^{2d},$$

can be regarded as an element of $M\Gamma_\mathcal{P}^1$. The $\mathcal{P}$-ellipticity assumption in (2.7.25),

$$|p(z)| \gtrsim \Lambda_\mathcal{P}(z), \quad |z| \geq R,$$

is then equivalent to the multi-quasi-ellipticity of $p(z)$ in the sense of Definition 2.6.8.

Concerning the $\mathcal{P}$-elliptic symbols we have the following proposition; since it has a key role in the following, we shall give the details of the proof.

Proposition 2.7.17. *Consider* $a(z) \in E\Gamma_{\rho,\mathcal{P}}^m$ *and let* $\psi(z)$ *be a function in* $C^\infty(\mathbb{R}^{2d})$ *which is identically zero for* $|z| \leq R'$ *and identically 1 for* $|z| \geq R''$, *with* $0 < R' < R''$ *sufficiently large. Then* $\frac{\psi(z)}{a(z)} \in E\Gamma_{\rho,\mathcal{P}}^{-m}$. *If* $a(z) \in EM\Gamma_{\rho,\mathcal{P}}^m$, *then* $\frac{\psi(z)}{a(z)} \in EM\Gamma_{\rho,\mathcal{P}}^{-m}$.

Proof. Since $a(z)$ fulfills estimate (2.7.25) with a suitable positive R, taking any $\psi(z)$ satisfying the prescribed assumptions with $R'' > R' > R$, we have $\frac{\psi(z)}{a(z)} \in C^\infty(\mathbb{R}^{2d})$. The first statement then follows from Lemma 1.3.5.

For the second one, it suffices to prove that $z^\gamma \partial^\gamma \frac{\psi(z)}{a(z)} \in \Gamma_{\rho,\mathcal{P}}^{-m}$ for any non-zero $\gamma \in \mathbb{K}$. By use of Leibniz' rule and by induction we see that for $\gamma \neq 0$, $\gamma \in \mathbb{K}$,

$$z^\gamma \partial^\gamma \frac{\psi(z)}{a(z)} = \frac{z^\gamma \partial^\gamma \psi(z)}{a(z)} + \sum_{0 \neq \nu \leq \gamma} \sum_{k=1}^{|\nu|} \sum C_{k,\nu,\nu^1,\ldots,\nu^k} \frac{z^{\gamma-\nu} \partial^{\gamma-\nu} \psi(z)}{a(z)^{k+1}}$$

$$\times z^{\nu^1} \partial^{\nu^1} a(z) \ldots z^{\nu^k} \partial^{\nu^k} a(z), \qquad (2.7.26)$$

where, for any $\nu \leq \gamma$ and $1 \leq k \leq |\nu|$, the last sum in the right-hand side is taken over all the k-tuples $(\nu^1, \ldots, \nu^k) \in \mathbb{K} \times \ldots \times \mathbb{K}$ such that $\nu^1 + \ldots + \nu^k = \nu$, while $C_{k,\nu,\nu^1,\ldots,\nu^k}$ are suitable constants depending on $\nu \leq \gamma$, $\nu^1, \ldots, \nu^k$ and k.

Since $a(z) \in E\Gamma_{\rho,\mathcal{P}}^m$, then $\frac{z^{\gamma-\nu} \partial^{\gamma-\nu} \psi(z)}{a(z)^{k+1}} \in \Gamma_{\rho,\mathcal{P}}^{-m(k+1)}$ for every $\nu \leq \gamma$, $k \geq 0$ (it belongs in particular to $\Gamma_{\rho,\mathcal{P}}^{-\infty} = \mathcal{S}(\mathbb{R}^{2d})$ when $\nu < \gamma$); on the other hand, $z^{\nu^J} \partial^{\nu^J} a(z) \in \Gamma_{\rho,\mathcal{P}}^m$ for $J = 1, \ldots, k$.

This just shows that $z^\gamma \partial^\gamma \frac{\psi(z)}{a(z)} \in \Gamma_{\rho,\mathcal{P}}^{-m}$ and completes the proof. $\qquad\square$

Remark 2.7.18. As an immediate consequence of Propositions 2.7.2 and 2.7.10, we see that any real power of a weight function $\Lambda_\mathcal{P}(z)^m$ is a $\mathcal{P}$-elliptic symbol of order m; more precisely $\Lambda_\mathcal{P}(z)^m \in EM\Gamma_{\frac{1}{\mu},\mathcal{P}}^m$ for any $m \in \mathbb{R}$.

We pass then to consider the related classes of pseudo-differential operators. From now on we split $z = (x, \xi)$, $x \in \mathbb{R}^d$, $\xi \in \mathbb{R}^d$. Firstly, we list some notions and related results which will be useful in the sequel.

We shall associate to each symbol $a(x, \xi) \in \Gamma_{\rho,\mathcal{P}}^m$ the pseudo-differential operator of τ-type, cf. (1.2.1):

$$Au(x) = \int e^{i(x-y)\xi} a((1-\tau)x + \tau y, \xi) u(y) \, dy \, d\xi, \quad u \in \mathcal{S}(\mathbb{R}^d), \qquad (2.7.27)$$

where $\tau \in \mathbb{R}$ is fixed. According to the results of Section 1.2, different choices of τ define the same class of operators, which we shall denote by $\mathrm{OP}\Gamma^m_{\rho,\mathcal{P}}$. We recall that if $A \in \mathrm{OP}\Gamma^m_{\rho,\mathcal{P}}$ and $a_{\tau_1}(x,\xi)$ and $a_{\tau_2}(x,\xi)$ are the τ_1 and the τ_2-symbol of A respectively, then

$$a_{\tau_2}(x,\xi) \sim \sum_\alpha \frac{1}{\alpha!} (\tau_1 - \tau_2)^{|\alpha|} \partial_\xi^\alpha D_x^\alpha a_{\tau_1}(x,\xi), \qquad (2.7.28)$$

see Remark 1.2.6. According to the notations of Section 1.2, $a_0(x,\xi)$ corresponding to the standard quantization $\tau = 0$ is called the *left symbol* of A, $a_1(x,\xi)$ corresponding to $\tau = 1$ the *right symbol* of A and $a_{\frac{1}{2}}(x,\xi)$ corresponding to $\tau = \frac{1}{2}$ the *Weyl symbol* of A. We refer to Section 1.2 for details concerning the pseudo-differential calculus in classes $\mathrm{OP}\Gamma^m_{\rho,\mathcal{P}}$.

Now we are interested in the operators in $\mathrm{OP}\Gamma^m_{\rho,\mathcal{P}}$ with τ-symbol $a(x,\xi) \in M\Gamma^m_{\rho,\mathcal{P}}$.

Proposition 2.7.19. *Let $A \in \mathrm{OP}\Gamma^m_{\rho,\mathcal{P}}$ and $\tau_1 \in \mathbb{R}$. If the τ_1-symbol $a_{\tau_1}(x,\xi)$ of A belongs to $M\Gamma^m_{\rho,\mathcal{P}}$, then the τ_2-symbol $a_{\tau_2}(x,\xi)$ also belongs to $M\Gamma^m_{\rho,\mathcal{P}}$, for every $\tau_2 \in \mathbb{R}$.*

Proof. For any $\tau_2 \in \mathbb{R}$, the τ_2-symbol of A can be expressed in terms of its τ_1-symbol by means of the asymptotic expansion (2.7.28). By Proposition 2.7.11, $\partial_\xi^\alpha D_x^\alpha a_{\tau_1}(x,\xi) \in M\Gamma^{m-2\rho N}_{\rho,\mathcal{P}}$ for $|\alpha| = N$ and then $a_{\tau_2}(x,\xi) \in M\Gamma^m_{\rho,\mathcal{P}}$ by Proposition 2.7.12. $\qquad \square$

We set $\mathrm{OP}M\Gamma^m_{\rho,\mathcal{P}}$ for the subclass of $\mathrm{OP}\Gamma^m_{\rho,\mathcal{P}}$ consisting of the operators with τ-symbol in $M\Gamma^m_{\rho,\mathcal{P}}$. Thanks to the above proposition, $\mathrm{OP}M\Gamma^m_{\rho,\mathcal{P}}$ is independent of τ.

Proposition 2.7.20. *If $A \in \mathrm{OP}M\Gamma^m_{\rho,\mathcal{P}}$ and $B \in \mathrm{OP}M\Gamma^{m'}_{\rho,\mathcal{P}}$, then*

(1) $AB \in \mathrm{OP}M\Gamma^{m+m'}_{\rho,\mathcal{P}}$;

(2) ${}^tA \in \mathrm{OP}M\Gamma^m_{\rho,\mathcal{P}}$;

(3) $A^* \in \mathrm{OP}M\Gamma^m_{\rho,\mathcal{P}}$.

Proof. Let $a(x,\xi) \in M\Gamma^m_{\rho,\mathcal{P}}$ and $b(x,\xi) \in M\Gamma^{m'}_{\rho,\mathcal{P}}$ be the Weyl symbols of A and B respectively. Then for the Weyl symbol $c(x,\xi) \in \Gamma^{m+m'}_{\rho,\mathcal{P}}$ of $C = AB$ we have

$$c(x,\xi) \sim \sum_{\alpha,\beta} \frac{(-1)^{|\beta|}}{\alpha!\beta!} 2^{-|\alpha+\beta|} \partial_\xi^\alpha D_x^\beta a(x,\xi) \partial_\xi^\beta D_x^\alpha b(x,\xi), \qquad (2.7.29)$$

cf. Theorem 1.2.17. In view of Proposition 2.7.11,

$$\partial_\xi^\alpha D_x^\beta a(x,\xi) \partial_\xi^\beta D_x^\alpha b(x,\xi) \in M\Gamma^{m+m'-2\rho N}_{\rho,\mathcal{P}},$$

for $|\alpha + \beta| = N$, and then $c(x, \xi) \in M\Gamma_{\rho,\mathcal{P}}^{m+m'}$ by Proposition 2.7.12. This proves the first statement.

For statements (2) and (3), it suffices to observe that the Weyl symbols ${}^t a(x, \xi)$ and $a^*(x, \xi)$ of ${}^t A$ and A^* respectively are related to the Weyl symbol $a(x, \xi)$ of A by the formulas

$$ {}^t a(x, \xi) = a(x, -\xi), \tag{2.7.30} $$

$$ a^*(x, \xi) = \overline{a(x, \xi)}, \tag{2.7.31} $$

cf. (1.2.14), (1.2.16). Thus ${}^t a(x, \xi), a^*(x, \xi) \in M\Gamma_{\rho,\mathcal{P}}^m$. $\qquad\square$

Given $a(x, \xi) \in \Gamma_{\rho,\mathcal{P}}^m$, we may also define the operator A_a with Anti-Wick symbol $a(x, \xi)$, cf. Section 1.7. As an easy consequence of the formula (1.7.17) for the asymptotic expansion of the Weyl symbol of A, we obtain the following result.

Proposition 2.7.21. *If $a(x, \xi) \in M\Gamma_{\rho,\mathcal{P}}^m$, then $A_a \in OPM\Gamma_{\rho,\mathcal{P}}^m$.*

We also recall from Section 1.3.1 that the notion of ellipticity does not depend on the quantization, cf. Proposition 1.3.4.

Proposition 2.7.22. *Let $\tau_1, \tau_2 \in \mathbb{R}$ be given. If the τ_1-symbol $a_{\tau_1}(x, \xi)$ of $A \in OP\Gamma_{\rho,\mathcal{P}}^m$ belongs to $E\Gamma_{\rho,\mathcal{P}}^m$, the same holds for the τ_2-symbol $a_{\tau_2}(x, \xi)$.*

The parametrix of an operator with symbol $a(x, \xi) \in E\Gamma_{\rho,\mathcal{P}}^m$ can be constructed as in Section 1.3.1. For the parametrix of an operator with symbol in $EM\Gamma_{\rho,\mathcal{P}}^m$ the following holds:

Proposition 2.7.23. *Let $P \in OPM\Gamma_{\rho,\mathcal{P}}^m$ be given with $\mathcal{P}$-elliptic symbol. Then there exists an operator Q with symbol in $EM\Gamma_{\rho,\mathcal{P}}^{-m}$ such that*

$$ PQ = I + R, \qquad QP = I + S, \tag{2.7.32} $$

where R, S are regularizing and I is the identity operator.

Proof. We know already from Section 1.3.1 that there exists $Q \in OP\Gamma_{\rho,\mathcal{P}}^{-m}$ with $\mathcal{P}$-elliptic symbol satisfying the identities (2.7.32).

Moreover, if $p(x, \xi) \in EM\Gamma_{\rho,\mathcal{P}}^m$ is the Weyl symbol of P and $\psi(x, \xi)$ is a C^∞ function as in Proposition 2.7.17 (with $z = (x, \xi)$) we may write $Q = RB_1$, where B_1 is the operator with Weyl symbol $b_1(x, \xi) = \frac{\psi(x,\xi)}{p(x,\xi)}$ while R is an operator in $OP\Gamma_{\rho,\mathcal{P}}^0$ whose Weyl symbol $r(x, \xi)$ has the following asymptotic expansion:

$$ r(x, \xi) \sim \sum_{j \geq 0} (-1)^j r_j(x, \xi), \tag{2.7.33} $$

where $r_1(x, \xi)$ is the Weyl symbol of the operator $R_1 := B_1 P - I$ and, for any $j \geq 0$, $r_j(x, \xi)$ is the Weyl symbol of the j-th power R_1^j of R_1 (see the proof of

Theorem 1.3.6). By Proposition 2.7.17, we know that $b_1(x,\xi) \in M\Gamma^{-m}_{\rho,\mathcal{P}}$.
Since moreover $r_1(x,\xi)$ has the asymptotic expansion

$$r_1(x,\xi) \sim \sum_{|\alpha+\beta|>0} \frac{(-1)^{|\beta|}}{\alpha!\beta!} 2^{-|\alpha+\beta|} \partial_\xi^\alpha D_x^\beta \frac{\psi(x,\xi)}{p(x,\xi)} \partial_\xi^\beta D_x^\alpha p(x,\xi) \qquad (2.7.34)$$

and, for all α,β, $\partial_\xi^\alpha D_x^\beta \frac{\psi(x,\xi)}{p(x,\xi)} \partial_\xi^\beta D_x^\alpha p(x,\xi) \in M\Gamma^{-2\rho|\alpha+\beta|}_{\rho,\mathcal{P}}$, in view of Propositions
2.7.11 and 2.7.12, it follows that $r_1(x,\xi) \in M\Gamma^{-2\rho}_{\rho,\mathcal{P}}$. From Proposition 2.7.20, we
deduce $r_j(x,\xi) \in M\Gamma^{-2\rho j}_{\rho,\mathcal{P}}$ and, by virtue of (2.7.33), $r(x,\xi) \in M\Gamma^0_{\rho,\mathcal{P}}$.
So $Q = RB_1 \in \text{OP}M\Gamma^{-m}_{\rho,\mathcal{P}}$ is the required parametrix. $\qquad\qquad\square$

Proposition 2.7.23 gives in particular the existence of a parametrix with sym-
bol in $EM\Gamma^{-1}_{\mathcal{P}} = EM\Gamma^{-1}_{1/\mu,\mathcal{P}}$ and the global regularity for the operators with
polynomial coefficients

$$P = \sum_{(\beta,\alpha)\in\mathcal{P}} c_{\alpha\beta} x^\beta D^\alpha,$$

obtained by standard quantization from a multi-quasi-elliptic polynomial

$$p(z) = \sum_{\gamma\in\mathcal{P}} c_\gamma z^\gamma, \quad z = (x,\xi),$$

cf. Examples 2.7.9 and 2.7.16 (actually, in view of Proposition 2.7.22, the multi-
quasi-ellipticity of the symbol classes does not depend on the quantization).
On the other hand, global regularity for such operators follows from Theorem
2.1.16 and Theorem 2.6.9. The advantage of providing a parametrix of P with
symbol in $EM\Gamma^{-1}_{\mathcal{P}}$ will be evident in the next section.

2.8 L^p-Estimates

As in the preceding section, $\mathcal{P}$ is a complete polyhedron in $\mathbb{R}^{2d}$ with formal order
μ and $\Lambda_{\mathcal{P}}(z)$ is the corresponding weight function. For $0 < \rho \leq 1/\mu$, $m \in \mathbb{R}$, the
classes of symbols $M\Gamma^m_{\rho,\mathcal{P}}$ and the classes of operators $\text{OP}M\Gamma^m_{\rho,\mathcal{P}}$ are defined as
before. Namely, $a(z) \in M\Gamma^m_{\rho,\mathcal{P}}$ means that for every $\alpha \in \mathbb{N}^{2d}$ and $\gamma \in \mathbb{K}_{2d} = \{\gamma \in \mathbb{N}^{2d} : \gamma_j \in \{0,1\}, j = 1,\dots,2d\}$,

$$|z^\gamma \partial^{\alpha+\gamma} a(z)| \lesssim \Lambda_{\mathcal{P}}(z)^{m-\rho|\alpha|}, \quad z = (x,\xi) \in \mathbb{R}^{2d}. \qquad (2.8.1)$$

Theorem 2.8.1. *Any operator $A \in \text{OP}M\Gamma^0_{\rho,\mathcal{P}}$ extends to a bounded operator from
$L^p(\mathbb{R}^d)$ to itself, for all $1 < p < \infty$.*

Actually, to obtain L^p-boundedness, we shall use much weaker assumptions
than $a \in M\Gamma^0_{\rho,\mathcal{P}}$ on the symbol of A. Namely it will be sufficient to assume for
$a(x,\xi) \in C^\infty(\mathbb{R}^{2d})$:

$$|\xi^\gamma \partial_x^\lambda \partial_\xi^{\nu+\gamma} a(x,\xi)| \lesssim \langle\xi\rangle^{-\epsilon|\nu|}, \quad (x,\xi) \in \mathbb{R}^{2d}, \qquad (2.8.2)$$

for some $\epsilon > 0$, all $\lambda \in \mathbb{N}^d$, $\nu \in \mathbb{N}^d$, $\gamma \in \mathbb{K}_d = \{\gamma \in \mathbb{N}^d : \gamma_j \in \{0,1\}, j = 1, \ldots, d\}$. Note that if (2.8.1) is satisfied with $m = 0$, then (2.8.2) is valid with $\epsilon = \rho\mu_0$, where μ_0 is the minimum order of $\mathcal{P}$, cf. (2.7.2).

The proof of Theorem 2.8.1 will use the classical result of Lizorkin and Marcinkiewicz concerning Fourier multipliers, that we begin to recall here below.

Theorem 2.8.2. *Let the function $m(\xi)$ be continuous together with its derivatives $\partial_\xi^\gamma m(\xi)$, for any $\gamma \in \mathbb{K}_d$. If there is a constant $B > 0$ such that*

$$|\xi^\gamma \partial_\xi^\gamma m(\xi)| \le B, \quad \xi \in \mathbb{R}^d, \ \gamma \in \mathbb{K}_d, \tag{2.8.3}$$

then for every $1 < p < \infty$ we can find a constant $A_p > 0$, depending only on p, B and the dimension d, such that

$$\|m(D)u\|_{L^p} \le A_p \|u\|_{L^p}$$

for all $u \in S(\mathbb{R}^d)$.

Proof of Theorem 2.8.1. We use for A the left-symbol representation

$$A\varphi(x) = a(x,D)\varphi(x) = \int e^{ix\xi} a(x,\xi)\widehat{\varphi}(\xi)\,d\xi, \quad \varphi \in S(\mathbb{R}^d), \tag{2.8.4}$$

with $a(x,\xi) \in M\Gamma^0_{\rho,\mathcal{P}}$. For $m = (m_1, \ldots, m_d) \in \mathbb{Z}^d$ let us consider

$$Q_m = \left\{ x \in \mathbb{R}^d : |x_j - m_j| \le \frac{1}{2}, j = 1,2,\ldots,d \right\},$$

$$Q_m^* = \left\{ x \in \mathbb{R}^d : |x_j - m_j| \le \frac{2}{3}, j = 1,2,\ldots,d \right\},$$

$$Q_m^{**} = \left\{ x \in \mathbb{R}^d : |x_j - m_j| \le 1, j = 1,2,\ldots,d \right\},$$

and cut-off functions $\psi_m(x) \in C_0^\infty(\mathbb{R}^d)$ such that $0 \le \psi_m(x) \le 1$, $\operatorname{supp}\psi_m \subset Q_m^{**}$ and $\psi_m = 1$ in Q_m^*. Setting $\varphi_{1,m} = \psi_m\varphi$ and $\varphi_{2,m} = (1 - \psi_m)\varphi$, we have

$$\int |a(x,D)\varphi(x)|^p\,dx \le 2^{p-1}\sum_{m \in \mathbb{Z}^d} \{I_{1,m} + I_{2,m}\}, \tag{2.8.5}$$

where

$$I_{i,m} = \int_{Q_m} |a(x,D)\varphi_{i,m}(x)|^p\,dx, \quad i = 1,2. \tag{2.8.6}$$

In fact, since $\mathbb{R}^d = \cup_{m \in \mathbb{Z}^d} Q_m$ and the measure of $Q_m \cap Q_n$ vanishes when $n \ne m$, we can write for $\varphi(x) \in S(\mathbb{R}^d)$:

$$\|a(x,D)\varphi\|_p^p = \sum_{m \in \mathbb{Z}^d} \int_{Q_m} |a(x,D)\varphi(x)|^p\,dx. \tag{2.8.7}$$

Moreover, for any $m \in \mathbb{Z}^d$,

$$\int_{Q_m} |a(x,D)\varphi(x)|^p \, dx = \int_{Q_m} |a(x,D)(\varphi_{1,m} + \varphi_{2,m})(x)|^p \, dx$$

$$\leq 2^{p-1}\{I_{1,m} + I_{2,m}\}. \tag{2.8.8}$$

Therefore (2.8.5) is proved. To continue the proof, we shall use the following lemma.

Lemma 2.8.3. *Let $a(x,\xi) \in M\Gamma^0_{\rho,\mathcal{P}}, \chi(x) \in C^\infty_0(\mathbb{R}^d)$ and set $a_\chi(x,\xi) = \chi(x)a(x,\xi)$; we can then consider the Fourier transform of $a_\chi(x,\xi)$ with respect to the x variable*

$$\widehat{a}_\chi(\eta,\xi) = \int e^{-ix\eta} a_\chi(x,\xi) \, dx. \tag{2.8.9}$$

For any $N > 0$ there is a positive constant $C_{N,\chi}$ such that

$$\left| \xi^\gamma \partial^\gamma_\xi \widehat{a}_\chi(\eta,\xi) \right| \leq C_{N,\chi}(1 + |\eta|)^{-N}, \tag{2.8.10}$$

for all $\eta, \xi \in \mathbb{R}^d$ and $\gamma \in \mathbb{K}_d$.

Proof. Since $a_\chi(x,\xi)$ has compact support with respect to the x variable, differentiation and integration by parts in (2.8.9) give

$$\eta^\beta \partial^\gamma_\xi \widehat{a}_\chi(\eta,\xi) = (-i)^{|\beta|} \int e^{-ix\eta} \partial^\beta_x \partial^\gamma_\xi a_\chi(x,\xi) \, dx, \tag{2.8.11}$$

where β is an arbitrary multi-index. Then by Leibniz' formula we get

$$\left| \eta^\beta \xi^\gamma \partial^\gamma_\xi \widehat{a}_\chi(\eta,\xi) \right| \leq \sum_{\nu \leq \beta} \binom{\beta}{\nu} \int |\partial^{\beta-\nu}_x \chi(x)| \, \left| \xi^\gamma \partial^\nu_x \partial^\gamma_\xi a(x,\xi) \right| \, dx. \tag{2.8.12}$$

Since $\left| \xi^\gamma \partial^\nu_x \partial^\gamma_\xi a(x,\xi) \right| \leq C_{\gamma,\nu} \Lambda_\mathcal{P}(x,\xi)^{-\rho|\nu|} \leq C'_{\gamma,\nu}$ (cf. (2.8.1) or (2.8.2)), we obtain

$$\left| \eta^\beta \xi^\gamma \partial^\gamma_\xi \widehat{a}_\chi(\eta,\xi) \right| \leq C_{\beta,\chi}, \tag{2.8.13}$$

for every $\eta, \xi \in \mathbb{R}^d$ and $\gamma \in \mathbb{K}_d$, where the positive constant on the right-hand side is given by

$$C_{\beta,\chi} = \max_{\gamma \in \mathbb{K}_d} \sum_{\nu \leq \beta} C'_{\gamma,\nu} \binom{\beta}{\nu} \int |\partial^{\beta-\nu}_x \chi(x)| \, dx. \tag{2.8.14}$$

Thus by summing on $|\beta| \leq N$ we obtain the desired conclusion. $\square$

Continuing the proof of Theorem 2.8.1. In order to estimate $I_{1,m}$ in (2.8.5), (2.8.6), let us consider a cut-off function $\theta(x) \in C^\infty_0(\mathbb{R}^d)$ such that $\theta(x) = 1$ for $x \in Q_0$

and define $\theta_m(x) = \theta(x - m)$ for any $m \in \mathbb{Z}^d$. By setting now $a_m(x, \xi) := \theta_m(x)a(x, \xi) \in M\Gamma^0_{\rho,\mathcal{P}}$ we get

$$I_{1,m} = \int_{Q_m} |a_m(x, D)\varphi_{1,m}|^p \, dx \leq \int_{\mathbb{R}^d} |a_m(x, D)\varphi_{1,m}|^p \, dx. \qquad (2.8.15)$$

If we set now $\chi(x) = \theta_m(x)$ in Lemma 2.8.3, we can show that for every $N > 0$ there exists a positive constant C_N such that

$$\left| \xi^\gamma \partial_\xi^\gamma \widehat{a}_m(\eta, \xi) \right| \leq C_N (1 + |\eta|)^{-N} \quad \eta, \xi \in \mathbb{R}^d, \ \gamma \in \mathbb{K}_d. \qquad (2.8.16)$$

Notice that, from the definition of the functions $\theta_m(x)$ and (2.8.14), the constant C_N in (2.8.16) depends on $\theta(x)$ but not on $m \in \mathbb{Z}^d$. Then by Theorem 2.8.2 for every $N > 0$ there exists $M_N > 0$, which does not depend on m, such that

$$\|\widehat{a}_m(\eta, D)u\|_{L^p} \leq M_N (1 + |\eta|)^{-N} \|u\|_{L^p}, \qquad (2.8.17)$$

for every $u \in S(\mathbb{R}^d)$, $\eta \in \mathbb{R}^d$ and $m \in \mathbb{Z}^d$. On the other hand, from Fubini's theorem we have

$$a_m(x, D)u(x) = \int e^{ix\xi} a_m(x, \xi)\widehat{u}(\xi) \, d\xi = \int e^{ix\eta} \widehat{a}_m(\eta, D)u(x) \, d\eta. \qquad (2.8.18)$$

So by using (2.8.17) and Minkowski's inequality in integral form, we have

$$I_{1,m} \leq \int \left| \int e^{ix\eta} \widehat{a}_m(\eta, D)\varphi_{1,m}(x) \, d\eta \right|^p dx \leq \left\{ \int \|\widehat{a}_m(\eta, D)\varphi_{1,m}\|_{L^p} \, d\eta \right\}^p$$

$$\leq M_N^p \|\varphi_{1,m}\|_{L^p}^p \left\{ \int (1 + |\eta|)^{-N} \, d\eta \right\}^p = M_N^p C_N^p \|\varphi_{1,m}\|_{L^p}^p, \qquad (2.8.19)$$

where, for $N > d$, $C_N := \int (1 + |\eta|)^{-N} \, d\eta < \infty$. To estimate the second term $I_{2,m}$ we use the kernel representation, cf. (1.2.2), (1.2.3), (1.2.4):

$$a(x, D)\varphi(x) = \int_{\mathbb{R}^d} K_0(x, y)\varphi(y) \, dy \qquad (2.8.20)$$

with $K_0(x, y) = K(x, x - y)$, where, in the distribution sense,

$$K(x, t) = (2\pi)^{-d} \int_{\mathbb{R}^d} e^{it\xi} a(x, \xi) \, d\xi. \qquad (2.8.21)$$

We need the following lemma.

Lemma 2.8.4. *If $a \in M\Gamma^0_{\rho,\mathcal{P}}$, then for every sufficiently large positive integer N, there is a positive constant C_N such that*

$$|K(x, t)| \leq C_N |t|^{-N}, \quad t \neq 0. \qquad (2.8.22)$$

Proof. Let $\nu \in \mathbb{N}^d$ be an arbitrary multi-index. We obtain from (2.8.21) in the distribution sense

$$(-it)^\nu K(x,t) = (2\pi)^{-d} \int_{\mathbb{R}^d} e^{it\xi} \partial_\xi^\nu a(x,\xi)\, d\xi.$$

Since $a \in M\Gamma^0_{\rho,\wp}$, then (2.8.2) is valid. Hence for sufficiently large ν we have

$$|t^\nu|\,|K(x,t)| \leq C_\nu, \quad x \in \mathbb{R}^d,\ t \in \mathbb{R}^d,$$

for a positive constant C_ν. The estimates (2.8.22) follow immediately. $\qquad\square$

End of the proof of Theorem 2.8.1. By (2.8.20), (2.8.21) and Lemma 2.8.4, there is a constant C_{2N} such that, for all $x \in Q_m$,

$$\begin{aligned}
|a(x,D)\varphi_{2,m}(x)| &= \left| \int_{\mathbb{R}^d} K(x,x-y)\varphi_{2,m}(y)dy \right| \\
&= \left| \int_{\mathbb{R}^d \setminus Q_m^*} K(x,x-y)\varphi_{2,m}(y)dy \right| \\
&\leq C_{2N} \int_{\mathbb{R}^d \setminus Q_m^*} |x-y|^{-2N} |\varphi_{2,m}(y)|\, dy.
\end{aligned}$$

We may further estimate under the integral, for $x \in Q_m$ and $y \notin Q_m^*$:

$$|x-y|^{-2N} \leq C_N'(\lambda + |x-y|)^{-N}(\lambda + |m-y|)^{-N},$$

for any fixed $\lambda > 0$ and for a suitable positive constant C_N' (depending on λ). Take in particular $\lambda = 1 + \frac{\sqrt{d}}{2}$. By Minkowski's inequality in integral form we obtain

$$\begin{aligned}
I_{2,m}^{1/p} &= \left(\int_{Q_m} |a(x,D)\varphi_{2,m}(x)|^p\, dx \right)^{1/p} \\
&\leq C_{2N} C_N' \left\{ \int_{Q_m} \left| \int_{\mathbb{R}^d \setminus Q_m^*} \frac{(\lambda + |x-y|)^{-N} |\varphi_{2,m}(y)|}{(\lambda + |m-y|)^N} dy \right|^p dx \right\}^{1/p} \\
&\leq C_{2N} C_N' \int_{\mathbb{R}^d \setminus Q_m^*} \left\{ \int_{Q_m} \frac{(\lambda + |x-y|)^{-Np} |\varphi_{2,m}(y)|^p}{(\lambda + |m-y|)^{Np}} dx \right\}^{1/p} dy \\
&= C_{2N} C_N' \int_{\mathbb{R}^d \setminus Q_m^*} \frac{|\varphi_{2,m}(y)|}{(\lambda + |m-y|)^N} \left\{ \int_{Q_m} (\lambda + |x-y|)^{-Np} dx \right\}^{1/p} dy,
\end{aligned}$$

so that for N sufficiently large and new constants $\tilde{C}_N > 0$,

$$I_{2,m}^{1/p} \leq \tilde{C}_N \int_{\mathbb{R}^d \setminus Q_m^*} (\lambda + |m-y|)^{-N} |\varphi_{2,m}(y)|\, dy.$$

Applying Hölder's inequality we easily deduce the following estimate:

$$I_{2,m} \leq H_{N,p} \int_{\mathbb{R}^d \setminus Q_m^*} \frac{|\varphi_{2,m}(y)|^p}{(\lambda + |m - y|)^{\frac{Np}{2}}} \, dy, \qquad (2.8.23)$$

where $H_{N,p}$ is a positive constant depending only on N, p and the dimension d. From the definition of $\varphi_{1,m}$ and Q_m^{**}, $m \in \mathbb{Z}^d$, we obtain

$$\sum_{m \in \mathbb{Z}^d} \|\varphi_{1,m}\|_{L^p}^p = \sum_{m \in \mathbb{Z}^d} \int_{Q_m^{**}} |\varphi_{1,m}(x)|^p \, dx \leq C_d \|\varphi\|_{L^p}^p, \qquad (2.8.24)$$

where the constant $C_d > 0$ depends only on the dimension d. Moreover

$$\sum_{m \in \mathbb{Z}^d} \int_{\mathbb{R}^d \setminus Q_m^*} \frac{|\varphi_{2,m}(y)|^p}{(\lambda + |m - y|)^{\frac{Np}{2}}} dy \leq \sum_{m \in \mathbb{Z}^d} \sum_{l \neq m} \int_{Q_l} \frac{|\varphi_{2,m}(y)|^p}{(\lambda + |m - y|)^{\frac{Np}{2}}} dy$$

$$\leq \sum_{m \in \mathbb{Z}^d} \sum_{l \in \mathbb{Z}^d} \frac{1}{(1 + |m - l|)^{\frac{Np}{2}}} \int_{Q_l} |\varphi(y)|^p \, dy$$

$$= \|\varphi\|_{L^p}^p \sum_{m \in \mathbb{Z}^d} \frac{1}{(1 + |m|)^{\frac{Np}{2}}} \qquad (2.8.25)$$

and $\sum_{m \in \mathbb{Z}^d} \frac{1}{(1+|m|)^{\frac{Np}{2}}} < \infty$ for suitably large N. By (2.8.6) and the estimates (2.8.5), (2.8.19), (2.8.23), (2.8.24), (2.8.25) we then get:

$$\|a(x, D)\varphi\|_{L^p}^p \leq C_{p,d}'' \|\varphi\|_{L^p}^p, \qquad (2.8.26)$$

which ends the proof. $\qquad \qquad \qquad \qquad \qquad \qquad \qquad \qquad \qquad \qquad \qquad \square$

We may now define L^p-Sobolev spaces related to a complete polyhedron $\mathcal{P}$.

Definition 2.8.5. For $s \in \mathbb{R}$ and $1 < p < \infty$, $H_{\mathcal{P}}^{s,p}$ is the space of all the temperate distributions $u \in \mathcal{S}'(\mathbb{R}^d)$ such that $\Lambda_{\mathcal{P}}^s(x, D)u \in L^p(\mathbb{R}^d)$. Here $\Lambda_{\mathcal{P}}^s(x, D)$ is the pseudo-differential operator with left symbol $\Lambda_{\mathcal{P}}(x, \xi)^s \in M\Gamma_{\frac{1}{\mu}, \mathcal{P}}^s$, that is

$$\Lambda_{\mathcal{P}}^s(x, D)u(x) = \int e^{ix\xi} \Lambda_{\mathcal{P}}(x, \xi)^s \hat{u}(\xi) \, d\xi, \quad u \in \mathcal{S}(\mathbb{R}^d).$$

Set $H_{\mathcal{P}}^p = H_{\mathcal{P}}^{1,p}$. Since $\Lambda_{\mathcal{P}}(x, \xi)^s$ is $\mathcal{P}$-elliptic of order s (see Remark 2.7.18), in view of Proposition 2.7.23, there exists an operator $\bar{\Lambda}_{-s}(x, D) \in OPM\Gamma_{\rho, \mathcal{P}}^{-s}$ such that

$$\bar{\Lambda}_{-s}(x, D)\Lambda_{\mathcal{P}}^s(x, D) = I + R_s, \qquad (2.8.27)$$

where R_s is regularizing. When $\Lambda_{\mathcal{P}}(x, \xi) = (1 + |x|^2 + |\xi|^2)^{1/2}$ and $p = 2$ the spaces $H_{\mathcal{P}}^{s,p}$ coincide with the spaces $H_{\Gamma}^s(\mathbb{R}^d)$ of Section 2.1. We impose a Banach topology on $H_{\mathcal{P}}^{s,p}$ by setting, for any $u \in H_{\mathcal{P}}^{s,p}$,

$$\|u\|_{s,p,\mathcal{P}} = \|\Lambda_{\mathcal{P}}^s(x, D)u\|_{L^p} + \|R_s u\|_{L^p}, \qquad (2.8.28)$$

where R_s is defined by (2.8.27).

Proposition 2.8.6. $H_{\mathcal{P}}^{s,p}$ is a Banach space with respect to the norm $\|\cdot\|_{s,p,\mathcal{P}}$ defined by (2.8.28).

Proof. In the following we write for short $\Lambda^s = \Lambda_{\mathcal{P}}^s(x,D)$ and $\overline{\Lambda}_{-s} = \overline{\Lambda}_{-s}(x,D)$. Of course, $\|\cdot\|_{s,p,\mathcal{P}}$ is a semi-norm in $H_{\mathcal{P}}^{s,p}$. Let us suppose $\|u\|_{s,p,\mathcal{P}} = 0$. By (2.8.28) it follows that $\Lambda^s u = 0$ and $R_s u = 0$; then $u = \overline{\Lambda}_{-s}(\Lambda^s u) - R_s u = 0$, in view of (2.8.27).

To prove the completeness, let $\{u_\nu\}_{\nu \in \mathbb{N}}$ be a Cauchy sequence in $H_{\mathcal{P}}^{s,p}$ with respect to $\|\cdot\|_{s,p,\mathcal{P}}$. Then it follows that $\{\Lambda^s u_\nu\}_{\nu \in \mathbb{N}}$ and $\{R_s u_\nu\}_{\nu \in \mathbb{N}}$ are Cauchy sequences in $L^p(\mathbb{R}^d)$; let v and w be their limits in $L^p(\mathbb{R}^d)$, respectively.

We prove now that $\{u_\nu\}_{\nu \in \mathbb{N}}$ converges to $u := \overline{\Lambda}_{-s}v - w$ in $H_{\mathcal{P}}^{s,p}$. Setting $v_\nu := \Lambda^s u_\nu$ and $w_\nu := R_s u_\nu$, we get $\Lambda^s u_\nu = \Lambda^s \overline{\Lambda}_{-s}v_\nu - \Lambda^s w_\nu$. Since $v_\nu \to v$ in $L^p(\mathbb{R}^d)$ and $\Lambda^s \overline{\Lambda}_{-s} \in OPM\Gamma_{\frac{1}{\mu},\mathcal{P}}^0$, $\Lambda^s \overline{\Lambda}_{-s}v_\nu \to \Lambda^s \overline{\Lambda}_{-s}v$ in $L^p(\mathbb{R}^d)$ in view of Theorem 2.8.1. On the other hand, since R_s is a regularizing operator, $u_\nu \to \overline{\Lambda}_{-s}v - w$ in $\mathcal{S}'(\mathbb{R}^d)$ yields $w_\nu = R_s u_\nu \to R_s(\overline{\Lambda}_{-s}v - w)$ in $\mathcal{S}(\mathbb{R}^d)$. Thus, for the uniqueness of the limit, $w_\nu \to w$ in $\mathcal{S}(\mathbb{R}^d)$ and then $\Lambda^s w_\nu \to \Lambda^s w$ in $\mathcal{S}(\mathbb{R}^d)$.

This shows that $\Lambda^s u \in L^p(\mathbb{R}^d)$ and that $\Lambda^s u_\nu = \Lambda^s \overline{\Lambda}_{-s}v_\nu - \Lambda^s w_\nu \to \Lambda^s \overline{\Lambda}_{-s}v - \Lambda^s w = \Lambda^s u$ in $L^p(\mathbb{R}^d)$. The proof is therefore concluded. $\square$

In fact the previous Sobolev spaces can be better described by means of any pseudo-differential operator with positive Anti-Wick elliptic symbol. Firstly we recall the following version of Proposition 1.7.12, cf. Theorem 2.1.7.

Lemma 2.8.7. Let $a(x,\xi) \in \Gamma_{\rho,\mathcal{P}}^m$ be $\mathcal{P}$-elliptic of order m and assume that $a(x,\xi) > 0$, for any $x,\xi \in \mathbb{R}^d$. Then the operator A_a with Anti-Wick symbol a is an isomorphism of $\mathcal{S}(\mathbb{R}^d)$ extending to an isomorphism of $\mathcal{S}'(\mathbb{R}^d)$. Moreover $A_a^{-1} \in OP\Gamma_{\rho,\mathcal{P}}^{-m}$ is $\mathcal{P}$-elliptic of order $-m$.

Proposition 2.8.8. Let $a(x,\xi) \in M\Gamma_{\rho,\mathcal{P}}^s$ be an arbitrary $\mathcal{P}$-elliptic symbol of order s such that $a(x,\xi) > 0$, for any $x,\xi \in \mathbb{R}^d$. Then for any $1 < p < \infty$,

$$H_{\mathcal{P}}^{s,p} = A_a^{-1}(L^p(\mathbb{R}^d)) = \{u \in \mathcal{S}'(\mathbb{R}^d) : A_a u \in L^p(\mathbb{R}^d)\},$$

where A_a is the operator with Anti-Wick symbol a. Moreover a norm on $H_{\mathcal{P}}^{s,p}$ equivalent to (2.8.28) is given by $\|A_a u\|_{L^p}$.

Proof. Firstly suppose that $A_a u \in L^p(\mathbb{R}^d)$ and write $\Lambda^s u = \Lambda^s A_a^{-1}(A_a u)$, with $\Lambda^s = \Lambda_{\mathcal{P}}^s(x,D)$. Then $\Lambda^s u \in L^p(\mathbb{R}^d)$ since $\Lambda^s A_a^{-1} \in OPM\Gamma_{\rho,\mathcal{P}}^0$.

Conversely let us take $u \in \mathcal{S}'(\mathbb{R}^d)$ such that $\Lambda^s u \in L^p(\mathbb{R}^d)$. By (2.8.27) we may write

$$A_a u = A_a \overline{\Lambda}_{-s}(\Lambda^s u) - A_a R_s \overline{\Lambda}_{-s}(\Lambda^s u) + A_a R_s(R_s u). \tag{2.8.29}$$

So using Theorem 2.8.1, we conclude that $A_a u \in L^p(\mathbb{R}^d)$ and there is a positive constant C independent of u such that

$$\|A_a u\|_{L^p} \leq C(\|\Lambda^s u\|_{L^p} + \|R_s u\|_{L^p}), \tag{2.8.30}$$

since $A_a \overline{\Lambda}_{-s} \in \mathrm{OPM}\Gamma^0_{\rho,\mathcal{P}}$ and $A_a R_s \overline{\Lambda}_{-s}, A_a R_s$ are regularizing.

On the other hand, since A_a is an isomorphism of $\mathcal{S}'(\mathbb{R}^d)$, $\|A_a u\|_{L^p}$ is a norm on $H^{s,p}_{\mathcal{P}}$ and it is easy to show that $H^{s,p}_{\mathcal{P}}$ is complete with respect to it; thus the equivalence with the norm (2.8.28) follows from the Open Mapping Theorem. $\quad\square$

Remark 2.8.9. We may as well reset Definition 2.8.5 by replacing $\Lambda^s_{\mathcal{P}}(x, D)$ with any elliptic pseudo-differential operator of order s. Namely, for any T with $\mathcal{P}$-elliptic symbol in $M\Gamma^s_{\rho,\mathcal{P}}$,

$$H^{s,p}_{\mathcal{P}} = \{u \in \mathcal{S}'(\mathbb{R}^d) : Tu \in L^p(\mathbb{R}^d)\}. \tag{2.8.31}$$

If moreover Q is a parametrix of T and $QT = I + R$, then $\|Tu\|_{L^p} + \|Ru\|_{L^p}$ is a norm in $H^{s,p}_{\mathcal{P}}$ equivalent to (2.8.28).

Remark 2.8.10. Notice also that for any $t < s$ and $1 < p < \infty$ the following inclusions hold:

$$\mathcal{S}(\mathbb{R}^d) \subset H^{s,p}_{\mathcal{P}} \subset H^{t,p}_{\mathcal{P}} \subset \mathcal{S}'(\mathbb{R}^d),$$

with compact embeddings.

Using the above description of the spaces $H^{s,p}_{\mathcal{P}}$ we can plainly deduce the action of pseudo-differential operators in classes $\mathrm{OPM}\Gamma^m_{\rho,\mathcal{P}}$. Namely the following holds.

Proposition 2.8.11. *If $A \in \mathrm{OPM}\Gamma^m_{\rho,\mathcal{P}}$, then*

$$A : H^{s+m,p}_{\mathcal{P}} \to H^{s,p}_{\mathcal{P}},$$

continuously, for all $s \in \mathbb{R}$ and $1 < p < \infty$.

Proof. In the following we set G_t for the operator with Anti-Wick symbol $\Lambda_{\mathcal{P}}(x, \xi)^t$ $\in M\Gamma^t_{\frac{1}{\mu},\mathcal{P}}$; since $\Lambda_{\mathcal{P}}(x, \xi)^t$ is positive and $\mathcal{P}$-elliptic of order t, in view of Proposition 2.8.8 we have $H^{t,p}_{\mathcal{P}} = G_t^{-1}(L^p(\mathbb{R}^d))$ for any $1 < p < \infty$.

Since $Q := G_s A G_{s+m}^{-1} \in \mathrm{OPM}\Gamma^0_{\rho,\mathcal{P}}$ and, for any $u \in H^{s+m,p}_{\mathcal{P}}$, $G_{s+m}u \in L^p(\mathbb{R}^d)$, then $G_s A u = Q(G_{s+m}u) \in L^p(\mathbb{R}^d)$; moreover there is a positive constant C such that

$$\|G_s A u\|_{L^p} = \|Q(G_{s+m}u)\|_{L^p} \leq C\|G_{s+m}u\|_{L^p}.$$

This shows the continuity of A and concludes the proof. $\quad\square$

Using the results of the preceding section, we may now obtain precise results of L^p-regularity for the solutions of the $\mathcal{P}$-elliptic equations.

Theorem 2.8.12. *Let $a(x, \xi) \in M\Gamma^m_{\rho,\mathcal{P}}$ be $\mathcal{P}$-elliptic, cf. Definition 2.7.15. Let A be the pseudo-differential operator with symbol a, in the standard left quantization. Let $1 < p < \infty$ and $s \in \mathbb{R}$. If $u \in \mathcal{S}'(\mathbb{R}^d)$ and $Au \in H^{s,p}_{\mathcal{P}}$, then $u \in H^{s+m,p}_{\mathcal{P}}$. Moreover for any $t < s + m$ there is a positive constant C, independent of u, for which*

$$\|u\|_{s+m,p,\mathcal{P}} \leq C(\|Au\|_{s,p,\mathcal{P}} + \|u\|_{t,p,\mathcal{P}}). \tag{2.8.32}$$

Proof. From Proposition 2.7.23 there exists an operator $B \in OPM\Gamma_{\rho,\mathcal{P}}^{-m}$ such that $BA = I + R$ with R regularizing. For any $u \in \mathcal{S}'(\mathbb{R}^d)$ we have then $u = BAu - Ru$. If $Au \in H_{\mathcal{P}}^{s,p}$ we get $BAu \in H_{\mathcal{P}}^{s+m,p}$ in view of Proposition 2.8.11, whereas $Ru \in \mathcal{S}(\mathbb{R}^d)$, hence $u \in H_{\mathcal{P}}^{s+m,p}$. The estimate (2.8.32) follows, because of the continuous embedding $\mathcal{S}(\mathbb{R}^d) \subset H_{\mathcal{P}}^{t,p}$.   □

From the theory of the Fredholm operators in the Banach spaces, cf. Remark 1.6.8, we have also readily the following theorem. To be precise, we shall denote here by A_s, $s \in \mathbb{R}$, the restriction of $A : \mathcal{S}'(\mathbb{R}^d) \to \mathcal{S}'(\mathbb{R}^d)$ to $H_{\mathcal{P}}^{s,p}$, or equivalently the extension of $A : \mathcal{S}(\mathbb{R}^d) \to \mathcal{S}(\mathbb{R}^d)$ to $H_{\mathcal{P}}^{s,p}$.

Theorem 2.8.13. *Consider $A \in OPM\Gamma_{\rho,\mathcal{P}}^{m}$ with $\mathcal{P}$-elliptic symbol. Then, for $1 < p < \infty$, $s \in \mathbb{R}$:*

(i) $A_s \in \text{Fred}\,(H_{\mathcal{P}}^{s,p}, H_{\mathcal{P}}^{s-m,p})$;

(ii) *ind A_s is independent of $s \in \mathbb{R}$ and $1 < p < \infty$;*

(iii) *If $T \in OPM\Gamma^{m'}$ with $m' < m$, then $A_s + T_s \in \text{Fred}\,(H_{\mathcal{P}}^{s,p}, H_{\mathcal{P}}^{s-m,p})$ and $\text{ind}\,(A_s + T_s) = \text{ind}\,A_s$.*

Similarly to Theorem 2.1.12, we have a more explicit definition of $H_{\mathcal{P}}^{s,p}$ when $s \in \mathbb{N}$; let us fix attention for simplicity on the case $s = 1$.

Theorem 2.8.14. *Writing $H_{\mathcal{P}}^p$ for the space $H_{\mathcal{P}}^{1,p}$, $1 < p < \infty$, we have:*

$$H_{\mathcal{P}}^p = \{u \in \mathcal{S}'(\mathbb{R}^d) : \, x^\beta D^\alpha u \in L^p(\mathbb{R}^d), \; (\beta,\alpha) \in V(\mathcal{P})\} \qquad (2.8.33)$$

with equivalent norm

$$\sum_{(\beta,\alpha)\in V(\mathcal{P})} \|x^\beta D^\alpha u\|_{L^p}, \qquad (2.8.34)$$

where $V(\mathcal{P})$ is the set of vertices of $\mathcal{P}$.

Proof. Let us assume that u belongs to $H_{\mathcal{P}}^{1,p}$, $1 < p < \infty$; since $x^\beta D_x^\alpha \in OPM\Gamma_{\frac{1}{\mu},\mathcal{P}}^1$ as long as $(\beta,\alpha) \in V(\mathcal{P})$, we have that

$$\|x^\beta D^\alpha u\|_{L^p} \leq C_{\alpha,\beta}\|u\|_{1,p,\mathcal{P}}$$

for a positive constant $C_{\alpha,\beta}$ independent of u ($\|\cdot\|_{1,p,\mathcal{P}}$ is the norm (2.8.28) in $H_{\mathcal{P}}^{1,p}$). Conversely, we argue similarly to the proof of Theorem 2.1.12. Namely, we consider the operator $T = W_{-1}P$ where W_{-1} is any $\mathcal{P}$-elliptic operator in $OPM\Gamma_{\rho,\mathcal{P}}^{-1}$, $0 < \rho \leq \frac{1}{\mu}$ (for instance the pseudo-differential operator with left symbol $\Lambda_{\mathcal{P}}(x,\xi)^{-1}$) and P is the differential operator $P = \sum_{(\beta,\alpha)\in V(\mathcal{P})} x^\beta D_x^\alpha(x^\beta D_x^\alpha)$. Since $P \in OPM\Gamma_{\frac{1}{\mu},\mathcal{P}}^2$ and is $\mathcal{P}$-elliptic, for some regularizing R we have that

$$\|Tu\|_{L^p} + \|Ru\|_{L^p} \leq C\left(\sum_{(\beta,\alpha)\in V(\mathcal{P})} \|W_{-1}x^\beta D^\alpha(x^\beta D^\alpha u)\|_{L^p} + \|Ru\|_{L^p}\right)$$

$$\leq C' \left(\sum_{(\beta,\alpha) \in V(\mathcal{P})} \|x^\beta D^\alpha u\|_{L^p} + \|u\|_{L^p} \right),$$

as $W_{-1} x^\beta D_x^\alpha$ are operators in $\text{OP}M\Gamma_{\rho,\mathcal{P}}^0$ whenever $(\beta,\alpha) \in V(\mathcal{P})$.

This ends the proof. □

Remark 2.8.15. The proof of Theorem 2.8.14 shows that we may take, as an equivalent norm in $H_{\mathcal{P}}^p$,

$$\|u\|_{H_{\mathcal{P}}^p} = \sum_{(\beta,\alpha) \in \mathcal{P}} \|x^\beta D^\alpha u\|_{L^p},$$

where the sum is now extended to all multi-indices $(\beta,\alpha) \in \mathcal{P}$.

In conclusion, we want to apply the preceding results to the operator with polynomial coefficients

$$P = \sum_{(\beta,\alpha) \in \mathcal{P}} c_{\alpha\beta} x^\beta D^\alpha \qquad (2.8.35)$$

obtained by standard left quantization from a multi-quasi-elliptic polynomial, cf. Examples 2.7.9 and 2.7.16. Let us refer to the space $H_{\mathcal{P}}^p$, $1 < p < \infty$, defined as in (2.8.33). We have that

$$P : H_{\mathcal{P}}^p \longrightarrow L^p \qquad (2.8.36)$$

is Fredholm, and the following a priori estimates are valid for a suitable $C > 0$:

$$\sum_{(\beta,\alpha) \in \mathcal{P}} \|x^\beta D^\alpha u\|_{L^p} \leq C \left(\|Pu\|_{L^p} + \|u\|_{L^p} \right). \qquad (2.8.37)$$

These L^p-estimates are in particular valid for the Γ-elliptic differential operators of Section 2.2, so for example for the harmonic oscillator $H = -\triangle + |x|^2$ in $\mathbb{R}^d$ we have

$$\sum_{j=1}^d \left(\|x_j^2 u\|_{L^p} + \|D_{x_j}^2 u\|_{L^p} \right) \leq C \left(\|Hu\|_{L^p} + \|u\|_{L^p} \right). \qquad (2.8.38)$$

Another relevant example is given by the more general Schrödinger operator

$$P = -\triangle + V(x), \qquad (2.8.39)$$

where $V(x)$ is a multi-quasi-elliptic polynomial

$$V(x) = \sum_{\beta \in \Omega} c_\beta x^\beta, \qquad (2.8.40)$$

with Ω a complete polyhedron in $\mathbb{R}^d$. The symbol of P in (2.8.39) is given by the polynomial in $\mathbb{R}^{2d}$,

$$|\xi|^2 + \sum_{\beta \in \Omega} c_\beta x^\beta. \qquad (2.8.41)$$

The corresponding Newton polyhedron $\mathcal{P}$ is easily computed, and turns out to be complete in $\mathbb{R}^{2d}$. Assuming further

$$V(x) > 0 \qquad (2.8.42)$$

we deduce multi-quasi-ellipticity of the polynomial (2.8.41) with respect to $\mathcal{P}$. The preceding results then apply and give in particular the estimates

$$\sum_{j=1}^{d} \|D_{x_j}^2 u\|_{L^p} + \sum_{\beta \in \Omega} \|x^\beta u\|_{L^p} \leq C\big(\|(-\triangle + V(x))u\|_{L^p} + \|u\|_{L^p}\big). \qquad (2.8.43)$$

Examples of potentials $V(x)$ satisfying the preceding assumptions are quasi-elliptic polynomials

$$V(x) = \sum_{j=1}^{d} x_j^{2N_j}, \qquad (2.8.44)$$

for positive integers N_j, and polynomials in $\mathbb{R}^2$ of the form

$$V(x_1, x_2) = x_1^{h_1} + x_1^{h_2} x_2^{k_2} + x_2^{k_1}, \qquad (2.8.45)$$

where the integers h_j, k_j, $j = 1, 2$, are even and satisfy the conditions in Example 2.6.10.

Notes

The classes $\Gamma^m(\mathbb{R}^d)$ and $\Gamma_\rho^m(\mathbb{R}^d)$, in Section 2.1, were studied by Shubin, see [183] and the references there to previous works. We also cite Helffer [109]. Such classes play an important role in semi-classical analysis, besides [183] see Robert [170] and references therein. Passing to consider the Γ-elliptic differential operators in Sections 2.2 and 2.3, we observe that they appeared earlier in Grushin [104], [105] as a tool for study of the local properties of partial differential equations with multiple characteristics. In this regard we mention also Helffer and Rodino [110], [111], Parenti and Rodino [158] and Mascarello and Rodino [142] for applications to the local regularity problem and to Parenti and Parmeggiani [157] and Mughetti and Nicola [149] for applications to lower bound estimates. In particular, the proofs of Theorems 2.3.3 and 2.3.5 were first given in [111], whereas we cite [142, Chapter 7] for details on the results of asymptotic integration used in Section 2.3; see also Gramchev and Popivanov [96].

The H-type and ρ-H-type polynomials of Section 2.4 were introduced by Hörmander [113], [114] in the context of the local regularity; see also Treves [189], Rodino [172]. We cite De Donno [70] for an alternative proof of Proposition 2.4.4. A main reference for Section 2.6, concerning multi-quasi-elliptic polynomials and generalizing Section 2.5, on the quasi-elliptic case, is the monograph of Boggiatto, Buzano and Rodino [19]; relevant previous works on the subject are Cattabriga

[46], Friberg [83], Gindikin and Volevich [91], Zanghirati [200]. For recent contributions on multi-quasi-elliptic polynomials see Gindikin and Volevich [92], Calvo [35], Bouzar and Chaili [25]. The source for Example 2.6.15 is Pini [163]; more general examples of H-polynomials, which are not multi-quasi-elliptic, can be found in De Donno and Oliaro [71].

The $\Gamma_{\mathcal{P}}$-pseudo-differential operators were first introduced in Boggiatto [17], and then studied in detail in [19]. The presentation in Sections 2.7, 2.8 follows Morando [148]; in the local context, the same L^p-estimates were proved by Garello and Morando [85], inspired by the original idea of Cattabriga [47]. Concerning the theorem of Lizorkin and Marcinkiewicz, we refer for the proof to Lizorkin [134] and Stein [184]. In our proof of the L^p-boundedness in Section 2.8, we also used some arguments from Wong [197]. For boundedness on L^p and on other function spaces arising in Fourier Analysis we also refer to Stein [185], the recent contribution by Cordero and Nicola [54] and the references therein.

Finally, we would like to cite Garetto [86], where the Γ-pseudo-differential calculus is recast in the context of Colombeau generalized functions, based on the space $\mathcal{S}(\mathbb{R}^d)$, see also Garetto, Gramchev and Oberguggenberger [87], Garetto and Hörmann [88], Hörmann, Oberguggenberger and Pilipovic [120].

Chapter 3

G-Pseudo-Differential Operators

Summary

As in Chapter 2, the basic example here is a partial differential operator with polynomial coefficients in $\mathbb{R}^d$, that is

$$P = \sum c_{\alpha\beta} x^\beta D^\alpha,$$

wherein the sum $(\alpha, \beta) \in \mathbb{N}^d \times \mathbb{N}^d$ runs over a finite subset of indices.

The symbol in the standard quantization is

$$p(x, \xi) = \sum c_{\alpha\beta} x^\beta \xi^\alpha = \sum c'_\gamma z^\gamma$$

with $z = (x, \xi) \in \mathbb{R}^{2d}$, $\gamma = (\beta, \alpha) \in \mathbb{N}^{2d}$.

As we have seen in Chapter 2, Hörmander's condition, i.e., Γ-hypoellipticity

$$|\partial_z^\gamma p(z)| \lesssim |p(z)| \langle z \rangle^{-\rho|\gamma|} \tag{3.0.1}$$

for $z \in \mathbb{R}^{2d}$, $|z| \geq R > 0$, $0 < \rho \leq 1$, is sufficient to obtain the global regularity of P. However, the condition is not necessary. In fact, the aim of the present chapter is to discuss classes of symbols p which do not satisfy (3.0.1) and nevertheless give rise to globally regular operators P.

As an elementary example in this connection, consider a polynomial $q(\xi)$ depending only on the ξ-variables and satisfying for some integer $m \geq 1$:

$$\langle \xi \rangle^m \lesssim |q(\xi)| \lesssim \langle \xi \rangle^m \quad \text{for } \xi \in \mathbb{R}^d. \tag{3.0.2}$$

It follows from (3.0.2) that $q(\xi)$ is elliptic of order m with respect to the ξ-variables, but the estimates (3.0.1) are not satisfied in the whole space $\mathbb{R}^{2d}$. On the other

hand, the global regularity of the partial differential operator with constant coefficients $q(D)$ is easily proved directly. To be definite, fix attention on the one-
dimensional case:

$$q(\xi) = \sum_{j=0}^{m} c_j \xi^j, \quad \xi \in \mathbb{R}, \ c_j \in \mathbb{C}, \ c_m \neq 0.$$

Then the lower bound (3.0.2) is satisfied if and only if the algebraic equation
$q(\xi) = 0$ has no real roots, which amounts to requiring that all the solutions of
the ordinary differential equation

$$q(D)u = \sum_{j=0}^{m} c_j D^j u = 0$$

have exponential decay/growth; it follows easily that $q(D)u \in \mathcal{S}(\mathbb{R})$, $u \in \mathcal{S}'(\mathbb{R})$
imply $u \in \mathcal{S}(\mathbb{R})$, i.e., $q(D)$ is globally regular. The lower bound in (3.0.2), corresponding to the absence of real roots, is essential to get the conclusion; for example,
if $q(\xi) = \xi$ and $q(D) = D$ in $\mathbb{R}$, then the equation $Du = 0$ admits $u(x) = \text{const.}$
as solution, with $u \in \mathcal{S}'(\mathbb{R})$, $u \notin \mathcal{S}(\mathbb{R})$. A relevant example of operator $q(D)$ in $\mathbb{R}^d$,
$d > 1$, with symbol satisfying (3.0.2), is the free particle Schrödinger operator

$$-\Delta - \lambda, \quad \lambda \notin \mathbb{R}_+ \cup \{0\}.$$

In Section 3.1 we re-consider the previous examples in the framework of the so-
called G-calculus. Namely, we define the class $G^{m,n}(\mathbb{R}^d)$, $m \in \mathbb{R}$, $n \in \mathbb{R}$, of all the
symbols $a(x,\xi)$ satisfying for $\alpha \in \mathbb{N}^d$, $\beta \in \mathbb{N}^d$,

$$|\partial_\xi^\alpha \partial_x^\beta a(x,\xi)| \lesssim \langle\xi\rangle^{m-|\alpha|}\langle x\rangle^{n-|\beta|}, \quad x \in \mathbb{R}^d, \ \xi \in \mathbb{R}^d. \tag{3.0.3}$$

The definition of G-ellipticity, generalizing (3.0.2), is given by

$$\langle x\rangle^n \langle\xi\rangle^m \lesssim |a(x,\xi)| \quad \text{for } |x|^2 + |\xi|^2 \geq R^2 > 0. \tag{3.0.4}$$

We refer to Section 3.1 for the corresponding pseudo-differential calculus. With
the notations in Chapter 1, we have $\Phi(x,\xi) = \langle x\rangle$, $\Psi(x,\xi) = \langle\xi\rangle$, that gives a
couple of sub-linear temperate weights, cf. (1.1.1) and (1.1.2).

In Section 3.2 we study classical G-symbols $a \in G_{\mathrm{cl}(\xi,x)}^{m,n}(\mathbb{R}^d) \subset G^{m,n}(\mathbb{R}^d)$,
possessing three principal parts: $\sigma_\psi^m(a)$, homogeneous with respect to ξ; $\sigma_e^n(a)$,
homogeneous with respect to x; $\sigma_{\psi,e}^{m,n}(a)$, homogeneous with respect to x and ξ
separately.

The definition of G-ellipticity can be reset in terms of principal parts. As an
example, consider

$$p(x,\xi) = \sum_{|\alpha|\leq m, \ |\beta|\leq n} c_{\alpha\beta} x^\beta \xi^\alpha, \tag{3.0.5}$$

which we may regard as an element of $G^{m,n}_{cl(\xi,x)}(\mathbb{R}^d)$, and assume

$$\sigma^{m,n}_{\psi,e}(p) = \sum_{\substack{|\alpha|=m \\ |\beta|=n}} c_{\alpha\beta}x^\beta\xi^\alpha \neq 0 \quad \text{for } x \neq 0,\ \xi \neq 0, \tag{3.0.6}$$

$$\sigma^m_\psi(p) = \sum_{\substack{|\alpha|=m \\ |\beta|\leq n}} c_{\alpha\beta}x^\beta\xi^\alpha \neq 0 \quad \text{for all } x \in \mathbb{R}^d \text{ and } \xi \neq 0, \tag{3.0.7}$$

$$\sigma^n_e(p) = \sum_{\substack{|\alpha|\leq m \\ |\beta|=n}} c_{\alpha\beta}x^\beta\xi^\alpha \neq 0 \quad \text{for } x \neq 0 \text{ and all } \xi \in \mathbb{R}^d. \tag{3.0.8}$$

Namely, (3.0.6), (3.0.7), (3.0.8) hold simultaneously if and only if the G-ellipticity condition (3.0.4) is satisfied. When $n = 0$, i.e., the case of an operator with constant coefficients, we recapture (3.0.2). Besides construction of parametrix and global regularity, the previous assumptions of G-ellipticity provide Fredholm property and a priori estimates in the Sobolev spaces $H^{m,n}_G(\mathbb{R}^d)$, defined for non-negative integers m, n by

$$\|u\|_{H^{m,n}_G} = \sum_{\substack{|\alpha|\leq m \\ |\beta|\leq n}} \|x^\beta D^\alpha u\|_{L^2} < \infty. \tag{3.0.9}$$

Section 3.3 is devoted to a detailed analysis of G-elliptic ordinary differential operators. Similarly to Section 2.3, we apply here the theory of Asymptotic Integration, and we deduce some exponential decay properties of the solutions, which will be extended later in the book to semi-linear equations, cf. Sections 6.3, 6.4, 6.5.

Section 3.4 concerns other applications of the general calculus of Chapter 1 to the problem of global regularity. Namely we take here, for some ρ with $0 < \rho < 1$, the couple of weights

$$\Phi(x,\xi) = 1, \quad \Psi(x,\xi) = (1 + |x|^2 + |\xi|^2)^{\rho/2} \tag{3.0.10}$$

or alternatively

$$\Phi(x,\xi) = (1 + |x|^2 + |\xi|^2)^{\rho/2}, \quad \Psi(x,\xi) = 1. \tag{3.0.11}$$

In the framework of the corresponding calculus, we prove global regularity of the operator

$$P = -\Delta + V(x), \tag{3.0.12}$$

where the potential $V(x)$ is a polynomial satisfying

$$\langle x\rangle^{2\rho} \lesssim V(x), \quad |\partial^\alpha V(x)| \lesssim V(x). \tag{3.0.13}$$

As an example, consider in $\mathbb{R}^2$,

$$V(x_1,x_2) = (1 + x_1^2)(1 + x_2^2). \tag{3.0.14}$$

Similarly, we have a global regularity for the operator

$$\widehat{P} = V(D) + |x|^2, \tag{3.0.15}$$

where the polynomial $V(\xi)$ satisfies, in the variable ξ, the same estimates (3.0.13). Notice that the assumption (3.0.13) for V is weaker than Hörmander's property in $\mathbb{R}^d$, cf. Example (3.0.14). Hence the operator $\widehat{P}$ is not locally regular in the Schwartz' sense, in general.

3.1 *G*-Pseudo-Differential Calculus

Other basic examples of the general classes of symbols considered in Chapter 1 are the *G*-classes defined below. With respect to the Γ-classes studied in Chapter 2, now the asymptotic behaviour in the (x, ξ)-space of the two weights Φ, Ψ is not optimal separately, nevertheless the strong uncertainty principle (1.1.10) is satisfied thanks to a favourable combination of decay properties.

Definition 3.1.1. We define $G^{m,n}(\mathbb{R}^d)$, $m \in \mathbb{R}$, $n \in \mathbb{R}$, as the set of all functions $a(x, \xi) \in C^\infty(\mathbb{R}^{2d})$ satisfying, for all $\alpha, \beta \in \mathbb{N}^d$, the estimates

$$|\partial_\xi^\alpha \partial_x^\beta a(x, \xi)| \lesssim \langle \xi \rangle^{m-|\alpha|} \langle x \rangle^{n-|\beta|}, \quad (x, \xi) \in \mathbb{R}^{2d}. \tag{3.1.1}$$

The weights $\Phi(x, \xi) = \langle x \rangle$, $\Psi(x, \xi) = \langle \xi \rangle$ are sub-linear (1.1.1) and temperate (1.1.2), and the strong uncertainty principle (1.1.10) is obviously satisfied, since

$$\langle x \rangle \langle \xi \rangle \gtrsim (1 + |x|^2 + |\xi|^2)^{1/2}.$$

Hence all the results of Chapter 1 apply for the corresponding pseudo-differential operators

$$Au(x) = a(x, D)u(x) = \int e^{ix\xi} a(x, \xi) \widehat{u}(\xi) \, d\xi, \tag{3.1.2}$$

with action $a(x, D) : \mathcal{S}(\mathbb{R}^d) \to \mathcal{S}(\mathbb{R}^d)$, $\mathcal{S}'(\mathbb{R}^d) \to \mathcal{S}'(\mathbb{R}^d)$. Namely, we shall denote by $\mathrm{OPG}^{m,n}(\mathbb{R}^d)$ the class of the operators in (3.1.2) with symbol $a \in G^{m,n}(\mathbb{R}^d)$.

The definition of $\mathrm{OPG}^{m,n}(\mathbb{R}^d)$ does not depend on the quantization. In the standard quantization, if $a \in G^{m,n}(\mathbb{R}^d)$, $b \in G^{m',n'}(\mathbb{R}^d)$, then the operator $c(x, D) = a(x, D)b(x, D)$ belongs to $\mathrm{OPG}^{m+m',n+n'}(\mathbb{R}^d)$, with symbol

$$c(x, \xi) \sim \sum_\alpha (\alpha!)^{-1} \partial_\xi^\alpha a(x, \xi) D_x^\alpha b(x, \xi), \tag{3.1.3}$$

where it is worth observing that the orders of the term $\partial_\xi^\alpha a(x, \xi) D_x^\alpha b(x, \xi)$ are given by $m + m' - |\alpha|, n + n' - |\alpha|$. Similarly we may argue for transposed operators and formal adjoints. The corresponding definition of ellipticity is the following.

Definition 3.1.2. Let $a \in G^{m,n}(\mathbb{R}^d)$, $m \in \mathbb{R}$, $n \in \mathbb{R}$. We say that a is *G*-elliptic if there exists $R > 0$ such that

$$\langle x \rangle^m \langle \xi \rangle^n \lesssim |a(x, \xi)| \quad \text{for } |x|^2 + |\xi|^2 \geq R^2. \tag{3.1.4}$$

We then repeat from Chapter 1 the results on the existence of parametrices, cf. Theorem 1.3.6, Corollary 1.3.9, particularized to G-elliptic symbols.

Theorem 3.1.3. *Let $a \in G^{m,n}(\mathbb{R}^d)$, $m \in \mathbb{R}$, $n \in \mathbb{R}$, be G-elliptic. Then there exists $b \in G^{-m,-n}(\mathbb{R}^d)$ such that $b(x, D)$ is a parametrix of $a(x, D)$, i.e.,*

$$a(x, D)b(x, D) = I + S_1, \quad b(x, D)a(x, D) = I + S_2$$

where S_1 and S_2 are regularizing operators. Hence $a(x, D)$ is globally regular, cf. Definition 1.3.8.

Passing to consider the corresponding Sobolev spaces, we may define them by means of the pseudo-differential operators $a_{s,t}(x, D) = \langle x \rangle^t \langle D \rangle^s$, $s \in \mathbb{R}$, $t \in \mathbb{R}$, that is the standard quantization of the symbol $a_{s,t}(x, \xi) = \langle x \rangle^t \langle \xi \rangle^s \in G^{s,t}(\mathbb{R}^d)$. Note that $a_{s,t}(x, \xi)$ is G-elliptic, and $a_{s,t}(x, D) : \mathcal{S}(\mathbb{R}^d) \to \mathcal{S}(\mathbb{R}^d)$ is an isomorphism, extending to an isomorphism on $\mathcal{S}'(\mathbb{R}^d)$, with inverse $b_{-s,-t}(x, D) = \langle D \rangle^{-s} \langle x \rangle^{-t}$; the asymptotic expansion of $b_{-s,-t}(x, \xi) \in G^{-s,-t}(\mathbb{R}^d)$ can be computed in terms of (3.1.3).

Definition 3.1.4. For $s \in \mathbb{R}$, $t \in \mathbb{R}$ we define the Hilbert space

$$H_G^{s,t}(\mathbb{R}^d) = \langle D \rangle^{-s} \langle x \rangle^{-t} \left(L^2(\mathbb{R}^d) \right) = \{ u \in \mathcal{S}'(\mathbb{R}^d) : \langle x \rangle^t \langle D \rangle^s u \in L^2(\mathbb{R}^d) \} \quad (3.1.5)$$

with the scalar product

$$(u, v)_{H_G^{s,t}} = \left(\langle x \rangle^t \langle D \rangle^s u, \langle x \rangle^t \langle D \rangle^s v \right)_{L^2} \quad (3.1.6)$$

and corresponding norm

$$\| u \|_{H_G^{s,t}} = \| \langle x \rangle^t \langle D \rangle^s u \|_{L^2}. \quad (3.1.7)$$

According to Proposition 1.5.3, for every $T \in OPG^{m,n}(\mathbb{R}^d)$ with G-elliptic symbol we have

$$H_G^{s,t}(\mathbb{R}^d) = \{ u \in \mathcal{S}'(\mathbb{R}^d) : Tu \in L^2(\mathbb{R}^d) \} \quad (3.1.8)$$

with scalar product equivalent to (3.1.6)

$$(u, v)_T = (Tu, Tv)_{L^2} + (Ru, Rv)_{L^2} \quad (3.1.9)$$

where R is a regularizing operator associated to a parametrix of T.

From Proposition 1.5.5, we then obtain the following boundedness result.

Theorem 3.1.5. *Every $A \in OPG^{m,n}(\mathbb{R}^d)$, $m \in \mathbb{R}$, $n \in \mathbb{R}$, defines for all $s \in \mathbb{R}$, $t \in \mathbb{R}$ a continuous operator*

$$A : H_G^{s,t}(\mathbb{R}^d) \to H_G^{s-m,t-n}(\mathbb{R}^d). \quad (3.1.10)$$

We have compact immersion $H_G^{s,t}(\mathbb{R}^d) \to H_G^{s',t'}(\mathbb{R}^d)$ for $s > s'$, $t > t'$, and compactness of the map $A : H_G^{s,t}(\mathbb{R}^d) \to H_G^{s',t'}(\mathbb{R}^d)$ for $A \in \mathrm{OPG}^{m,n}(\mathbb{R}^d)$ whenever $s - s' > m$, $t - t' > n$. Moreover

$$\bigcap_{s \in \mathbb{R}, \, t \in \mathbb{R}} H_G^{s,t}(\mathbb{R}^d) = \mathcal{S}(\mathbb{R}^d), \qquad \bigcup_{s \in \mathbb{R}, \, t \in \mathbb{R}} H_G^{s,t}(\mathbb{R}^d) = \mathcal{S}'(\mathbb{R}^d). \tag{3.1.11}$$

For a couple m, n of non-negative integers, an equivalent definition of the space $H_G^{s,t}(\mathbb{R}^d)$ is given by

$$H_G^{m,n}(\mathbb{R}^d) = \{u \in \mathcal{S}'(\mathbb{R}^d) : \; x^\beta D^\alpha u \in L^2(\mathbb{R}^d) \text{ for } |\alpha| \le m, \, |\beta| \le n\}, \tag{3.1.12}$$

with equivalent norm

$$\|u\|_{H_G^{m,n}} = \sum_{\substack{|\alpha| \le m \\ |\beta| \le n}} \|x^\beta D^\alpha u\|_{L^2}. \tag{3.1.13}$$

Observe also that for $t = 0$ the space $H_G^{s,0}(\mathbb{R}^d)$ gives the standard Sobolev space $H^s(\mathbb{R}^d)$.

We may now improve Theorem 3.1.3 adding precise Sobolev estimates.

Theorem 3.1.6. *Consider* $A \in \mathrm{OPG}^{m,n}(\mathbb{R}^d)$, $m \in \mathbb{R}$, $n \in \mathbb{R}$, *with G-elliptic symbol and assume* $u \in \mathcal{S}'(\mathbb{R}^d)$, $Au \in H_G^{s,t}(\mathbb{R}^d)$, $s \in \mathbb{R}$, $t \in \mathbb{R}$. *Then* $u \in H_G^{s+m,t+n}(\mathbb{R}^d)$ *and for every* $s' < s + m$, $t' < t + n$,

$$\|u\|_{H_G^{s+m,t+n}} \le C\big(\|Au\|_{H_G^{s,t}} + \|u\|_{H_G^{s',t'}}\big), \tag{3.1.14}$$

for a positive constant C *depending on* s *and* t. *Moreover, the operator* A, *considered as a map* $H_G^{s+m,t+n}(\mathbb{R}^d) \to H_G^{s,t}(\mathbb{R}^d)$ *is Fredholm, cf. Theorem 1.6.9.*

If m, n is a couple of positive integers, we may refer to the equivalent norm (3.1.13) and re-write (3.1.14) for $s = s' = 0$, $t = t' = 0$ as

$$\sum_{\substack{|\alpha| \le m \\ |\beta| \le n}} \|x^\beta D^\alpha u\|_{L^2} \le C \left(\|Au\|_{L^2} + \|u\|_{L^2}\right). \tag{3.1.15}$$

We may as well consider G-hypoelliptic symbols and related operators.

Theorem 3.1.7. *Assume* $a \in G_\rho^{m,n}(\mathbb{R}^d)$, $m \in \mathbb{R}$, $n \in \mathbb{R}$, $0 < \rho \le 1$, *i.e., for all* $(\alpha, \beta) \in \mathbb{N}^{2d}$,

$$|\partial_\xi^\alpha \partial_x^\beta a(x,\xi)| \lesssim \langle\xi\rangle^{m-\rho|\alpha|} \langle x\rangle^{n-\rho|\beta|}, \quad (x,\xi) \in \mathbb{R}^{2d}. \tag{3.1.16}$$

Assume that a *is hypoelliptic, i.e., there exist* m', n' *with* $m' \le m$, $n' \le n$ *and* $R > 0$ *such that*

$$\langle\xi\rangle^{m'} \langle x\rangle^{n'} \lesssim |a(x,\xi)| \quad \text{for } |x|^2 + |\xi|^2 \ge R^2 \tag{3.1.17}$$

and for every $(\alpha, \beta) \in \mathbb{N}^{2d}$,

$$|\partial_\xi^\alpha \partial_x^\beta a(x, \xi)| \lesssim |a(x, \xi)| \langle \xi \rangle^{-\rho|\alpha|} \langle x \rangle^{-\rho|\beta|}. \tag{3.1.18}$$

Then, denoted by $A = a(x, D)$ *the pseudo-differential operator with symbol* a, *there exists* $B = b(x, D)$ *with symbol* $b \in G^{-m', -n'}(\mathbb{R}^d)$ *such that*

$$AB = I + S_1, \quad BA = I + S_2,$$

with S_1 *and* S_2 *regularizing operators. The estimate* (3.1.14) *remains valid with* m, n *replaced by* m', n'.

Let us now compare G-classes with Γ-classes. We may say that G-classes are more general, in the sense that $\Gamma^M(\mathbb{R}^d)$, $M \in \mathbb{R}$, is included in $G^{m,n}(\mathbb{R}^d)$ for a suitable couple m, n, take for example $m = n = M$, if $M > 0$, and $m = n = 0$ for $M < 0$, and use the obvious estimates $\langle x \rangle \leq (1 + |x|^2 + |\xi|^2)^{1/2}$, $\langle \xi \rangle \leq (1 + |x|^2 + |\xi|^2)^{1/2}$, $(1 + |x|^2 + |\xi|^2)^{1/2} \leq \langle x \rangle \langle \xi \rangle$. Also, the generalized classes Γ_ρ^m, $\Gamma_{\rho,\mathcal{P}}^m$, $M\Gamma_{\rho,\mathcal{P}}^m$ in Chapter 2 are included in the classes defined by (3.1.16) and Γ-hypoellipticity implies G-hypoellipticity, cf. (3.1.18). Summing up, all the results of global regularity in Chapter 2 are recaptured by the above Theorem 3.1.7. However, Γ-elliptic symbols and $\mathcal{P}$-elliptic symbols are not G-elliptic, and we cannot identify in the G-frame the peculiar properties of the corresponding operators, in particular the sharp Sobolev estimates in Sections 2.1, 2.7 and 2.8.

In the opposite direction, the G-symbols do not satisfy in general the estimates required for Γ-symbols. Relevant examples are given by Fourier multipliers, i.e., operators with symbols depending only on the ξ-variables. Namely, consider $a(\xi) \in C^\infty(\mathbb{R}^d)$ satisfying for all $\alpha \in \mathbb{N}^d$ and some $m \in \mathbb{R}$,

$$|\partial^\alpha a(\xi)| \lesssim \langle \xi \rangle^{m - |\alpha|}, \quad \xi \in \mathbb{R}^d. \tag{3.1.19}$$

We have $a \in G^{m,0}(\mathbb{R}^d)$ and a is G-elliptic when

$$\langle \xi \rangle^m \lesssim |a(\xi)|, \quad \xi \in \mathbb{R}^d. \tag{3.1.20}$$

It is worth emphasizing that estimate (3.1.20) is assumed to be valid in the whole $\mathbb{R}^d$, which implies $a(\xi) \neq 0$ for all $\xi \in \mathbb{R}^d$. In fact, the vanishing of $a(\xi)$ at a point ξ_0 would contradict (3.1.4) for any choice of the constant $R > 0$. Note also that a symbol $a(\xi)$ satisfying (3.1.19) does not belong to any of the Γ-classes, but when $a(\xi)$ is a polynomial, and in this case Γ-ellipticity fails anyhow for $m \geq 1$.

If (3.1.20) is satisfied, we may apply Theorems 3.1.3 and 3.1.6. The parametrix of $a(D)$ is actually an inverse, with symbol $b(\xi) = \frac{1}{a(\xi)} \in G^{-m,0}(\mathbb{R}^d)$. Then $a(D) : \mathcal{S}(\mathbb{R}^d) \to \mathcal{S}(\mathbb{R}^d)$ is an isomorphism extending to an isomorphism $a(D) : \mathcal{S}'(\mathbb{R}^d) \to \mathcal{S}'(\mathbb{R}^d)$ and

$$a(D) : H_G^{s,t}(\mathbb{R}^d) \to H_G^{s-m,t}(\mathbb{R}^d) \tag{3.1.21}$$

where $s \in \mathbb{R}$, $t \in \mathbb{R}$, with inverse

$$b(D): \ H_G^{s-m,t}(\mathbb{R}^d) \to H_G^{s,t}(\mathbb{R}^d). \tag{3.1.22}$$

Let us discuss in detail some examples of Fourier multipliers, considering first differential operators with constant coefficients.

Example 3.1.8. *Partial differential operators.* For m a positive integer, consider

$$p(\xi) = \sum_{|\alpha| \leq m} c_\alpha \xi^\alpha, \quad c_\alpha \in \mathbb{C}, \tag{3.1.23}$$

giving the operator

$$P = p(D) = \sum_{|\alpha| \leq m} c_\alpha D^\alpha. \tag{3.1.24}$$

Define

$$p_m(\xi) = \sum_{|\alpha| = m} c_\alpha \xi^\alpha. \tag{3.1.25}$$

Then the G-ellipticity (3.1.20) is equivalent to assume simultaneously:

(i) $p_m(\xi) \neq 0$ for $\xi \neq 0$;

(ii) $p(\xi) \neq 0$ for all $\xi \in \mathbb{R}^d$.

In fact, from the proof of Proposition 2.1.5 we know that (i) is equivalent to the ellipticity of $p(\xi)$ as a polynomial in ξ-variables:

$$|\xi|^m \lesssim |p(\xi)| \ \text{ for } |\xi| \geq R, \tag{3.1.26}$$

with $R > 0$ sufficiently large. On the other hand (3.1.26) and (ii), jointly, are equivalent to

$$\langle \xi \rangle^m \lesssim |p(\xi)|, \quad \xi \in \mathbb{R}^d. \tag{3.1.27}$$

Under the G-ellipticity assumption (3.1.27), the previous conclusions hold, in particular P in (3.1.24) is globally regular. Note that if $p(\xi_0) = 0$ for some $\xi_0 \in \mathbb{R}^d$, then global regularity fails, because $u(x) = e^{i\xi_0 x}$ satisfies $u \in \mathcal{S}'(\mathbb{R}^d)$, $u \notin \mathcal{S}(\mathbb{R}^d)$ and $P(e^{i\xi_0 x}) = p(\xi_0)e^{i\xi_0 x} = 0$.

Example 3.1.9. *Ordinary differential operators.* In the one-dimensional case, consider the symbol

$$p(\xi) = \xi^m + \sum_{j=0}^{m-1} c_j \xi^j, \quad c_j \in \mathbb{C}, \ \xi \in \mathbb{R},$$

with corresponding ordinary differential operator

$$P = p(D) = D_x^m + \sum_{j=0}^{m-1} c_j D_x^j.$$

The G-ellipticity reduces to assume $p(\xi) \neq 0$ for all $\xi \in \mathbb{R}$, that is all roots $r_j \in \mathbb{C}$, $j = 1, ..., m$ of the algebraic equation $p(\xi) = 0$ satisfy $\operatorname{Im} r_j \neq 0$. Consequently a basis for the classical solutions of $Pu = 0$ is given by functions of the form

$$u(x) = Q(x)e^{ir_j x}$$

for some polynomial $Q(x)$, with $\operatorname{Re} ir_j \neq 0$, giving exponential growth in $\mathbb{R}_+$ or in $\mathbb{R}_-$. Hence the only solution $u \in \mathcal{S}(\mathbb{R})$ of $Pu = 0$ is the trivial solution, as expected from the previous arguments. We get also that $Pu = f \in \mathcal{S}(\mathbb{R})$ admits a unique solution $u \in \mathcal{S}(\mathbb{R})$. More precisely, $P : H_G^{s+m,t}(\mathbb{R}) \to H_G^{s,t}(\mathbb{R})$ is an isomorphism for all $s \in \mathbb{R}$, $t \in \mathbb{R}$.

Example 3.1.10. *The free particle.* As a particular case of the above examples, consider in $\mathbb{R}^d$ the operator of the free particle

$$P = -\Delta - \lambda$$

with symbol

$$p(\xi) = |\xi|^2 - \lambda.$$

The symbol is G-elliptic and the operator globally regular, if and only if $\lambda \notin \mathbb{R}_+ \cup \{0\}$.

Example 3.1.11. *The intermediate-long-wave operator.* The following non-polynomial symbol appears in the study of solitary waves, see later in this book, Section 6.4:

$$p(\xi) = \xi \operatorname{Ctgh} \xi + \gamma, \quad \xi \in \mathbb{R},$$

with $\operatorname{Ctgh} \xi = \operatorname{Ch} \xi / \operatorname{Sh} \xi = (e^\xi + e^{-\xi})/(e^\xi - e^{-\xi})$ and $\gamma \in \mathbb{R}$. We have $p \in G^{1,0}(\mathbb{R})$ and the G-ellipticity condition is satisfied if $\gamma > -1$.

3.2 Polyhomogeneous G-Operators

It is natural to introduce the subclass of $G^{m,n}(\mathbb{R}^d)$ of the polyhomogeneous (or classical) symbols. The definition is somewhat more complicated than in the Γ-case, since homogeneity has to be considered separately in the x and ξ variables. Building blocks of the calculus are the following different types of homogeneous symbols.

Type 1 symbols: orders m, n; bi-homogeneity with respect to x, ξ. We assume that $a(x, \xi) \in C^\infty((\mathbb{R}^d \setminus \{0\}) \times (\mathbb{R}^d \setminus \{0\}))$ is positively homogeneous of degree $m \in \mathbb{R}$, $n \in \mathbb{R}$ with respect to x and ξ separately, i.e.,

$$a(sx, t\xi) = t^m s^n a(x, \xi) \quad \text{for all } t > 0, s > 0.$$

We shall also denote by $\mathcal{H}_{\xi,x}^{m,n}$ the class of such functions.

If we take $\chi \in C^\infty(\mathbb{R}^d)$, $\chi(x) = 1$ for $|x| \geq 2$, $\chi(x) = 0$ for $|x| \leq 1$, then $\chi(x)\chi(\xi)a(x,\xi) \in G^{m,n}(\mathbb{R}^d)$.

Type 2 symbols:

(i) *orders m, n; homogeneity m with respect to ξ.* We assume that $a(x,\xi) \in C^\infty(\mathbb{R}^d \times (\mathbb{R}^d \setminus \{0\}))$ is positively homogeneous of degree m with respect to ξ, i.e.,

$$a(x, t\xi) = t^m a(x,\xi), \quad \text{for all } s > 0.$$

We assume moreover for all $\alpha, \beta \in \mathbb{N}^d$,

$$|\partial_\xi^\alpha \partial_x^\beta a(x,\xi)| \lesssim \langle x \rangle^{n-|\beta|} \quad \text{for } x \in \mathbb{R}^d, \ \xi \in \mathbb{S}^{d-1},$$

where $\mathbb{S}^{d-1} = \{\xi \in \mathbb{R}^d : |\xi| = 1\}$. Taking χ as before, we have $\chi(\xi)a(x,\xi) \in G^{m,n}(\mathbb{R}^d)$.

(ii) *orders m, n; homogeneity n with respect to x.* Same definition as in (i), interchanging the role of x and ξ.

Type 3 symbols:

(i) *orders m, n; homogeneity m with respect to ξ, polyhomogeneity with respect to x.* Let $a(x,\xi)$ satisfy the same conditions as in Type 2, (i), and assume the existence of $b_{n-j}(x,\xi) \in \mathcal{H}_{\xi,x}^{m,n-j}$, cf. Type 1, $j = 0, 1, ...,$ such that $a(x,\xi) \sim \sum_{j=0}^\infty b_{n-j}(x,\xi)$, in the sense that, for all $\alpha, \beta \in \mathbb{N}^d$,

$$\left| \partial_\xi^\alpha \partial_x^\beta \left(a(x,\xi) - \sum_{0 \leq j < N} \chi(x)b_{n-j}(x,\xi) \right) \right| \lesssim \langle x \rangle^{n-N-|\beta|}, \quad x \in \mathbb{R}^d, \ \xi \in \mathbb{S}^{d-1}.$$

As before we have $\chi(\xi)a(x,\xi) \in G^{m,n}(\mathbb{R}^d)$.

(ii) *orders m, n; homogeneity n with respect to x, polyhomogeneity with respect to ξ.* Same definition as in (i), interchanging the role of x and ξ.

We may now define the classes of the G-polyhomogeneous symbols. Again, we distinguish three types: symbols which are classical with respect to ξ, with respect to x, with respect to ξ, x simultaneously.

Definition 3.2.1. We define $G_{\mathrm{cl}(\xi)}^{m,n}(\mathbb{R}^d)$ as the class of all $a \in G^{m,n}(\mathbb{R}^d)$ admitting asymptotic expansion

$$a(x,\xi) \sim \sum_{k=0}^\infty a_{m-k}(x,\xi) \tag{3.2.1}$$

where a_{m-k} are Type 2, (i), namely of orders $m - k$, n and homogeneity $m - k$ with respect to ξ, for $k = 0, 1, \ldots$. The asymptotic expansion (3.2.1) is in the G-sense, i.e., for χ as before:

$$a(x,\xi) - \sum_{0 \leq k < N} \chi(\xi)a_{m-k}(x,\xi) \in G^{m-N,n}(\mathbb{R}^d).$$

We define consequently by standard quantization $\text{OPG}_{\text{cl}(\xi)}^{m,n}(\mathbb{R}^d)$. We shall also write

$$\sigma_\psi^{m-k}(a) = a_{m-k} \tag{3.2.2}$$

and we call it the *interior symbol of order* $m - k$ of $a \in G_{\text{cl}(\xi)}^{m,n}(\mathbb{R}^d)$.

Example 3.2.2. Consider the partial differential operator

$$P = \sum_{|\alpha| \le m} c_\alpha(x) D^\alpha,$$

with symbol

$$p(x, \xi) = \sum_{|\alpha| \le m} c_\alpha(x) \xi^\alpha,$$

where we assume

$$|\partial_x^\beta c_\alpha(x)| \lesssim \langle x \rangle^{n - |\beta|} \quad \text{for } x \in \mathbb{R}^d.$$

We have then $p(x, \xi) \in G_{\text{cl}(\xi)}^{m,n}(\mathbb{R}^d)$, with

$$\sigma_\psi^{m-k}(p) = \sum_{|\alpha| = m-k} c_\alpha(x) \xi^\alpha, \quad k = 0, 1, \ldots, m.$$

Definition 3.2.3. We define $G_{\text{cl}(x)}^{m,n}(\mathbb{R}^d)$ as the class of all $a \in G^{m,n}(\mathbb{R}^d)$ with asymptotic expansion

$$a(x, \xi) \sim \sum_{j=0}^\infty b_{n-j}(x, \xi) \tag{3.2.3}$$

where b_{n-j} are Type 2, (ii), namely of orders $m, n - j$ and homogeneity $n - j$ with respect to x, for $j = 0, 1, \ldots$.

The classes $\text{OPG}_{\text{cl}(x)}^{m,n}(\mathbb{R}^d)$ are defined consequently. We shall write

$$\sigma_e^{n-j}(a) = b_{n-j} \tag{3.2.4}$$

and call it the *exit symbol of order* $n - j$ of $a \in G_{\text{cl}(x)}^{m,n}(\mathbb{R}^d)$.

Example 3.2.4. Consider the symbol

$$p(x, \xi) = \sum_{|\beta| \le n} c_\beta(\xi) x^\beta,$$

where

$$|\partial_\xi^\alpha c_\beta(\xi)| \lesssim \langle \xi \rangle^{m - |\alpha|} \quad \text{for } \xi \in \mathbb{R}^d.$$

Then $p(x, \xi) \in G_{\text{cl}(x)}^{m,n}(\mathbb{R}^d)$ with

$$\sigma_e^{n-j}(p) = \sum_{|\beta| = n-j} c_\beta(\xi) x^\beta, \quad j = 0, 1, \ldots, n.$$

The corresponding pseudo-differential operator, in the standard quantization, is given by

$$P = \sum_{|\beta| \leq n} x^\beta c_\beta(D),$$

where the Fourier multipliers $c_\beta(D)$ can be regarded as elements of $G^{m,0}(\mathbb{R}^d)$, cf. (3.1.19).

We finally consider the subclass of $G^{m,n}_{\mathrm{cl}(\xi)}(\mathbb{R}^d) \cap G^{m,n}_{\mathrm{cl}(x)}(\mathbb{R}^d)$, obtained by requiring further that a_{m-k} in (3.2.1) and b_{n-j} in (3.2.3) are Type 3 symbols.

Definition 3.2.5. We define $G^{m,n}_{\mathrm{cl}(\xi,x)}(\mathbb{R}^d)$ as the class of all $a \in G^{m,n}(\mathbb{R}^d)$ such that:

(I) We have

$$a(x,\xi) \sim \sum_{k=0}^{\infty} a_{m-k}(x,\xi),$$

where a_{m-k}, $k = 0, 1, \ldots$, are Type 3, (i) symbols, with orders $m - k, n$, and homogeneity $m - k$ with respect to ξ, polyhomogeneity with respect to x.

(II) We have

$$a(x,\xi) \sim \sum_{j=0}^{\infty} b_{n-j}(x,\xi),$$

where b_{n-j}, $j = 0, 1, \ldots$, are Type 3, (ii) symbols, with orders $m, n - j$, and homogeneity $n - j$ with respect to x, polyhomogeneity with respect to ξ.

Interior symbols $\sigma_\psi^{m-k}(a)(x,\xi)$, $k = 0, 1, \ldots$, and exit symbols $\sigma_e^{n-j}(a)(x,\xi)$, $j = 0, 1, \ldots$, are defined as before, cf. (3.2.2) and (3.2.4). Since now they are Type 3 symbols, we may actually proceed further. Namely, for $\sigma_\psi^{m-k}(a)(x,\xi)$ we know the existence of elements of $\mathcal{H}^{m-k,n-j}_{\xi,x}$, that we may denote by $\sigma_e^{n-j}(\sigma_\psi^{m-k}(a))$, $j = 0, 1, \ldots$, such that

$$\sigma_\psi^{m-k}(a) \sim \sum_{j=0}^{\infty} \sigma_e^{n-j}(\sigma_\psi^{m-k}(a)), \tag{3.2.5}$$

cf. the definition of Type 3, (i) symbols. Similarly we may introduce functions $\sigma_\psi^{m-k}(\sigma_e^{n-j}(a)) \in \mathcal{H}^{m-k,n-j}_{\xi,x}$ such that

$$\sigma_e^{n-j}(a) \sim \sum_{k=0}^{\infty} \sigma_\psi^{m-k}(\sigma_e^{n-j}(a)). \tag{3.2.6}$$

The information given by (I) and (II) in Definition 3.2.5 is overabundant, in the sense that the two procedures give actually the same result:

$$\sigma_{\psi,e}^{m-k,n-j}(a) = \sigma_e^{n-j}(\sigma_\psi^{m-k}(a)) = \sigma_\psi^{m-k}(\sigma_e^{n-j}(a)), \quad j, k = 0, 1, \ldots. \tag{3.2.7}$$

Definition 3.2.6. Let $a \in G^{m,n}_{\mathrm{cl}(\xi,x)}(\mathbb{R}^d)$. We call $\sigma^m_\psi(a)$ the *interior principal part* of a, and $\sigma^n_e(a)$ the *exit principal part*. They are Type 3 symbols, with homogeneity m in ξ and, respectively, homogeneity n in x. Fixing $k = j = 0$ in (3.2.7), we obtain $\sigma^{m,n}_{\psi,e}(a) \in \mathcal{H}^{m,n}_{\xi,x}$, which we call the *bi-homogeneous principal part* of a.

The class of pseudo-differential operators $\mathrm{OPG}^{m,n}_{\mathrm{cl}(\xi,x)}(\mathbb{R}^d)$ is defined consequently.

Example 3.2.7. Let

$$P = \sum_{\substack{|\alpha|\leq m \\ |\beta|\leq n}} c_{\alpha\beta} x^\beta D^\alpha. \tag{3.2.8}$$

We have $P \in \mathrm{OPG}^{m,n}_{\mathrm{cl}(\xi,x)}(\mathbb{R}^d)$ and the principal parts of the symbol $p(x,\xi)$ in the standard quantization are given by:

$$\sigma^m_\psi(p)(x,\xi) = \sum_{\substack{|\alpha|=m \\ |\beta|\leq n}} c_{\alpha\beta} x^\beta \xi^\alpha, \tag{3.2.9}$$

$$\sigma^n_e(p)(x,\xi) = \sum_{\substack{|\alpha|\leq m \\ |\beta|=n}} c_{\alpha\beta} x^\beta \xi^\alpha, \tag{3.2.10}$$

$$\sigma^{m,n}_{\psi,e}(p)(x,\xi) = \sum_{\substack{|\alpha|=m \\ |\beta|=n}} c_{\alpha\beta} x^\beta \xi^\alpha. \tag{3.2.11}$$

Example 3.2.8. Consider $P = \langle x \rangle^n \langle D \rangle^m$, $m \in \mathbb{R}$, $n \in \mathbb{R}$, with symbol

$$p(x,\xi) = \langle x \rangle^n \langle \xi \rangle^m. \tag{3.2.12}$$

We have $P \in \mathrm{OPG}^{m,n}_{\mathrm{cl}(\xi,\tau)}(\mathbb{R}^d)$ and $\sigma^m_\psi(p) = \langle x \rangle^n |\xi|^m$, $\sigma^n_e(p) = |x|^n \langle \xi \rangle^m$, $\sigma^{m,n}_{\psi,e}(p) = |x|^n |\xi|^m$.

Our attention in the sequel will be restricted to the classes $\mathrm{OPG}^{m,n}_{\mathrm{cl}(\xi,x)}(\mathbb{R}^d)$. The results in Chapter 1 and Section 3.1 obviously apply. In particular, the definition of the class, as well as the notions of the principal parts $\sigma^m_\psi(a)$, $\sigma^n_e(a)$, $\sigma^{m,n}_{\psi,e}(a)$, do not depend on the quantization.

If $a(x,\xi) \in G^{m,n}_{\mathrm{cl}(\xi,x)}(\mathbb{R}^d)$ and $b \in G^{m',n'}_{\mathrm{cl}(\xi,x)}(\mathbb{R}^d)$, then $c(x,D) = a(x,D)b(x,D) \in \mathrm{OPG}^{m+m',n+n'}_{\mathrm{cl}(\xi,x)}(\mathbb{R}^d)$, with asymptotic expansions for the symbol $c(x,\xi)$ as standard; in particular

$$\sigma^{m+m'}_\psi(c) = \sigma^m_\psi(a)\sigma^{m'}_\psi(b), \tag{3.2.13}$$

$$\sigma^{n+n'}_e(c) = \sigma^n_e(a)\sigma^{n'}_e(b), \tag{3.2.14}$$

$$\sigma^{m+m',n+n'}_{\psi,e}(c) = \sigma^{m,n}_{\psi,e}(a)\sigma^{m',n'}_{\psi,e}(b). \tag{3.2.15}$$

We may treat a similarly transposed and formal adjoint of $a(x,D) \in \mathrm{OPG}^{m,n}_{\mathrm{cl}(\xi,x)}(\mathbb{R}^d)$. The main novelty with respect to Section 3.1, and motivation for the present study of the polyhomogeneous symbols, is given by the following equivalent definition of G-ellipticity.

Theorem 3.2.9. *Let $a \in G_{\mathrm{cl}(\xi,x)}^{m,n}(\mathbb{R}^d)$, $m \in \mathbb{R}$, $n \in \mathbb{R}$. The symbol a is G-elliptic, cf. Definition 3.1.2, if and only if*

$$\sigma_\psi^m(a)(x,\xi) \neq 0 \quad \text{for all } x \in \mathbb{R}^d, \ \xi \neq 0, \tag{3.2.16}$$

$$\sigma_e^n(a)(x,\xi) \neq 0 \quad \text{for all } x \neq 0, \ \xi \in \mathbb{R}^d, \tag{3.2.17}$$

$$\sigma_{\psi,e}^{m,n}(a)(x,\xi) \neq 0 \quad \text{for all } x \neq 0, \ \xi \neq 0. \tag{3.2.18}$$

In order to prove Theorem 3.2.9 it will be convenient to define for any given $a \in G_{\mathrm{cl}(\xi,x)}^{m,n}(\mathbb{R}^d)$ the associated symbol

$$\tilde{a}(x,\xi) = \chi(\xi)\sigma_\psi^m(a)(x,\xi) + \chi(x)\sigma_e^n(a)(x,\xi) - \chi(x)\chi(\xi)\sigma_{\psi,e}^{m,n}(a)(x,\xi), \tag{3.2.19}$$

where χ is defined as before: $\chi \in C^\infty(\mathbb{R}^d)$, $\chi(x) = 1$ for $|x| \geq 2$, $\chi(x) = 0$ for $|x| \leq 1$. We have $\tilde{a} \in G_{\mathrm{cl}(\xi,x)}^{m,n}(\mathbb{R}^d)$, with the same principal parts as a:

$$\sigma_\psi^m(\tilde{a}) = \sigma_\psi^m(a), \ \sigma_e^n(\tilde{a}) = \sigma_e^n(a), \ \sigma_{\psi,e}^{m,n}(\tilde{a}) = \sigma_{\psi,e}^{m,n}(a). \tag{3.2.20}$$

It follows easily from Definition 3.2.5 that

$$a(x,\xi) - \tilde{a}(x,\xi) \in G_{\mathrm{cl}(\xi,x)}^{m-1,n-1}(\mathbb{R}^d). \tag{3.2.21}$$

We may also re-write (3.2.19) in the form:

$$\tilde{a}(x,\xi) = \chi(x)\chi(\xi)\sigma_{\psi,e}^{m,n}(a)(x,\xi) + r_\psi(x,\xi) + r_e(x,\xi), \tag{3.2.22}$$

where

$$r_\psi(x,\xi) = \chi(x)\sigma_e^n(a)(x,\xi) - \chi(x)\chi(\xi)\sigma_{\psi,e}^{m,n}(a)(x,\xi) \in G_{\mathrm{cl}(\xi,x)}^{m-1,n}(\mathbb{R}^d), \tag{3.2.23}$$

$$r_e(x,\xi) = \chi(\xi)\sigma_\psi^m(a)(x,\xi) - \chi(x)\chi(\xi)\sigma_{\psi,e}^{m,n}(a)(x,\xi) \in G_{\mathrm{cl}(\xi,x)}^{m,n-1}(\mathbb{R}^d). \tag{3.2.24}$$

Proof of Theorem 3.2.9. Multiplying by the symbol $\langle x \rangle^{-n} \langle \xi \rangle^{-m}$, cf. Example 3.2.8, we are reduced to proving the theorem in the case $m = 0$, $n = 0$. Define then $\tilde{a}$ according to (3.2.19), (3.2.22),

$$\tilde{a}(x,\xi) = \chi(\xi)\sigma_\psi^0(a)(x,\xi) + \chi(x)\sigma_e^0(a)(x,\xi) - \chi(x)\chi(\xi)\sigma_{\psi,e}^{0,0}(a)(x,\xi) \tag{3.2.25}$$

and set consequently $r_\psi(x,\xi) \in G_{\mathrm{cl}(\xi,x)}^{-1,0}(\mathbb{R}^d)$, $r_e \in G_{\mathrm{cl}(\xi,x)}^{0,-1}(\mathbb{R}^d)$. Since $a - \tilde{a} \in G_{\mathrm{cl}(\xi,x)}^{-1,-1}(\mathbb{R}^d) \subset G^{-1,-1}(\mathbb{R}^d)$ in view of (3.2.21), the G-ellipticity of a, cf. Definition 3.1.2, is equivalent to the G-ellipticity of $\tilde{a}$, i.e.,

$$|\tilde{a}(x,\xi)| \geq \text{const.} > 0, \quad |x|^2 + |\xi|^2 \geq R^2 \tag{3.2.26}$$

for a suitable positive constant R. We are therefore reduced to prove that (3.2.26) is equivalent to (3.2.16), (3.2.17), (3.2.18), that read now

$$\sigma_\psi^0(\tilde{a})(x,\xi) \neq 0 \quad \text{for all } x \in \mathbb{R}^d, \ \xi \neq 0, \tag{3.2.27}$$

$$\sigma_e^0(\tilde{a})(x,\xi) \neq 0 \quad \text{for all } x \neq 0, \ \xi \in \mathbb{R}^d, \tag{3.2.28}$$

$$\sigma_{\psi,e}^{0,0}(\tilde{a})(x,\xi) \neq 0 \quad \text{for all } x \neq 0, \ \xi \neq 0. \tag{3.2.29}$$

Let us first assume that (3.2.26) is valid, and deduce (3.2.27), (3.2.28), (3.2.29). Considering first $\sigma_{\psi,e}^{0,0}(\tilde{a}) \in \mathcal{H}_{\xi,x}^{0,0}$, we have from (3.2.22), (3.2.23), (3.2.24), for $x \neq 0$, $\xi \neq 0$:

$$\sigma_{\psi,e}^{0,0}(\tilde{a})(x,\xi) = \lim_{\lambda \to +\infty} \chi(\lambda x)\chi(\lambda\xi)\sigma_{\psi,e}^{0,0}(\tilde{a})(\lambda x, \lambda\xi)$$

$$= \lim_{\lambda \to +\infty} \left[\tilde{a}(\lambda x, \lambda\xi) - r_\psi(\lambda x, \lambda\xi) - r_e(\lambda x, \lambda\xi)\right]$$

$$= \lim_{\lambda \to +\infty} \tilde{a}(\lambda x, \lambda\xi) \neq 0$$

in view of (3.2.26). This proves (3.2.29). On the other hand, for $\xi \neq 0$ we have

$$\sigma_\psi^0(\tilde{a})(x,\xi) = \lim_{\lambda \to +\infty} \chi(\lambda\xi)\sigma_\psi^0(\tilde{a})(x, \lambda\xi)$$

$$= \lim_{\lambda \to +\infty} \left[\tilde{a}(x, \lambda\xi) - r_\psi(x, \lambda\xi)\right] = \lim_{\lambda \to +\infty} \tilde{a}(x, \lambda\xi) \neq 0.$$

This gives (3.2.27). Similarly we prove (3.2.28).

In the opposite direction, assume (3.2.27), (3.2.28), (3.2.29). Because of the bi-homogeneity, we have from (3.2.29)

$$|\sigma_{\psi,e}^{0,0}(\tilde{a})(x,\xi)| \geq C > 0 \quad \text{for } x \neq 0, \ \xi \neq 0.$$

Applying then (3.2.22), (3.2.23), (3.2.24), we obtain similar estimates for $\tilde{a}$:

$$|\tilde{a}(x,\xi)| \geq C > 0, \quad \text{for } |x| \geq T, \ |\xi| \geq T,$$

with suitable new constants C and $T > 0$. To handle $\tilde{a}(x,\xi)$ in the strip $|x| \leq T$, we write

$$\tilde{a}(x,\xi) = \chi(\xi)\sigma_\psi^0(\tilde{a})(x,\xi) + r_\psi(x,\xi).$$

From (3.2.27), in view of the homogeneity with respect to ξ we have

$$|\chi(\xi)\sigma_\psi^0(\tilde{a})(x,\xi)| \geq \epsilon > 0, \quad \text{for } |\xi| \geq 2, \ |x| \leq T,$$

for a suitable ϵ depending on T. Since $r_\psi \in G_{\text{cl}(\xi,x)}^{-1,0}(\mathbb{R}^d)$, the same estimates are valid for $\tilde{a}(x,\xi)$, if we shrink $\epsilon > 0$ and impose $|\xi| \geq R$ for a sufficiently large R. The same argument applies in the strip $|\xi| \leq T$, by writing

$$\tilde{a}(x,\xi) = \chi(x)\sigma_e^0(\tilde{a})(x,\xi) + r_e(x,\xi)$$

and using the assumption (3.2.28). This gives (3.2.26) and concludes the proof of Theorem 3.2.9. $\qquad\square$

Theorem 3.2.10. *Let $a \in G^{m,n}_{\mathrm{cl}(\xi,x)}(\mathbb{R}^d)$, $m \in \mathbb{R}$, $n \in \mathbb{R}$, satisfy (3.2.16), (3.2.17), (3.2.18). Then there exists $b \in G^{-m,-n}_{\mathrm{cl}(\xi,x)}(\mathbb{R}^d)$ such that $b(x, D)$ is a parametrix of $a(x, D)$. Hence $A = a(x, D)$ is globally regular, satisfies the estimates (3.1.14) and is Fredholm, considered as a map $H^{s+m,t+n}_G(\mathbb{R}^d) \to H^{s,t}_G(\mathbb{R}^d)$.*

Proof. In view of Theorem 3.2.9, the symbol a is G-elliptic, and we may apply Theorem 3.1.3 and Theorem 3.1.6. It remains to prove that the symbol $b(x, \xi)$ of the parametrix is polyhomogeneous. To this aim, set

$$b^{-m}_\psi(x, \xi) = \left(\sigma^m_\psi(a)(x, \xi)\right)^{-1}, \quad x \in \mathbb{R}^d, \ \xi \neq 0,$$

$$b^{-n}_e(x, \xi) = \left(\sigma^n_e(a)(x, \xi)\right)^{-1}, \quad x \neq 0, \ \xi \in \mathbb{R}^d,$$

which are symbols of Type 3, in view of (3.2.16) and (3.2.17), and

$$b^{-m,-n}_{\psi,e}(x, \xi) = \left(\sigma^{m,n}_{\psi,e}(a)(x, \xi)\right)^{-1}, \quad x \neq 0, \ \xi \neq 0,$$

well defined in $\mathcal{H}^{-m,-n}_{\xi,x}$ in view of (3.2.18). We then construct $\tilde{b} \in G^{-m,-n}_{\mathrm{cl}(\xi,x)}(\mathbb{R}^d)$ as

$$\tilde{b}(x, \xi) = \chi(\xi)b^{-m}_\psi(x, \xi) + \chi(x)b^{-n}_e(x, \xi) - \chi(x)\chi(\xi)b^{-m,-n}_{\psi,e}(x, \xi).$$

In view of (3.2.13), (3.2.14), (3.2.15), we obtain

$$\tilde{b}(x, D)a(x, D) = I + S_1, \quad a(x, D)\tilde{b}(x, D) = I + S_2,$$

where $S_1, S_2 \in OPG^{-1,-1}_{\mathrm{cl}(\xi,x)}(\mathbb{R}^d)$. The expression of the symbol $b(x, \xi) \in G^{-m,-n}_{\mathrm{cl}(\xi,x)}(\mathbb{R}^d)$ of the parametrix follows then from iterative arguments, as in the proof of Theorem 1.3.6. $\qquad\square$

Example 3.2.11. Reconsidering Example 3.2.7, we have from Theorem 3.2.9 that the symbol

$$p(x, \xi) = \sum_{\substack{|\alpha| \leq m \\ |\beta| \leq n}} c_{\alpha\beta} x^\beta \xi^\alpha$$

of the operator P in (3.2.8) is G-elliptic if

$$\sum_{\substack{|\alpha| = m \\ |\beta| \leq n}} c_{\alpha\beta} x^\beta \xi^\alpha \neq 0 \quad \text{for } x \in \mathbb{R}^d, \ \xi \neq 0, \tag{3.2.30}$$

that is, we have local ellipticity in the classical sense, and additionally

$$\sum_{\substack{|\alpha| \leq m \\ |\beta| = n}} c_{\alpha\beta} x^\beta \xi^\alpha \neq 0 \quad \text{for } x \neq 0, \ \xi \in \mathbb{R}^d, \tag{3.2.31}$$

$$\sum_{\substack{|\alpha| = m \\ |\beta| = n}} c_{\alpha\beta} x^\beta \xi^\alpha \neq 0 \quad \text{for } x \neq 0, \ \xi \neq 0. \tag{3.2.32}$$

In the case $n = 0$ we recapture the conclusions in Example 3.1.8.

Note that (3.2.32) is necessary to get G-ellipticity, and to conclude global regularity through Theorem 3.2.10. Consider in fact the operator in $\mathrm{OPG}_{\mathrm{cl}(\xi,x)}^{1,1}(\mathbb{R})$

$$P = D_x - x, \quad x \in \mathbb{R},$$

with symbol $\xi - x$. Conditions (3.2.30) and (3.2.31) are obviously satisfied, however $\sigma_{\psi,e}^{1,1} = 0$, hence we do not have G-ellipticity. A solution $u \in \mathcal{S}'(\mathbb{R})$ of $Pu = 0$ is given by $u(x) = e^{i\frac{x^2}{2}} \notin \mathcal{S}(\mathbb{R})$, and global regularity fails. We refer to Section 3.3 for a detailed discussion of G-ellipticity for ordinary differential operators.

3.3 G-Elliptic Ordinary Differential Operators

Let us first fix attention on ordinary differential operators with polynomial coefficients. We may re-write in this case the operator $P \in \mathrm{OPG}_{\mathrm{cl}(\xi,x)}^{m,n}(\mathbb{R})$, $m \in \mathbb{N}$, $n \in \mathbb{N}$, in the form

$$P = \sum_{j=0}^{m} Q_j(x) D_x^j, \quad x \in \mathbb{R}, \tag{3.3.1}$$

where Q_j, $j = 0, \ldots, m$, are polynomials of degree $\leq n$:

$$Q_j(x) = \sum_{k=0}^{n} c_{jk} x^k, \quad c_{jk} \in \mathbb{C}, \ x \in \mathbb{R}. \tag{3.3.2}$$

We may read the G-ellipticity of the symbol of P in terms of (3.2.30), (3.2.31), (3.2.32) in Example 3.2.11. Namely, (3.2.30) corresponds to assuming that

$$Q_m(x) \neq 0 \quad \text{for all } x \in \mathbb{R}. \tag{3.3.3}$$

Note that this implies analyticity in $\mathbb{R}$ of all the classical solutions of $Pu = 0$. As for condition (3.2.31), this is equivalent to assuming that

$$\sum_{j=0}^{m} c_{jn} \xi^j \neq 0 \quad \text{for all } \xi \in \mathbb{R}, \tag{3.3.4}$$

granting exponential asymptotic behaviour of the classical solutions, see below. Finally, (3.2.32) amounts to assuming that

$$c_{mn} \neq 0. \tag{3.3.5}$$

Theorem 3.3.1. *Under the conditions* (3.3.3), (3.3.4), (3.3.5), *we may regard the operator in* (3.3.1), (3.3.2) *as a Fredholm map*

$$P: H_G^{m,n}(\mathbb{R}) \to L^2(\mathbb{R}), \tag{3.3.6}$$

with $\mathrm{ind}\, P = 0$.

Proof. The Fredholm property of P follows from Theorem 3.2.10. It remains to prove the assertion on the index. Assume then $c_{mn} = 1$, without loss of generality, and consider the operator

$$P_0 = Q_m(x)P_{00}$$

with

$$P_{00} = D_x^m + \sum_{j=0}^{m-1} c_{jn} D_x^j.$$

The symbols of P and P_0 have the same principal parts, hence we have $P - P_0 \in \mathrm{OPG}_{\mathrm{cl}(\xi,x)}^{m-1,n-1}(\mathbb{R})$. Therefore, $P - P_0$ is a compact operator, so that $\mathrm{ind}\, P = \mathrm{ind}\, P_0$, cf. Theorems 3.1.5 and 1.6.6. On the other hand P_{00} belongs to $\mathrm{OPG}_{\mathrm{cl}(\xi,x)}^{m,0}(\mathbb{R})$ with G-elliptic symbol, in view of (3.3.4). As observed in Example 3.1.9, $P_{00} : H_G^{m,n}(\mathbb{R}) \to H_G^{0,n}(\mathbb{R})$ is then an isomorphism. The same conclusion is valid for the multiplication operator

$$Q_m : \ H_G^{0,n}(\mathbb{R}) \to L^2(\mathbb{R}),$$

because of (3.3.3), and in view of Corollary 1.6.5 we conclude $\mathrm{ind}\, P = \mathrm{ind}\, Q_m P_{00} = \mathrm{ind}\, Q_m + \mathrm{ind}\, P_{00} = 0.$ □

We may go further in the analysis of the solutions of the equation $Pu = 0$ by applying the theory of asymptotic integration. Let us first observe that the solutions $u(x)$ extend to the complex domain, with possibly a finite number of singularities at the points $x \in \mathbb{C}$ where $Q_m(x) = 0$. Consider then the roots $r_1, \dots, r_m$ of the equation

$$\sum_{j=0}^{m} c_{jn} \xi^j = 0. \tag{3.3.7}$$

In view of (3.3.4), we have $\mathrm{Im}\, r_j \neq 0$ for all $j = 1, \dots, m$. Assume for simplicity that all the r_j, $j = 1, \dots, m$, are distinct. We recall the following basic result on the asymptotic expansions of the solutions of $Pu = 0$ in sectors of the complex plane, see for example Wasow [194].

Proposition 3.3.2. *Given any sector* $\Lambda = \{x \in \mathbb{C} \setminus \{0\} : \ \varphi_1 < \arg x < \varphi_2\}$ *with* $\varphi_2 - \varphi_1 < \pi$, *there exist* $u_{\Lambda,j}(x)$, $j = 1, \dots, m$, *linearly independent solutions of* $Pu = 0$, *such that for* $x \in \Lambda$, $x \to \infty$,

$$u_{\Lambda,j}(x) \sim x^{s_j} \exp[ir_j x] \sum_{p=0}^{\infty} \beta_{-p,j} x^{-p}. \tag{3.3.8}$$

We can determine s_j, $\beta_{-p,j}$, $j = 1, \dots, m$, $p = 0, 1, \dots$, by letting the right-hand side of (3.3.8) be a formal solution of $Pu = 0$, cf. Section 2.3. The asymptotic meaning of (3.3.8) is then the same as in Proposition 2.3.1. Here we understand $x \in \mathbb{C}$, $|x| > R$, with R sufficiently large, so that $Q_m(x) \neq 0$ and the solutions $u_{\Lambda,j}$ are holomorphic in $\Lambda \cap \{|x| \geq R\}$.

Taking now two sectors Λ_+, Λ_- containing $\mathbb{R}_+$, $\mathbb{R}_-$ respectively, we may conclude the existence of u_j^+, u_j^-, $j = 1, \ldots, m$, satisfying (3.3.8) in $\mathbb{R}_+$, $\mathbb{R}_-$ respectively. Every classical solution $u(x)$ can be written as

$$u = \sum_{j=1}^{m} \mu_j^+ u_j^+ = \sum_{j=1}^{m} \mu_j^- u_j^-, \quad \mu_j^+, \mu_j^- \in \mathbb{C}.$$

Assume

$$\operatorname{Im} r_j > 0 \text{ for } j = 1, \ldots, m^+, \quad \operatorname{Im} r_j < 0 \text{ for } j = m^+ + 1, \ldots, m, \quad m^+ + m^- = m. \tag{3.3.9}$$

Then u belongs to $\mathcal{S}(\mathbb{R})$, or equivalently to $\mathcal{S}'(\mathbb{R})$, if and only if $\mu_j^+ = 0$ for $m^+ < j \leq m$ and $\mu_j^- = 0$ for $1 \leq j \leq m^+$, that is

$$u = \sum_{1 \leq j \leq m^+} \mu_j^+ u_j^+ = \sum_{m^+ < j \leq m} \mu_j^- u_j^-.$$

Summing up, we have obtained the following result.

Theorem 3.3.3. *Let $u \in \mathcal{S}'(\mathbb{R})$ be a solution of $Pu = 0$, with P as in (3.3.1), (3.3.2) satisfying (3.3.3), (3.3.4), (3.3.5). Then $u \in \mathcal{S}(\mathbb{R})$ and*

$$|u(x)| \lesssim e^{-\delta|x|}, \quad x \in \mathbb{R}, \tag{3.3.10}$$

for some constant $\delta > 0$. Moreover $u(x)$ has holomorphic extension for $x \in \mathbb{C}$ in the strip $\{|\operatorname{Im} x| < T\}$ for some $T > 0$. Letting m^+, m^- be defined as in (3.3.9), we have

$$\dim \left(\operatorname{Ker} P \cap \mathcal{S}(\mathbb{R}) \right) \leq \min\{m^+, m^-\}. \tag{3.3.11}$$

In particular, if $m^+ = 0$ or $m^- = 0$, then the equation $Pu = 0$, $u \in \mathcal{S}'(\mathbb{R}^d)$, admits only the trivial solution.

By a slight improvement of Proposition 3.3.2, cf. (2.3.4), we may see that Theorem 3.3.3 is valid regardless of the multiplicity of the roots r_j in (3.3.9).

Let us pass now to consider differential operators in $\operatorname{OPG}_{\mathrm{cl}(\xi,x)}^{m,n}(\mathbb{R})$, $m \in \mathbb{N}$, $n \in \mathbb{N}$, with C^∞-coefficients. Actually, for the sake of brevity, we shall limit the analysis to the first-order operator

$$L = D_x - r(x), \quad x \in \mathbb{R}. \tag{3.3.12}$$

We have $L \in \operatorname{OPG}_{\mathrm{cl}(\xi,x)}^{1,0}(\mathbb{R})$ if $r \in C^\infty(\mathbb{R})$ admits an asymptotic expansion

$$r(x) \sim \sum_{j=0}^{\infty} r_{-j}(x), \tag{3.3.13}$$

where $r_{-j}(x)$ is positively homogeneous of degree $-j$. In particular

$$r_0(x) = r_0^+ \text{ if } x > 0, \quad r_0(x) = r_0^- \text{ if } x < 0, \tag{3.3.14}$$

for constants $r_0^+, r_0^- \in \mathbb{C}$. To check (3.2.16), (3.2.17), (3.2.18), we compute $\sigma_\psi^1 = \sigma_{\psi,e}^{1,0} = \xi$ and $\sigma_e^0 = \xi - r_0^+$ if $x > 0$, $= \xi - r_0^-$ if $x < 0$. Hence G-ellipticity for the symbol of (3.3.12) amounts to requiring

$$\operatorname{Im} r_0^+ \neq 0, \quad \operatorname{Im} r_0^- \neq 0. \tag{3.3.15}$$

In fact, the classical solutions of $Lu = 0$ are the functions

$$u(x) = C \exp\left[i \int_0^x r(t)\, dt\right], \quad C \in \mathbb{C}. \tag{3.3.16}$$

Since $\int_0^x r(t)\, dt = r_0^+ x + O(\log x)$ for $x \to +\infty$, $= r_0^- x + O(\log |x|)$ for $x \to -\infty$, we have that the solutions in (3.3.16) with $C \neq 0$ belong to $\mathcal{S}(\mathbb{R})$, or equivalently to $\mathcal{S}'(\mathbb{R})$, if and only if

$$\operatorname{Im} r_0^+ > 0, \quad \operatorname{Im} r_0^- < 0.$$

So for example $u(x) = e^{-\langle x \rangle}$ solves $u' + \frac{x}{\langle x \rangle} u = 0$, which is an equation of the type $Lu = 0$ with $r_0^+ = i$, $r_0^- = -i$.

Applying the results of the previous sections, we may then summarize as follows:

Theorem 3.3.4. *Under the condition* (3.3.15), *we may regard the operator* L *in* (3.3.12), (3.3.13), (3.3.14) *as a Fredholm map*

$$L: H_G^{1,0}(\mathbb{R}) \to L^2(\mathbb{R}) \tag{3.3.17}$$

with

$$\operatorname{ind} L = \frac{1}{2}\left(\operatorname{sign} \operatorname{Im} r_0^+ - \operatorname{sign} \operatorname{Im} r_0^-\right). \tag{3.3.18}$$

Here we used the fact that $\operatorname{ind} L = \dim \operatorname{Ker} L - \dim \operatorname{Ker} L^*$, and $L^* = D_x - \overline{r(x)}$, cf. Theorem 1.6.9.

Note that $H_G^{1,0}(\mathbb{R})$ in (3.3.17) coincides with the standard Sobolev space $H^1(\mathbb{R})$, and that the exponential bound (3.3.10) remains valid for the solutions $u \in \mathcal{S}'(\mathbb{R})$ of $Lu = 0$.

3.4   Other Classes of Globally Regular Operators

A couple Φ, Ψ of sub-linear and temperate weights can satisfy the strong uncertainty principle (1.1.10) even in the case when $\Phi(x, \xi) = 1$, provided good estimates in the whole (x, ξ) variables are valid for the other weight $\Psi(x, \xi)$. A basic example is given by

$$\Psi(x, \xi) = (1 + |x|^2 + |\xi|^2)^{\rho/2}, \quad \Phi(x, \xi) = 1, \tag{3.4.1}$$

with $0 < \rho \leq 1$. We may as well interchange the role of Φ and Ψ in the above argument. Rather than detailing the calculus for the corresponding classes $S(M; \Phi, \Psi)$, we immediately propose some examples of operators with polynomial coefficients, to which this calculus applies.

Theorem 3.4.1. *Consider the operator*

$$P = -\Delta + V(x) \tag{3.4.2}$$

where $V(x)$ is a positive polynomial in $\mathbb{R}^d$ satisfying, for some $\rho > 0$, $N > 0$ and every $\beta \in \mathbb{N}$,

$$\langle x \rangle^{2\rho} \lesssim V(x) \lesssim \langle x \rangle^N, \quad x \in \mathbb{R}^d, \tag{3.4.3}$$

$$|\partial^\beta V(x)| \lesssim V(x), \quad x \in \mathbb{R}^d. \tag{3.4.4}$$

The operator P is globally regular.

Example 3.4.2. As an example of potential $V(x)$ satisfying (3.4.3), (3.4.4) consider in $\mathbb{R}^2$,

$$V(x_1, x_2) = (1 + x_1^2)(1 + x_2^2). \tag{3.4.5}$$

The estimates (3.4.3) are valid for $\rho = 1$, $N = 4$. Note that also (3.4.4) is satisfied, however $V(x)$ has not the Hörmander property with respect to $x \in \mathbb{R}^d$.

Proof of Theorem 3.4.1. We refer directly to the general calculus of Chapter 1, by taking Ψ, Φ as in (3.4.1) with ρ as in (3.4.3), where we may assume $\rho \leq 1$. Let $M(x, \xi)$ be defined as the symbol of P:

$$M(x, \xi) = |\xi|^2 + V(x), \tag{3.4.6}$$

which is a temperate weight. We now prove that $M \in S(M; \Phi, \Psi)$, according to Definition 1.1.1. In fact

$$|\partial_{\xi_j} M(x, \xi)| = |2\xi_j| \lesssim M(x, \xi)\Psi(x, \xi)^{-1},$$
$$|\partial_{\xi_j}^2 M(x, \xi)| = 2 \lesssim M(x, \xi)\Psi(x, \xi)^{-2},$$

in view of (3.4.3) and (3.4.1). Moreover

$$|\partial_x^\beta M(x, \xi)| = |\partial_x^\beta V(x)| \lesssim V(x) \leq M(x, \xi)$$

in view of (3.4.4). Other derivatives of the symbol vanish, hence (1.1.4) is satisfied.

On the other hand, $M(x, \xi)$ is obviously elliptic in $S(M; \Phi, \Psi)$ and, applying Corollary 1.3.9, we deduce the global regularity of P. $\qquad \square$

Theorem 3.4.3. *Consider the operator*

$$Q = V(D) + |x|^2 \tag{3.4.7}$$

where $V(D)$ is a partial differential operator with constant coefficients and symbol $V(\xi)$ satisfying the estimates (3.4.3), (3.4.4) with x replaced by ξ. The operator Q in (3.4.7) is globally regular.

Example 3.4.4. Rephrasing the example in Theorem 3.4.1, we have that

$$Q = (1 + D_{x_1}^2)(1 + D_{x_2}^2) + |x|^2 \qquad (3.4.8)$$

is globally regular in $\mathbb{R}^2$.

As for the proof of Theorem 3.4.3, we may repeat the arguments in the proof of Theorem 3.4.1, by taking

$$\Psi(x, \xi) = 1, \quad \Phi(x, \xi) = \langle (x, \xi) \rangle^\rho = (1 + |x|^2 + |\xi|^2)^{\rho/2}, \qquad (3.4.9)$$

$$M(x, \xi) = V(\xi) + |x|^2. \qquad (3.4.10)$$

As alternative proof, we may observe that Q in (3.4.7) is obtained from P in (3.4.2) by Fourier conjugation, $Q = \widehat{P} = \mathcal{F}^{-1} P \mathcal{F}$. Since global regularity is invariant under Fourier conjugation, Theorem 3.4.3 follows from Theorem 3.4.1.

Notes

G-pseudo-differential operators were introduced by Parenti [156] and then studied in detail by Cordes [59]. An exhaustive treatment of polyhomogeneous G-operators is in Schulze [178, Section 1.4], which we followed closely in the previous Section 3.2. Concerning the results on asymptotic integration used in Section 3.3, besides Wasow [194], see also Mascarello and Rodino [142], Gramchev and Popivanov [96]. Finally, the examples in Section 3.4 are included in Buzano [27], Buzano and Ziggioto [33].

In the following we list some other important aspects of G-theory, which are not treated in the present book. First, the G–calculus can be extended to a suitable class of non-compact manifolds, see Schrohe [175]. In this setting, G-operators appear in the study of boundary value problems, see for example Kapanadze and Schulze [125], Harutyunyan and Schulze [108], as well as in the analysis of equations on manifolds with edges that have conical exits at infinity, see for example Calvo and Schulze [37]. For a polyhomogeneous G-calculus on manifolds, we refer also to Melrose [144], [145]. Other important topics are given by the G-hyperbolic equations, i.e., hyperbolic equations with coefficients globally defined in space variables, having G-type asymptotic behaviour. Their study was initiated in [59] and carried on by Coriasco [61], [62], Coriasco and Panarese [64], presenting the theory of the G-Fourier integral operators; see also Coriasco and Maniccia [63] for the related notion of G-wave front set, Coriasco and Rodino [65] for G-hyperbolic equations with multiple characteristics. Finally, we refer to Dasgupta and Wong [68] for the L^p-theory of the G-pseudo-differential operators.

Let us return, in conclusion, to the problem of the characterization of the globally regular operators with polynomial coefficients:

$$P = \sum c_{\alpha\beta} x^\beta D^\alpha.$$

Namely, one would like to give necessary and sufficient conditions on the symbol

$$p(x, \xi) = \sum c_{\alpha\beta} x^\beta \xi^\alpha$$

for the global regularity of P. The results in Chapters 1,2,3, give sufficient conditions, but do not provide a complete answer to the problem. We refer to Camperi [38] for generalized G-operators included in the calculus of Hörmander [119, Vol. III], leading to further classes of globally regular operators with polynomial coefficients. As striking examples, giving evidence of the difficulty of the problem in general, we quote the operators in $\mathbb{R}^2$,

$$R = D_{x_1}^2 + x_1^2 - \lambda,$$

i.e., the harmonic oscillator with respect to x_1, which is globally regular in the (x_1, x_2)-variables when $\lambda \neq 2n + 1$, $n = 0, 1, \ldots$, cf. Gramchev, Pilipovic and Rodino [95], and the twisted Laplacian

$$Q = -\Delta + \frac{1}{4}(x_1^2 + x_2^2) + x_1 D_{x_2} - x_2 D_{x_1},$$

also globally regular, cf. Wong [199] and also Dasgupta and Wong [69]. The symbols of R and Q fail to be elliptic or hypoelliptic in Γ and G classes, neither enter Hörmander's calculus [119, Vol. III].

Chapter 4

Spectral Theory

Summary

In this chapter we study some problems of Spectral Theory for pseudo-differential operators with hypoelliptic symbols in the classes $S(M; \Phi, \Psi)$ considered in Chapter 1; see in particular Sections 1.1 and 1.3.1.

For a symbol $a \in S(M; \Phi, \Psi)$ we denote by a^w its Weyl quantization, as a continuous operator on $\mathcal{S}'(\mathbb{R}^d)$, cf. Section 1.2, Proposition 1.2.13. Consider then the unbounded operator A in $L^2(\mathbb{R}^d)$, with domain $\mathcal{S}(\mathbb{R}^d)$, defined as $Au = a^w u$, for $u \in \mathcal{S}(\mathbb{R}^d)$. Clearly, the restriction of a^w to the subspace

$$\{u \in L^2(\mathbb{R}^d) : a^w u \in L^2(\mathbb{R}^d)\} \tag{4.0.1}$$

defines a closed extension of A. It is called the *maximal realization* of A. In particular, the operator A is closable, i.e., the closure of its graph in $L^2(\mathbb{R}^d) \times L^2(\mathbb{R}^d)$ is still a graph of a linear operator, which is called *closure*, or *minimal realization*, of A, and denoted by $\overline{A}$. Moreover, the maximal realization of A extends $\overline{A}$.

In Section 4.1 we recall some basic facts about unbounded operators in Hilbert spaces. Then, assuming the strong uncertainty principle (1.1.10), in Section 4.2 we will prove that when a is hypoelliptic, cf. Definition 1.3.2, the above minimal and maximal realizations coincide, i.e., the domain of $\overline{A}$ is the space in (4.0.1).

In Section 4.2 we also study the spectrum of operators with a hypoelliptic Weyl symbol $a \in \text{Hypo}(M, M_0; \Phi, \Psi)$. The main result in this connection states that, if a is real-valued and $M_0(x, \xi) \to +\infty$ at infinity, then the spectrum of $\overline{A}$ is given by a sequence of real eigenvalues either diverging to $+\infty$ or $-\infty$. The eigenvalues are all of finite multiplicity and the eigenfunctions belong to $\mathcal{S}(\mathbb{R}^d)$. Moreover $L^2(\mathbb{R}^d)$ has an orthonormal basis made of eigenfunctions of $\overline{A}$. This generalizes the results obtained for the harmonic oscillator in Section 2.2. Basic examples of operators which this result applies to are the generalized harmonic oscillator

$$A = -\Delta + |x|^{2k}, \quad k \in \mathbb{N}, \ k \neq 0,$$

or

$$A = \langle x \rangle^s (-\Delta + I)^k \langle x \rangle^s, \quad s > 0, \ k \in \mathbb{N}, \ k \neq 0.$$

Section 4.3 is devoted to complex powers of pseudo-differential operators with a hypoelliptic symbol $a(x, \xi)$ taking values in a closed angle $\operatorname{Re} z \geq -R \operatorname{Im} z$, when $|x| + |\xi| \geq R$, for some $R > 0$. Here the basic assumption at the operator level is that the operator $\overline{A}$ is non-negative in the sense of Komatsu, i.e., $(-\infty, 0)$ is contained in the resolvent set of $\overline{A}$ and

$$\sup_{\lambda \in \mathbb{R}_+} \lambda \left\| (\overline{A} + \lambda I)^{-1} \right\|_{\mathcal{B}(L^2(\mathbb{R}^d))} < \infty.$$

This guarantees that the complex powers $\overline{A}^z$, $\operatorname{Re} z > 0$, are well defined, (in the sense of Balakhrishnan), as densely defined operators on $L^2(\mathbb{R}^d)$. We will show that $\overline{A}^z$ is in fact a pseudo-differential operator with hypoelliptic symbol given essentially by a^z plus a remainder satisfying suitable estimates. Notice that we allow the symbol to tend to zero and infinity in different directions and we also allow the spectrum of $\overline{A}$ to have zero as an accumulation point.

We are also interested in the asymptotic behaviour of the eigenvalues of pseudo-differential operators $A = a^w$, for a real elliptic symbol a diverging to $+\infty$ at infinity at least algebraically. A classical strategy to attack this problem consists in studying the trace (that is the sum of the eigenvalues) of some parameter dependent function $F_t(A)$ of the operator A. Then one deduces the asymptotic behaviour of the eigenvalues of A via a suitable Tauberian theorem. In Sections 4.5, 4.6 we will apply the above program with the function $F_t(x) = e^{-tx}$, where $t \geq 0$ is here regarded as a parameter (this is known as the *heat method*). Namely, denote by λ_j, $j \in \mathbb{N}$, the sequence of eigenvalues (counted according to their multiplicity) of A. We have $\lambda_j \to +\infty$ as $j \to \infty$. We will show that the heat semigroup e^{-tA} is in fact a regularizing operator for $t > 0$, and its symbol $u(t, x, \xi)$ is given by $e^{-ta(x,\xi)}$, plus a remainder which will be conveniently estimated. Moreover the trace of e^{-tA} can be expressed in terms of its symbol by the formula

$$\sum_{j=0}^{\infty} e^{-t\lambda_j} = \int u(t, x, \xi) \, dx \, d\xi.$$

An explicit computation of this integral allows us to deduce the asymptotic behaviour of the above trace as $t \to 0$. Karamata's Tauberian Theorem finally yields the asymptotic behaviour for the so-called counting function $N(\lambda) = \{ j \in \mathbb{N} : \lambda_j \leq \lambda \}$, and therefore that of the λ_j's.

In fact, for simplicity we will develop this program only for the classes $\mathrm{OP\Gamma}^m_\rho(\mathbb{R}^d)$, $m > 0$, $0 < \rho \leq 1$, studied in Chapter 2 and the classes $\mathrm{OPG}^{m,n}_{\mathrm{cl}(\xi,x)}(\mathbb{R}^d)$,

$m > 0$, $n > 0$ of Chapter 3. For operators in the former class or even in the second class when $m \neq n$ we will obtain a formula of the type $N(\lambda) \sim C\lambda^\alpha$ as $\lambda \to +\infty$, for certain explicit values of $\alpha, C > 0$, whereas operators in $\mathrm{OPG}^{m,n}_{\mathrm{cl}(\xi,x)}(\mathbb{R}^d)$, $m = n > 0$, will fulfill $N(\lambda) \sim C\lambda^\alpha \log(\lambda)$. More refined methods yield an estimate of the remainder term. We will discuss this briefly in the Notes at the end of the present chapter.

Preliminary results about trace-class operators are also recalled in detail in Section 4.4.

4.1 Unbounded Operators in Hilbert spaces

In this section we recall some basic results for unbounded operators in Hilbert spaces.

Let H be a separable complex Hilbert space of infinite dimension. Let A be a linear operator, from a *dense* linear subspace $\mathrm{Dom}(A) \subset H$, into H. We refer to $\mathrm{Dom}(A)$ as the *domain* of A. We do not require it to be closed, nor the operator A to be bounded.

The graph of A is the linear subspace $\{(u, Au) \in H \times H : u \in \mathrm{Dom}(A)\}$ of $H \times H$. A is called *closed* if its graph is a closed subspace, namely, if $u_n \in \mathrm{Dom}(A) \to u \in H$ and $Au_n \to v \in H$ imply $u \in \mathrm{Dom}(A)$ and $Au = v$.

A is called *closable* if the closure of its graph is still a graph of a (linear) operator, which is then denoted by $\overline{A}$. Clearly, it is the smallest closed extension of A. It is easy to see that, in fact, it suffices that A admits a closed extension for A to be closable.

We also recall that the *adjoint operator* A^* has domain $\mathrm{Dom}(A^*)$ given by all $v \in H$ such that the map $\mathrm{Dom}(A) \subset H \to H$, $u \mapsto (Au, v)_H$, is bounded. Then A^*v is defined as the unique $w \in H$ such that

$$(Au, v)_H = (u, w)_H, \quad \text{for all } u \in \mathrm{Dom}(A).$$

Thus we have

$$(Au, v)_H = (u, A^*v)_H, \quad \text{for all } u \in \mathrm{Dom}(A) \text{ and } v \in \mathrm{Dom}(A^*).$$

One can prove that A^* is a closed operator. Moreover it is densely defined if A is closable.

A is called *self-adjoint* if $A = A^*$. A weaker definition than self-adjointness is that of symmetric operator: A is called *symmetric* if

$$(Au, v)_H = (u, Av)_H, \quad \text{for } u, v \in \mathrm{Dom}(A),$$

this is, if A^* is an extension of A.

Proposition 4.1.1. *A densely defined symmetric operator has symmetric closure.*

Proof. Let $u_n \in \mathcal{D}om(A)$ be a sequence such that $u_n \to 0$ and $Au_n \to v$. Let $w \in \mathcal{D}om(A)$. We have

$$(v, w)_H = \lim_{n \to \infty} (Au_n, w)_H = \lim_{n \to \infty} (u_n, Aw)_H = 0.$$

Thus $v = 0$, because $\mathcal{D}om(A)$ is dense in H. This implies that if (u, v_1), (u, v_2) are two elements in the closure of the graph of A, then $v_1 = v_2$, and therefore A is closable.

Let $u, v \in \mathcal{D}om(\overline{A})$. Then there exist two sequences $u_n, v_n \in \mathcal{D}om(A)$ such that $u_n \to u$, $v_n \to v$, $Au_n \to \overline{A}u$, $Av_n \to \overline{A}v$. Hence

$$(\overline{A}u, v)_H = \lim_{n \to \infty} (Au_n, v_n)_H = \lim_{n \to \infty} (u_n, Av_n)_H = (u, \overline{A}v)_H.$$

This proves that $\overline{A}$ is symmetric. $\qquad\qquad\qquad\qquad\qquad\qquad\qquad\qquad\Box$

A densely defined symmetric operator A is called *essentially self-adjoint* if $\overline{A}$ is self-adjoint. It is equivalent to saying that $A^* = \overline{A}$, because for any closable operator A we have $\overline{A}^* = A^*$.

Given a closed densely defined operator A on a *complex* Hilbert space H we define the *resolvent set* of A as the set $\rho(A)$ of complex numbers λ such that $A - \lambda I$ is a bijection $\mathcal{D}om(A) \to H$, with a bounded inverse $R_A(\lambda) = (A - \lambda I)^{-1}$ (which has domain H). $R_A(\lambda)$ is called the *resolvent (operator)* of A. The *spectrum* of A is by definition the complementary set of $\rho(A)$ in $\mathbb{C}$:

$$\sigma(A) = \mathbb{C} \setminus \rho(A).$$

Let us prove that self-adjoint operators have real spectrum.

Proposition 4.1.2. *If A is self-adjoint, then $\sigma(A) \subset \mathbb{R}$.*

Proof. Let $\lambda \in \mathbb{C} \setminus \mathbb{R}$. Then $\lambda = \xi + i\eta$ for real ξ, η, with $\eta \neq 0$. Because A is self-adjoint we have, for $u \in \mathcal{D}om(A)$,

$$\left\| (A - (\xi \pm i\eta)I)u \right\|_{L^2}^2 = \left\| (A - \xi I)u \right\|_{L^2}^2 + \eta^2 \|u\|_{L^2}^2. \qquad (4.1.1)$$

This implies that $A - \lambda I$ is one-to-one and with closed range. In fact, if u_n is a sequence in $\mathcal{D}om(A)$ such that

$$(A - \lambda)u_n \to v,$$

then by (4.1.1) we have that u_n and Au_n are convergent. But A is closed, hence

$$u_n \to u \in \mathcal{D}om(A) \quad \text{and} \quad (A - \lambda I)u_n \to (A - \lambda I)u.$$

This implies that $v = (A - \lambda I)u$ belongs to the range of $A - \lambda I$, which is therefore closed. Since A is self-adjoint, the range of $A - \lambda I$ is then the orthogonal space of $\mathrm{Ker}(A - \overline{\lambda}I) = \{0\}$, so that $A - \lambda I$ is onto. Hence $A - \lambda I : \mathcal{D}om(A) \to H$ is invertible. Moreover its inverse is bounded, because from (4.1.1) we obtain, for $u = (A - \lambda I)^{-1}v$,

$$\|v\|_{L^2} \geq |\eta| \|(A - \lambda I)^{-1}v\|_{L^2}.$$

This means that $\lambda \in \rho(A)$. $\qquad\square$

We say that A has *compact resolvent* if there exists $\lambda \in \rho(A)$ such that $R_A(\lambda)$ is compact.

Proposition 4.1.3. *If A has compact resolvent, then $R_A(\lambda)$ is compact for all $\lambda \in \rho(A)$.*

Proof. Let $\lambda_0 \in \rho(A)$ be such that $R_A(\lambda_0)$ is compact. Consider $\lambda \in \rho(A)$ and let

$$A - \lambda I = A - \lambda_0 I - (\lambda - \lambda_0) I. \tag{4.1.2}$$

Multiplication of (4.1.2) from the left by $R_A(\lambda_0)$ and from the right by $R_A(\lambda)$ yields

$$R_A(\lambda) - R_A(\lambda_0) = (\lambda - \lambda_0) R_A(\lambda) R_A(\lambda_0).$$

This identity shows that $R_A(\lambda)$ is compact. $\qquad\square$

Self-adjoint operators with compact resolvent have very simple spectrum. In order to show this, we use the following well-known result about the spectrum of compact operators, combined with the subsequent lemma.

Theorem 4.1.4. *Let T be a compact operator on a complex Hilbert space H. Then $\sigma(T)$ is an at most countable set with no accumulation point different from 0. Each non-zero $\lambda \in \sigma(T)$ is an eigenvalue with finite multiplicity. If T is also self-adjoint, then all the eigenvalues are real and H has an orthonormal basis made of eigenvectors of T.*

Lemma 4.1.5. *Let A be a closed densely defined operator on a Hilbert space H such that $\rho(A) \neq \emptyset$. Then for any $\lambda_0 \in \rho(A)$ we have*

$$\rho(A) = \{\lambda_0\} \cup \{\lambda \in \mathbb{C}: \ \lambda \neq \lambda_0 \ \text{and} \ (\lambda - \lambda_0)^{-1} \in \rho(R_A(\lambda_0))\}.$$

Proof. Let $B = A - \lambda_0 I$. Then we have

$$R_A(\lambda_0) = B^{-1}$$

and

$$\lambda \in \rho(A) \iff \lambda - \lambda_0 \in \rho(B).$$

We have to prove that

$$\lambda \in \rho(B) \iff \lambda = 0, \ \text{or} \ \lambda \neq 0 \ \text{and} \ \lambda^{-1} \in \rho(B^{-1}).$$

Let $\lambda \in \rho(B)$, $\lambda \neq 0$. Then

$$S := BR_B(\lambda) = I + \lambda R_B(\lambda)$$

is bounded, and

$$B^{-1}S = R_B(\lambda) = \lambda^{-1}(S - I).$$

This means that

$$(B^{-1} - \lambda^{-1}I)S = -\lambda^{-1}I. \tag{4.1.3}$$

Hence $B^{-1} - \lambda^{-1}I$ is onto. But $B^{-1} - \lambda^{-1}I$ is also one-to-one, because from $(B^{-1} - \lambda^{-1}I)u = 0$ we obtain $\lambda u = Bu$, which implies $u = 0$ because $\lambda \in \rho(B)$. Then $B^{-1} - \lambda^{-1}I$ is invertible. Moreover (4.1.3) shows that

$$\left(B^{-1} - \lambda^{-1}I\right)^{-1} = -\lambda S,$$

which is bounded.

Let now $\lambda \neq 0$, $\lambda^{-1} \in \rho(B^{-1})$, and set

$$S' = B^{-1}R_{B^{-1}}(\lambda^{-1}) = I + \lambda^{-1}R_{B^{-1}}(\lambda^{-1}).$$

Then S' is bounded and

$$BS' = \lambda(S' - I), \tag{4.1.4}$$

so

$$(B - \lambda I)S' = -\lambda I.$$

This implies that $B - \lambda I$ is onto. Moreover it is one-to-one, because from $(B - \lambda I)u = 0$ we obtain $B^{-1}u = \lambda^{-1}u$, which implies $u = 0$ because $\lambda^{-1} \in \rho(B^{-1})$. Then we have that $B - \lambda I$ is invertible. Moreover (4.1.4) shows that $(B - \lambda I)^{-1} = -\lambda^{-1}S'$, which is bounded. $\qquad\square$

Now we can describe the spectrum of a self-adjoint operator with compact resolvent.

Theorem 4.1.6. *Let A be a densely defined self-adjoint operator on a complex Hilbert space H (of infinite dimension). If A has compact resolvent, then $\sigma(A)$ is a sequence of real isolated eigenvalues, diverging to ∞. Each eigenvalue has finite multiplicity and H has an orthonormal basis made of eigenfunctions of A.*

Proof. Fix any $\lambda_0 \in \rho(A)$, then $R_A(\lambda_0)$ is compact. By Theorem 4.1.4 and Lemma 4.1.5, $\sigma(A)$ must be at most countable. This means that there exists $\mu \in \rho(A) \cap \mathbb{R}$. Then $R_A(\mu)$ is compact self-adjoint. Moreover $\mathrm{Ker}\big(R_A(\mu)\big) = 0$ because $R_A(\mu)$ is one-to-one. The result now follows from Theorem 4.1.4 and Lemma 4.1.5 (applied now with μ in place of λ_0). $\qquad\square$

4.2   Pseudo-Differential Operators in L^2: Realization and Spectrum

We deal now with pseudo-differential operators having symbols in the classes $S(M) = S(M; \Phi, \Psi)$ and $\mathrm{Hypo}(M, M_0) = \mathrm{Hypo}(M, M_0; \Phi, \Psi)$; see Definitions 1.1.1 and 1.3.2. Moreover, we use here the Weyl quantization, cf. (1.0.10).

Given two symbols $a \in S(M_1)$ and $b \in S(M_2)$, the Weyl symbol of the composition $a^w b^w$ will be denoted by $a \# b$. Hence we recall from Theorem 1.2.17 that $a \# b \in S(M_1 M_2)$ and the remainder term can be rewritten as

$$\mathcal{R}_N(a, b) = a \# b - \sum_{j=0}^{N} \frac{\{a, b\}_j}{(2i)^j \, j!} \in S(M_1 M_2 h^{N+1}), \qquad (4.2.1)$$

for all $N \in \mathbb{N}$, where $\{a, b\}_0 := ab$,

$$\{a, b\}_j := \left[\left(\sum_{i=1}^{n} \left(\frac{\partial}{\partial \eta_i} \frac{\partial}{\partial y_i} - \frac{\partial}{\partial x_i} \frac{\partial}{\partial \xi_i} \right) \right)^j a(x, \eta) b(y, \xi) \right]_{\substack{y=x \\ \eta=\xi}}$$

$$= j! \sum_{|\alpha + \beta| = j} (-1)^{|\beta|} (\alpha! \beta!)^{-1} \partial_\xi^\alpha \partial_x^\beta a(x, \xi) \partial_\xi^\beta \partial_x^\alpha b(x, \xi), \qquad (4.2.2)$$

for $j > 0$, and h is the Planck function in (1.1.8). More precisely, for each $N, k \in \mathbb{N}$ there exists an integer $l_{N,k}$ such that

$$\|\mathcal{R}_N(a, b)\|_{k, S(M_1 M_2 h^{N+1})} \lesssim \|a\|_{l_{N,k}, S(M_1)} \|b\|_{l_{N,k}, S(M_2)} \qquad (4.2.3)$$

for all $a \in S(M_1)$ and all $b \in S(M_2)$. This is easily seen, e.g., from (4.2.1) as a consequence of the Closed Graph Theorem.

In the sequel we will also use the following notation: for $k \in \mathbb{N}$, $a \in C^\infty(\mathbb{R}^{2d})$ we set

$$|a|_k(x, \xi) = \sup_{|\alpha| + |\beta| \le k} |\partial_\xi^\alpha \partial_x^\beta a(x, \xi)| \Psi(x, \xi)^{-|\alpha|} \Phi(x, \xi)^{-|\beta|}. \qquad (4.2.4)$$

Hence, a symbol $a \in \mathrm{Hypo}(M, M_0)$ satisfies $|a|_k \lesssim |a|$ for every $k \in \mathbb{N}$, when $|x| + |\xi| \ge R$, with R large enough.

Now, given $a \in S(M)$, we regard its Weyl quantization as an unbounded operator in $L^2(\mathbb{R}^d)$ with dense domain $\mathcal{S}(\mathbb{R}^d)$, namely, we consider the operator A with domain $\mathcal{S}(\mathbb{R}^d)$ defined as $Au = a^w u$, for $u \in \mathcal{S}(\mathbb{R}^d)$. When not specified, a^w is understood as a continuous operator on $\mathcal{S}'(\mathbb{R}^d)$. As we saw in the summary of the present chapter, the restriction of a^w to the linear subspace (4.0.1) is a closed extension of A, the maximal realization of A. Hence A is closable. As we observed there, the closure is also called minimal realization.

We now study the relationship between the formal adjoint and adjoint in $L^2(\mathbb{R}^d)$ for a pseudo-differential operator with symbol $a \in S(M)$.

Given $a \in S(M)$, let A be as above, and let A^* be its adjoint in $L^2(\mathbb{R}^d)$, as defined in the previous section. Consider moreover the formal adjoint $(a^w)^*$ ($= \bar{a}^w$) defined in Section 1.2.2, namely the operator $(a^w)^* : \mathcal{S}(\mathbb{R}^d) \to \mathcal{S}(\mathbb{R}^d)$ such that

$$((a^w)^* u, v)_{L^2} = (u, a^w v)_{L^2} \quad \text{for } u, v \in \mathcal{S}(\mathbb{R}^d). \qquad (4.2.5)$$

Observe that A is symmetric if and only if it is formally self-adjoint, which is equivalent to saying that a is real-valued. We now show that A^* coincides with the maximal realization of the formal adjoint $(a^w)^*$, when regarded as an operator in $L^2(\mathbb{R}^d)$ with dense domain $\mathcal{S}(\mathbb{R}^d)$.

Denote by $A^+ : \mathcal{S}'(\mathbb{R}^d) \to \mathcal{S}'(\mathbb{R}^d)$ the continuous extension to $\mathcal{S}'(\mathbb{R}^d)$ of the pseudo-differential operator $(a^w)^*$.

Proposition 4.2.1. *Let $a \in S(M; \Phi, \Psi)$ and let A, A^+ be as above. Then the adjoint A^* in $L^2(\mathbb{R}^d)$ has domain*

$$\mathrm{Dom}(A^*) = \{u \in L^2(\mathbb{R}^d) : A^+ u \in L^2(\mathbb{R}^d)\}, \tag{4.2.6}$$

and coincides with the restriction of $A^+ : \mathcal{S}'(\mathbb{R}^d) \to \mathcal{S}'(\mathbb{R}^d)$ to such a linear subspace.

Proof. Given $u \in \mathrm{Dom}(A^*)$, $A^* u$ is the unique element of $L^2(\mathbb{R}^d)$ such that

$$(A^* u, v)_{L^2} = (u, Av)_{L^2} = (A^+ u, v), \quad \text{for } v \in \mathcal{S}(\mathbb{R}^d).$$

This means that

$$A^* = A^+|_{\mathrm{Dom}(A^*)}$$

and that

$$\mathrm{Dom}(A^*) \subset \{u \in L^2(\mathbb{R}^d) : A^+ u \in L^2(\mathbb{R}^d)\}.$$

Vice-versa, let $u \in L^2(\mathbb{R}^d)$ be such that $A^+ u \in L^2(\mathbb{R}^d)$. Then we have

$$(A^+ u, v)_{L^2} = (u, Av)_{L^2}, \quad \text{for } v \in \mathcal{S}(\mathbb{R}^d).$$

This implies that $v \mapsto (u, Av)_{L^2}$ is continuous and therefore $u \in \mathrm{Dom}(A^*)$. $\qquad\square$

We next result shows that, for operators with hypoelliptic symbols, the minimal and maximal realizations coincide. We need the following easy preliminary results.

Lemma 4.2.2. *Given two smooth functions a, $b \in C^\infty(\mathbb{R}^{2d})$, we have*

$$\left| \{a, b\}_j \right|_k \lesssim \sum_{l=0}^{k} |a|_{j+l} \, |b|_{j+k-l} \, h^j,$$

for all $k, j \in \mathbb{N}$.

Proof. This follows from the very definition (4.2.2) and Leibniz' formula. $\qquad\square$

Lemma 4.2.3. *Let ϕ_j be a bounded sequence in $S(1; \Phi, \Psi)$, converging in some symbol class $S(M; \Phi, \Psi)$ to $\phi \in S(1; \Phi, \Psi)$. Then $\phi_j^w u \to \phi^w u$ in $L^2(\mathbb{R}^d)$ for all $u \in L^2(\mathbb{R}^d)$.*

Proof. In view of the uniform estimate (1.4.1), it suffices to prove the desired result when $u \in \mathcal{S}(\mathbb{R}^d)$. In that case we have $\phi_j^w u \to \phi^w u$ in $\mathcal{S}(\mathbb{R}^d)$ by Remark 1.2.8. $\qquad\square$

Theorem 4.2.4. *Consider a pseudo-differential operator A with Weyl symbol $a \in$ Hypo$(M, M_0; \Phi, \Psi)$. Assume there exists $N_0 \in \mathbb{N}$ such that*

$$h^{N_0} \lesssim \frac{\inf\{1, M_0\}}{M}. \tag{4.2.7}$$

Then $\overline{A}$ has domain $\{u \in L^2(\mathbb{R}^d) : a^w u \in L^2(\mathbb{R}^d)\}$ and coincides with the restriction of a^w to this subspace.

Proof. We have to show that for any $u \in L^2(\mathbb{R}^d)$ such that $a^w u \in L^2(\mathbb{R}^d)$, there exists a sequence u_j in $\mathcal{S}(\mathbb{R}^d)$ such that

$$u_j \to u \quad \text{and} \quad a^w u_j \to a^w u \quad \text{in } L^2(\mathbb{R}^d).$$

By arguing as in the proof of Theorem 1.3.6 one sees that there exists $q \in$ Hypo(M_0^{-1}, M^{-1}) such that

$$r := 1 - q \# a \in S(h^{N_0+1}).$$

By Proposition 1.1.5 there exists a sequence of symbols $\chi_j \in \mathcal{S}(\mathbb{R}^{2d})$ which is bounded in $S(1)$ and converges to 1, say, in $S(1 + |x| + |\xi|)$.

Then we set

$$u_j = \chi_j^w u.$$

We have that $u_j \in \mathcal{S}(\mathbb{R}^d)$, because $\chi_j \in \mathcal{S}(\mathbb{R}^{2d})$.

From the definition of r we have

$$
\begin{aligned}
a^w u_j &= a^w \chi_j^w q^w a^w u + a^w \chi_j^w r^w u \\
&= (a \# \chi_j \# q)^w a^w u + (a \# \chi_j \# r)^w u.
\end{aligned} \tag{4.2.8}
$$

Thanks to (4.2.3) and Lemma 4.2.2, and using the fact that a and q are hypoelliptic symbols, it follows that for all $k \in \mathbb{N}$ there exist $l, l' \in \mathbb{N}$ such that the following estimates hold true:

$$
\begin{aligned}
|a \# \chi_j \# q|_k &\leq \sum_{n=0}^{N_0} \frac{1}{2^n \, n!} |\{a \# \chi_j, q\}_n|_k + |R_{N_0}(a \# \chi_j, q)|_k \\
&\lesssim \sup_{m \leq l} |a \# \chi_j|_m \sup_{m \leq l} |q|_m + \frac{M h^{N_0+1}}{M_0} \\
&\lesssim \left(\sup_{m \leq l'} |a|_m \sup_{m \leq l'} |\chi_j|_m + M h^{N_0+1} \right) \sup_{m \leq l} |q|_m + \frac{M h^{N_0+1}}{M_0} \\
&\lesssim (|a| + M h^{N_0+1}) |a|^{-1} + \frac{M h^{N_0+1}}{M_0} \\
&\lesssim 1 + \frac{M h^{N_0+1}}{M_0} \lesssim 1,
\end{aligned}
$$

for all $|x| + |\xi| \geq R$ and $j \in \mathbb{N}$, with R large enough.

As a consequence the symbols $a\#\chi_j\#q$ belong to a bounded subset of $S(1)$. Moreover from (4.2.7) and (4.2.3) we have also

$$\|a\#\chi_j\#r\|_{k,S(1)} \lesssim \|a\#\chi_j\#r\|_{k,S(Mh^{N_0+1})}$$

$$\lesssim \|a\|_{l,S(M)} \|\chi_j\|_{l,S(1)} \|r\|_{l,S(h^{N_0+1})},$$

for some $l = l_k \in \mathbb{N}$ and all $j \in \mathbb{N}$. This means that also the sequence $a\#\chi_j\#r$ belongs to a bounded set of $S(1)$.

On the other hand, $a\#\chi_j\#q$ and $a\#\chi_j\#r$ converge to $a\#q$ and $a\#r$ respectively, in suitable symbol classes, by (4.2.1), (4.2.2) and (4.2.3).

From Lemma 4.2.3 we obtain that $u_j = \chi_j^w u \to u$ in $L^2(\mathbb{R}^d)$, whereas $a^w u_j$ in (4.2.8) converges to

$$(a\#q)^w a^w u + (a\#r)^w u = (a\#(q\#a + r))^w u = a^w u$$

in $L^2(\mathbb{R}^d)$, which concludes the proof. $\square$

We now study the spectrum of operators with hypoelliptic symbols.

Proposition 4.2.5. *Assume the strong uncertainty principle (1.1.10). Consider a pseudo-differential operator A with Weyl symbol in $\mathrm{Hypo}(M, M_0; \Phi, \Psi)$, with $M_0(x, \xi) \to +\infty$ at infinity. Then its closure $\overline{A}$ in L^2 has either spectrum $\sigma(\overline{A}) = \mathbb{C}$ or has compact resolvent.*

Proof. Let $\sigma(\overline{A}) \neq \mathbb{C}$. Then there exists $\lambda_0 \in \rho(\overline{A})$ such that $\overline{A} - \lambda_0 I$ has a bounded inverse $R_{\overline{A}}(\lambda_0)$. By Theorem 1.3.6 there exists a parametrix $B \in S(M_0^{-1})$ of A. This means that $S = BA - I$ is regularizing. In particular, by Theorem 1.4.2, B and S extend to compact operators on $L^2(\mathbb{R}^d)$. This implies that

$$R_{\overline{A}}(\lambda_0) = B + (\lambda_0 B - S)R_{\overline{A}}(\lambda_0)$$

is compact. $\square$

Remark 4.2.6. It may happen that $\sigma(\overline{A}) = \mathbb{C}$. For example the ordinary differential operator $A = iD + x$ has Weyl symbol $i\xi + x$, which is elliptic in the class $S(M; \Phi, \Psi)$, with $M(x, \xi) = \Phi(x, \xi) = \Psi(x, \xi) = (1 + |x|^2 + |\xi|^2)^{1/2}$, and has spectrum equal to $\mathbb{C}$. Indeed, any $\lambda \in \mathbb{C}$ is an eigenvalue, with eigenfunction $u(x) = e^{-(x-\lambda)^2/2}$.

Hence, we get the following result.

Corollary 4.2.7. *Assume the strong uncertainty principle. Consider a pseudo-differential operator A having real-valued Weyl symbol in $\mathrm{Hypo}(M, M_0; \Phi, \Psi)$, with $M_0(x, \xi) \to +\infty$ at infinity. Then A is essentially self-adjoint, and its closure $\overline{A}$ in L^2 has compact resolvent.*

Proof. The fact that A is essentially self-adjoint follows from Theorem 4.2.4 and Proposition 4.2.1. The remaining part of the statement is a consequence of Propositions 4.1.2 and 4.2.5. $\qquad\square$

Bounds on the spectrum of pseudo-differential operators will be obtained in the next theorem, by using the following result.

Lemma 4.2.8. *Assume the strong uncertainty principle. Consider a real-valued positive symbol $a \in \mathrm{Hypo}(M, M_0; \Phi, \Psi)$. Then a^w is bounded from below, i.e., there exists a positive constant C such that*

$$(a^w u, u)_{L^2(\mathbb{R}^d)} \geq -C\|u\|^2_{L^2(\mathbb{R}^d)} \quad for\ u \in \mathcal{S}(\mathbb{R}^d).$$

Proof. For every $N \in \mathbb{N}$, we will prove the existence of a symbol $c \in S(M^{1/2})$ such that

$$c^w (c^w)^* - a^w = r_N^w, \quad r_N \in S(h^N M), \tag{4.2.9}$$

which by Theorem 1.4.1 implies the desired lower bound, if N is so large that $h^N M \lesssim 1$.

Now, it follows from Lemma 1.3.5 that $b := a^{1/2} \in \mathrm{Hypo}(M^{1/2}, M_0^{1/2})$. As a consequence of (4.2.3) and Lemma 4.2.2 we obtain that

$$I - b^w a^w b^w = r^w,$$

where $r \in S(h)$ is real; cf. the proof of Lemma 4.2.3.

Denote now by $T_N(x)$ the first N terms in the power series expansion of $(1 - x)^{1/2}$ at 0. Since $1 - x - T_N(x)^2$ is a polynomial divisible by x^N, it turns out that $I - r^w - T_N(r^w)^2$ has symbol in $S(h^N)$, so that

$$T_N(r^w)^2 - b^w a^w b^w$$

has symbol in $S(h^N; \Phi, \Psi)$. Let B a parametrix of b^w, with real-valued Weyl symbol. If we set $c^w = BT_N(r^w)$, since B and $T_N(r^w)$ are formally self-adjoint we obtain (4.2.9). $\qquad\square$

Notice that, in comparison with the Sharp Gårding Inequality in Theorem 1.7.15, the conclusion in Lemma 4.2.8 is much stronger, but we have here the additional hypothesis that a is hypoelliptic.

Now we can prove the spectral theorem for operators with hypoelliptic symbols.

Theorem 4.2.9. *Assume the strong uncertainty principle (1.1.10). Consider a pseudo-differential operator A, with real-valued Weyl symbol in the class $\mathrm{Hypo}(M, M_0; \Phi, \Psi)$, and assume $M_0(x, \xi) \to +\infty$ at infinity. Its closure $\overline{A}$ in $L^2(\mathbb{R}^d)$ has spectrum given by a sequence of real eigenvalues either diverging to $+\infty$ or $-\infty$. The eigenvalues have all finite multiplicity and the eigenfunctions belong to $\mathcal{S}(\mathbb{R}^d)$. Moreover $L^2(\mathbb{R}^d)$ has an orthonormal basis made of eigenfunctions of $\overline{A}$.*

Proof. The closure $\overline{A}$ of A is self-adjoint and has a compact resolvent by Corollary 4.2.7. Then we can apply Theorem 4.1.6. Hence there are infinite eigenvalues diverging to ∞. The corresponding eigenfunctions must belong to $S(\mathbb{R}^d)$ by Corollary 1.3.9. If we show that A is semi-bounded we have that its eigenvalues diverge either to $+$ or $-\infty$.

Since
$$|a(x,\xi)| \gtrsim M_0(x,\xi) > 0, \quad \text{for } |x| + |\xi| \geq R,$$

and $a(x,\xi)$ is real-valued, $a(x,\xi)$ has always the same sign on the connected set $|x| + |\xi| \geq R$, i.e., it is either bounded from below or from above. Lemma 4.2.8 then implies that A is semibounded. So are its eigenvalues, which must diverge to $+\infty$ or $-\infty$ according to the sign of a.

The proof therefore is complete. $\square$

4.3 Complex Powers

This section is devoted to the study of the complex powers of hypoelliptic pseudo-differential operators. We begin by recalling some results on complex powers of a non-negative operator in the sense of Komatsu.

Definition 4.3.1. A closed operator A on a Banach space X is called *non-negative* if

(a) $(-\infty, 0)$ is contained in the resolvent set of A;

(b) $\sup\limits_{\lambda \in \mathbb{R}_+} \lambda \left\| (A + \lambda I)^{-1} \right\|_{\mathcal{B}(X)} < \infty.$

Remark 4.3.2. If A is a densely defined self-adjoint operator in a Hilbert space H, then A is non-negative if and only if $(Au, u)_H \geq 0$ for all $u \in Dom(H)$ (see, e.g., Martínez Carracedo and Sanz Alix [141, Proposition 1.3.6]).

Set $\mathbb{C}_+ = \{z \in \mathbb{C} : \operatorname{Re} z > 0\}$ and

$$\gamma_k(z) = \frac{\Gamma(k)}{\Gamma(z)\Gamma(k-z)} = \frac{(k-1)!\sin \pi z}{(k-1-z)\cdots(1-z)\pi}, \tag{4.3.1}$$

for $k \in \mathbb{N}$, $k \neq 0$ and $z \in \mathbb{C} \setminus \mathbb{Z}$.

Proposition 4.3.3. *Consider a non-negative operator A on a Banach space X. Given $z \in \mathbb{C}_+$, and $u \in Dom(A^{[\operatorname{Re} z]+1})$ we have that the integral*

$$I_{A,k}^z u = \gamma_k(z) \int_0^\infty \lambda^{z-1} \left[A \left(A + \lambda I \right)^{-1} \right]^k u \, d\lambda \tag{4.3.2}$$

is absolutely convergent for all integers $k > \operatorname{Re} z$, as an improper Riemann integral taking values in X.

Moreover these integrals are independent of k:

$$I_{A,k+1}^z u = I_{A,k}^z u, \quad k > \operatorname{Re} z.$$

Proof. See [141, Proposition 3.1.3].　　　　　　　　　　　　□

Following Balakhrishnan, we now define the complex powers of a non-negative operator.

Definition 4.3.4. Given a non-negative operator A on a Banach space X and a complex number $z \in \mathbb{C}_+$, define a new operator J_A^z on X as

$$\begin{cases} \mathcal{D}om\left(J_A^z\right) = \mathcal{D}om\left(A^{[\operatorname{Re} z]+1}\right), \\ J_A^z u = I_{A,k}^z u, \quad \text{for any } k > \operatorname{Re} z. \end{cases}$$

Theorem 4.3.5. *Assume that A is a non-negative, densely defined operator on a Banach space X, then*

$$A^z = \overline{J_A^z}, \quad z \in \mathbb{C}_+,$$

is the unique *family of operators which enjoys the following set of properties:*

(a) A^z *extends* J_A^z;

(b) A^z *is closed;*

(c) $A^1 = A$;

(d) $A^z A^w = A^{z+w}$ *for all $z, w \in \mathbb{C}_+$.*

In particular, $A^k = \underbrace{AA \cdots A}_{k-times}$, with $k \in \mathbb{N}$. Moreover we have:

(e) *(Spectral Mapping Theorem) the spectrum of A^z is given by*[1]

$$\sigma(A^z) = \{\lambda^z : \lambda \in \sigma(A)\};$$

(f) *For all $u \in \mathcal{D}om(A^n)$, with $n \in \mathbb{N}$, $n \neq 0$, the mapping*

$$z \mapsto A^z u$$

is analytic in the strip $\{z \in \mathbb{C} : 0 < \operatorname{Re} z < n\}$.

Proof. See [141, Theorems 3.1.5 and 3.1.8, Corollary 5.1.12 and Section 6.2] .　□

Let now a be a symbol in some class $S(M; \Phi, \Psi)$ and consider the unbounded operator A in $L^2(\mathbb{R}^d)$ with domain $\mathcal{S}(\mathbb{R}^d)$, defined by $Au = a^w u$, $u \in \mathcal{S}(\mathbb{R}^d)$. In Theorem 4.3.6 we show that under suitable hypotheses $\overline{A}^z$ is pseudo-differential.

Theorem 4.3.6. *Assume the strong uncertainty principle (1.1.10) and consider a hypoelliptic symbol $a \in \text{Hypo}(M, M_0; \Phi, \Psi)$ such that*

$$\operatorname{Re} a(x, \xi) \geq -R \left|\operatorname{Im} a(x, \xi)\right|, \tag{4.3.3}$$

---

[1] The complex power λ^z is the principal branch $\lambda^z = \exp(z (\log |\lambda| + i \arg \lambda))$, with $-\pi < \arg \lambda \leq \pi$.

for $|x| + |\xi| \geq R$, where R is a positive constant such that estimates (1.3.2) and (1.3.3) are satisfied. Let A be as above and suppose that $\overline{A}$ is non-negative. Let

$$a_0 = a + \chi, \qquad (4.3.4)$$

where $\chi \in C_0^\infty(\mathbb{R}^{2d})$, $\chi \geq 0$ everywhere, and $\chi(x,\xi) > 0$ for $|x| + |\xi| \leq R$. Then for all $z \in \mathbb{C}_+$ there exists a hypoelliptic symbol

$$a^{\#z} \in \mathrm{Hypo}\left(M^{\mathrm{Re}\,z}, M_0^{\mathrm{Re}\,z}; \Phi, \Psi\right)$$

such that

(i) *for all $k \in \mathbb{N}$ and $z \in \mathbb{C}_+$ we have*

$$\left|a^{\#z} - a_0^z\right|_k (x,\xi) \lesssim |a_0(x,\xi)|^{\mathrm{Re}\,z}\, h(x,\xi), \quad \text{for } (x,\xi) \in \mathbb{R}^{2d}, \qquad (4.3.5)$$

   where h is the Planck function defined in (1.1.8), (see (4.2.4) for the notation $|\cdot|_k$);

(ii) *for all $z \in \mathbb{C}_+$ we have*

$$\overline{A}^z = \overline{(a^{\#z})^{\mathrm{w}}|_{\mathcal{S}(\mathbb{R}^d)}}. \qquad (4.3.6)$$

We shall prove this theorem in Section 4.3.2.

We observe that the power a_0^z is well defined by taking the principal branch of the logarithm, by (4.3.7) below. Notice also that from Theorems 4.3.5 and 4.3.6 it follows that $a^{\#k} = \underbrace{a\# \cdots \#a}_{k-\text{times}}$, when $k \in \mathbb{N}$, $k \neq 0$.

We can also give a simple description of the domain $\mathcal{D}om\left(\overline{A}^z\right)$ of $\overline{A}^z$.

Corollary 4.3.7. *Under the hypotheses of Theorem 4.3.6, we have*

$$\mathcal{D}om\left(\overline{A}^z\right) = \left\{u \in L^2(\mathbb{R}^d) : \left(a^{\#z}\right)^{\mathrm{w}} u \in L^2(\mathbb{R}^d)\right\}, \quad z \in \mathbb{C}_+,$$

where $\left(a^{\#z}\right)^{\mathrm{w}}$ is here regarded as an operator on $\mathcal{S}'(\mathbb{R}^d)$.

Proof. Because $a^{\#z}$ is hypoelliptic, the result follows from Theorems 4.3.6 and 4.2.4. $\qquad \square$

4.3.1 The Resolvent Operator

As formula (4.3.2) suggests, to study the complex powers $\overline{A}^z$ we have to understand the structure of the resolvent operator of $\overline{A}$.

To to this, we need some lemmata. The proof of the first lemma is elementary and is left to the reader.

Lemma 4.3.8. *Let a_0 be given in* (4.3.4). *Possibly for a greater R we have*

$$\operatorname{Re} a_0(x,\xi) > -R\,|\operatorname{Im} a_0(x,\xi)|, \quad \text{for all } (x,\xi) \in \mathbb{R}^{2d}. \tag{4.3.7}$$

As a consequence,

$$|a_0(x,\xi)| \leq \sqrt{1+R^2}\,|a_0(x,\xi) + \lambda|, \tag{4.3.8}$$

$$\lambda \leq \sqrt{1+R^2}\,|a_0(x,\xi) + \lambda|, \tag{4.3.9}$$

for all $(x,\xi) \in \mathbb{R}^{2d}$ and $\lambda \geq 0$.

Lemma 4.3.9. *Let a_0 be given in* (4.3.4). *Then, for all $k \in \mathbb{N}$, it turns out that*

$$|a_0|_k\,(x,\xi) \lesssim |a_0(x,\xi)|, \quad \text{for all } (x,\xi) \in \mathbb{R}^{2d}. \tag{4.3.10}$$

Moreover, for all $k \in \mathbb{N}$ we have

$$\left|(a_0+\lambda)^{-1}\right|_k (x,\xi) \lesssim \frac{1}{|a_0(x,\xi) + \lambda|}, \quad \text{for all } (x,\xi) \in \mathbb{R}^{2d} \text{ and } \lambda > 0. \tag{4.3.11}$$

Proof. Since a is hypoelliptic, $a_0 - a$ has compact support and a_0 never vanishes, it follows that (4.3.10) holds. This fact, together with (4.3.8) shows that $a_0 + \lambda$ satisfies the estimates $|a_0 + \lambda|_k \lesssim |a_0 + \lambda|$ uniformly with respect to λ. Hence, arguing as in the proof of Lemma 1.3.5 gives (4.3.11), as desired. $\square$

From the strong uncertainty principle and Lemma 4.3.8 one easily obtains the following result.

Lemma 4.3.10. *Assume the strong uncertainty principle* (1.1.10) *and let a_0 be given in* (4.3.4). *Consider two real numbers ν, ν_0 such that*

$$(1+|x|+|\xi|)^{-\nu_0} \lesssim |a_0(x,\xi)| \lesssim (1+|x|+|\xi|)^{\nu}. \tag{4.3.12}$$

Then we have

$$h(x,\xi)^{\nu_0/\delta}\,(1+\lambda) \lesssim |a_0(x,\xi) + \lambda| \lesssim h(x,\xi)^{-\nu/\delta}\,(1+\lambda), \tag{4.3.13}$$

for all $(x,\xi) \in \mathbb{R}^{2d}$ and $\lambda \geq 0$.

Lemma 4.3.11. *Assume the strong uncertainty principle, and let a_0 be given in* (4.3.4). *Assume that for all $\lambda \in \mathbb{R}_+$ two symbols ϕ_λ and ψ_λ are given, such that for all $k \in \mathbb{N}$ we have*

$$|\phi_\lambda|_k \lesssim |a_0|^{L'}\,|a_0 + \lambda|^{J'}\,h^{N'}, \qquad |\psi_\lambda|_k \lesssim |a_0|^{L''}\,|a_0 + \lambda|^{J''}\,h^{N''}, \tag{4.3.14}$$

for $\lambda > 0$, with L', L'', J', J'', N', $N'' \in \mathbb{R}$.

Then for all N, $k \in \mathbb{N}$ we have

$$|\phi_\lambda \# \psi_\lambda|_k \lesssim |a_0|^{L'+L''}\,|a_0 + \lambda|^{J'+J''}\,h^{N'+N''}, \tag{4.3.15}$$

and

$$|\mathcal{R}_N(\phi_\lambda, \psi_\lambda)|_k \lesssim |a_0|^{L'+L''}\,|a_0 + \lambda|^{J'+J''}\,h^{N'+N''+N+1}, \tag{4.3.16}$$

for $\lambda > 0$, where $\mathcal{R}_N(\phi_\lambda, \psi_\lambda)$ is defined in (4.2.1).

Proof. From (4.3.14) and (4.3.13), for all $k \in \mathbb{N}$ we obtain

$$\|\phi_\lambda\|_{k,S(h^{\gamma'},g)} \lesssim (1+\lambda)^{J'}, \quad \lambda > 0,$$

$$\|\psi_\lambda\|_{k,S(h^{\gamma''},g)} \lesssim (1+\lambda)^{J''}, \quad \lambda > 0,$$

with [2]

$$\gamma' = \left(L'_- + J'_-\right)\nu_0/\delta - \left(L'_+ + J'_+\right)\nu/\delta + N',$$
$$\gamma'' = \left(L''_- + J''_-\right)\nu_0/\delta - \left(L''_+ + J''_+\right)\nu/\delta + N''.$$

From Lemma 4.2.2 and (4.2.3) for all N_0, $k \in \mathbb{N}$ there exists an integer $l = l_{N_0,k} \geq N_0 + k$ such that

$$
\begin{aligned}
|\phi_\lambda \# \psi_\lambda|_k &\lesssim \sup_{0 \leq j \leq l} |\phi_\lambda|_j \sup_{0 \leq j \leq l} |\psi_\lambda|_j + |\mathcal{R}_{N_0}(\phi_\lambda, \psi_\lambda)|_k \\
&\lesssim \sup_{0 \leq j \leq l} |\phi_\lambda|_j \sup_{0 \leq j \leq l} |\psi_\lambda|_j + \|\mathcal{R}_{N_0}(\phi_\lambda, \psi_\lambda)\|_{k,S(h^{\gamma'+\gamma''+N_0+1})} h^{\gamma'+\gamma''+N_0+1} \\
&\lesssim \sup_{0 \leq j \leq l} |\phi_\lambda|_j \sup_{0 \leq j \leq l} |\psi_\lambda|_j + \|\phi_\lambda\|_{l,S(h^{\gamma'})} \|\psi_\lambda\|_{l,S(h^{\gamma''})} h^{\gamma'+\gamma''+N_0+1} \\
&\lesssim |a_0|^{L'+L''} |a_0 + \lambda|^{J'+J''} h^{N'+N''} + (1+\lambda)^{J'+J''} h^{\gamma'+\gamma''+N_0+1},
\end{aligned}
$$

for $\lambda > 0$.

Then (4.3.15) follows from (4.3.13), when we choose

$$
\begin{aligned}
N_0 &\geq \left(L'_+ + L''_+\right)\nu_0/\delta - \left(L'_- + L''_-\right)\nu/\delta + \\
&\quad + \left(J'_+ + J''_+\right)\nu_0/\delta - \left(J'_- + J''_-\right)\nu/\delta + N' + N'' - \gamma' - \gamma'' - 1 \\
&= \left(|L'| + |L''| + |J'| + |J''|\right)\left(\nu_0 + \nu\right)/\delta - 1.
\end{aligned}
$$

The proof of (4.3.16) is similar. $\square$

Theorem 4.3.12. *Assume the strong uncertainty principle. Let $a \in \mathrm{Hypo}(M, M_0; \Phi, \Psi)$ satisfying (4.3.3). Then for each $\lambda > 0$ there exists a hypoelliptic symbol*

$$q_\lambda \in \mathrm{Hypo}\left(M_0^{-1}, (M+\lambda)^{-1}; \Phi, \Psi\right),$$

such that

(a) for all $k \in \mathbb{N}$ we have

$$\left|q_\lambda - \frac{1}{a_0 + \lambda}\right|_k \lesssim \frac{h}{|a_0 + \lambda|}, \quad \text{for } \lambda > 0, \tag{4.3.17}$$

where a_0 is defined in (4.3.4);

---

[2] As usual we set $x_- = \min\{x, 0\}$ and $x_+ = \max\{x, 0\}$.

(b) *for all* k, $N \in \mathbb{N}$ *we have*

$$\begin{cases} |1 - q_\lambda \# (a + \lambda)|_k \lesssim h^N, \\ |1 - (a + \lambda) \# q_\lambda|_k \lesssim h^N, \end{cases} \tag{4.3.18}$$

for $\lambda > 0$.

Proof. Define

$$r_{1,\lambda} = 1 - (a_0 + \lambda)^{-1} \# (a + \lambda),$$

$$a_1 = a - a_0. \tag{4.3.19}$$

Because a_1 has compact support, by (4.3.12) and (4.3.8), for all $k \in \mathbb{N}$ we have

$$|a_1|_k (x, \xi) \lesssim (1 + |x| + |\xi|)^{-\nu} h(x, \xi) \lesssim |a_0(x, \xi)| \, h(x, \xi) \lesssim |a_0(x, \xi) + \lambda| \, h(x, \xi), \tag{4.3.20}$$

and

$$|a + \lambda|_k \lesssim |a_0 + \lambda|,$$

for $\lambda > 0$. Then from Lemma 4.3.11 and (4.3.11) we may conclude that for all $k \in \mathbb{N}$ we have

$$|r_{1,\lambda}|_k \lesssim \left| \frac{a_1}{a_0 + \lambda} \right|_k + \left| \mathcal{R}_0 \left((a_0 + \lambda)^{-1}, a + \lambda \right) \right|_k \lesssim h, \tag{4.3.21}$$

for $\lambda > 0$, where $\mathcal{R}_0$ is defined in (4.2.1).

Define $r_{0,\lambda} = 1$ and $r_{j,\lambda} = r_{1,\lambda} \# \cdots \# r_{1,\lambda}$, j times, when $j \geq 1$. Then from Lemma 4.3.11, (4.3.21) and (4.3.11), for all j, $k \in \mathbb{N}$ we have

$$|r_{j,\lambda}|_k \lesssim h^j, \tag{4.3.22}$$

$$\left| r_{j,\lambda} \# (a_0 + \lambda)^{-1} \right|_k \lesssim \frac{h^j}{|a_0 + \lambda|}, \tag{4.3.23}$$

for $\lambda > 0$.

Now we consider the asymptotic sum of the symbols $r_{j,\lambda} \# (a_0 + \lambda)^{-1}$. An application of Proposition 1.1.6 (or better, of its proof[3]) shows that there exists q_λ such that

$$q_\lambda - \sum_{j=0}^{N} r_{j,\lambda} \# (a_0 + \lambda)^{-1} \in S(|a_0 + \lambda|^{-1} h^{N+1}), \tag{4.3.24}$$

for all $N \in \mathbb{N}$, uniformly with respect to $\lambda > 0$.

Estimate (4.3.17) follows from (4.3.24) with $N = 0$, which also gives

$$|q_\lambda|_k \lesssim \frac{1}{|a_0 + \lambda|}, \quad \text{for } \lambda > 0. \tag{4.3.25}$$

---

[3] Although $|a_0 + \lambda|$ need not be temperate, we can apply that result to the space $S(|a_0 + \lambda|^{-1} h^{N+1})$, because in the proof of Proposition 1.1.6 we did not use that hypothesis.

Since $h(x, \xi) \to 0$ as $|x| + |\xi| \to +\infty$, (4.3.17) and (4.3.25) imply that $q_\lambda \in$ Hypo$(M_0^{-1}, (M + \lambda)^{-1})$.

It remains to prove estimates (4.3.18). Setting

$$c_\lambda = q_\lambda - \sum_{j=0}^{N} r_{j,\lambda} \# (a_0 + \lambda)^{-1},$$

by (4.3.24) and Lemma 4.3.11 we have

$$1 - q_\lambda \# (a + \lambda) = 1 - \sum_{j=0}^{N} r_{j,\lambda} \# (a_0 + \lambda)^{-1} \# (a + \lambda) - c_\lambda \# (a + \lambda)$$

$$= 1 - \sum_{j=1}^{N} r_{j,\lambda} \# (1 - r_{1,\lambda}) - c_\lambda \# (a + \lambda)$$

$$= r_{N+1,\lambda} - c_\lambda \# (a + \lambda) \in S(h^{N+1}),$$

uniformly with respect to λ, which implies the first one of the estimates (4.3.18).

In order to prove the second estimate in (4.3.18), we observe that starting from

$$\tilde{r}_{1,\lambda} = 1 - (a + \lambda) \# (a_0 + \lambda)^{-1},$$

and

$$\tilde{r}_{j,\lambda} = \underbrace{\tilde{r}_{1,\lambda} \# \cdots \# \tilde{r}_{1,\lambda}}_{j-\text{times}}, \quad j \geq 1,$$

we may consider

$$\tilde{q}_\lambda \sim \sum_{j=0}^{\infty} (a_0 + \lambda)^{-1} \# \tilde{r}_{j,\lambda} \in S(|a_0 + \lambda|^{-1}),$$

such that for all $k, N \in \mathbb{N}$ we have

$$|1 - (a + \lambda) \# \tilde{q}_\lambda|_k \lesssim h^N, \quad \text{for } \lambda > 0. \tag{4.3.26}$$

Thus, from Lemma 4.3.11 and (4.3.25), for all $k, N \in \mathbb{N}$ we have

$$|q_\lambda - q_\lambda \# (a + \lambda) \# \tilde{q}_\lambda|_k \lesssim \frac{h^N}{|a_0 + \lambda|},$$

whereas from the first estimate in (4.3.18) and (4.3.26) we get

$$|\tilde{q}_\lambda - q_\lambda \# (a + \lambda) \# \tilde{q}_\lambda|_k \lesssim \frac{h^N}{|a_0 + \lambda|},$$

so that

$$|q_\lambda - \tilde{q}_\lambda|_k \lesssim \frac{h^N}{|a_0 + \lambda|}.$$

Hence

$$|1 - (a + \lambda)\#q_\lambda|_k \le |1 - (a + \lambda)\#\tilde{q}_\lambda|_k + |(a + \lambda) \# (q_\lambda - \tilde{q}_\lambda)|_k \lesssim h^N,$$

for $\lambda > 0$. $\qquad\qquad\qquad\qquad\qquad\qquad\qquad\qquad\qquad\qquad\qquad\qquad\square$

Theorem 4.3.13. *Under the hypotheses of Theorem 4.3.6, we have that for all $\lambda > 0$ the operator $a^w + \lambda I : \mathcal{S}(\mathbb{R}^d) \to \mathcal{S}(\mathbb{R}^d)$ is invertible and $a^w (a^w + \lambda I)^{-1} : \mathcal{S}(\mathbb{R}^d) \to \mathcal{S}(\mathbb{R}^d)$ is a hypoelliptic pseudo-differential operator with Weyl symbol*

$$b_\lambda \in \mathrm{Hypo}(1, M_0/(M + \lambda); \Phi, \Psi)$$

such that for all $k \in \mathbb{N}$ we have the estimates

$$\left| b_\lambda - \frac{a_0}{a_0 + \lambda} \right|_k \lesssim \left| \frac{a_0 h}{a_0 + \lambda} \right|, \tag{4.3.27}$$

for $\lambda > 0$.

Proof. We begin by observing that from Corollary 1.3.9 we have the following global regularity property: for all $\lambda > 0$ we have

$$u \in \mathcal{S}'(\mathbb{R}^d) \text{ and } (a^w + \lambda I) u \in \mathcal{S}(\mathbb{R}^d) \implies u \in \mathcal{S}(\mathbb{R}^d),$$

where a^w is here understood as an operator on $\mathcal{S}'(\mathbb{R}^d)$. Now we show that $a^w + \lambda I : \mathcal{S}(\mathbb{R}^d) \to \mathcal{S}(\mathbb{R}^d)$ is invertible for all $\lambda > 0$.

Let A be the unbounded operator in $L^2(\mathbb{R}^d)$ with domain $\mathcal{S}(\mathbb{R}^d)$ defined by $Au = a^w u$, $u \in \mathcal{S}(\mathbb{R}^d)$. Then $a^w + \lambda I$ is one-to-one because, by hypothesis, $\overline{A} + \lambda I$ is. On the other hand we know that the range of $\overline{A} + \lambda I$ is $L^2(\mathbb{R}^d)$. Therefore, given any $f \in \mathcal{S}(\mathbb{R}^d)$ there exists $u \in L^2(\mathbb{R}^d)$ such that $(\overline{A} + \lambda I) u = f$. But $u \in \mathcal{S}(\mathbb{R}^d)$ by the above mentioned global regularity, and therefore $(a^w + \lambda I) u = f$, that is $a^w + \lambda I$ is onto.

Then for all $\lambda > 0$ we may consider $a^w (a^w + \lambda I)^{-1} : \mathcal{S}(\mathbb{R}^d) \to \mathcal{S}(\mathbb{R}^d)$. We want to show that this operator is pseudo-differential. We have

$$a^w (a^w + \lambda I)^{-1} = a^w q_\lambda^w + q_\lambda^w a^w (1 - (a + \lambda) \# q_\lambda)^w$$
$$+ (1 - q_\lambda \# (a + \lambda))^w a^w (a^w + \lambda I)^{-1} (1 - (a + \lambda) \# q_\lambda)^w. \tag{4.3.28}$$

For all $r, s \in \mathbb{R}$, consider the Sobolev spaces

$$H^{r,s} := H(\Phi^r \Psi^s),$$

cf. Definition 1.5.2. We have

$$\mathcal{S}(\mathbb{R}^d) = \bigcap_{r,s} H^{r,s}, \qquad \mathcal{S}'(\mathbb{R}^d) = \bigcup_{r,s} H^{r,s},$$

with the topologies of $S(\mathbb{R}^d)$ and $S'(\mathbb{R}^d)$ equal to the initial and final topology of intersection and union. In particular it follows that an operator is continuous from S' into S if and only if it is continuous from $H^{r,s}$ into $H^{p,q}$ for all $r, s, p, q \in \mathbb{R}$.

From (4.3.18), the strong uncertainty principle and Proposition 1.5.5 (a) we obtain easily that for all $r, s, p, q \in \mathbb{R}$ we have

$$\left\| (1 - q_\lambda \# (a + \lambda))^{\text{w}} u \right\|_{H^{r,s}} \lesssim \| u \|_{H^{p,q}}, \quad \left\| (1 - (a + \lambda) \# q_\lambda)^{\text{w}} u \right\|_{H^{r,s}} \lesssim \| u \|_{H^{p,q}},$$
$$\left\| (1 - q_\lambda \# (a + \lambda))^{\text{w}} a^{\text{w}} u \right\|_{H^{r,s}} \lesssim \| u \|_{H^{p,q}},$$

for all $u \in S(\mathbb{R}^d)$ and $\lambda > 0$. Let

$$S_\lambda = (1 - q_\lambda \# (a + \lambda))^{\text{w}} a^{\text{w}} (a^{\text{w}} + \lambda I)^{-1} (1 - (a + \lambda) \# q_\lambda)^{\text{w}}. \qquad (4.3.29)$$

By hypothesis $\overline{A} + \lambda I$ is non-negative, so we have

$$\sup_{\lambda > 0} \left\| \overline{A} (\overline{A} + \lambda I)^{-1} \right\|_{\mathcal{B}(L^2)} \leq 1 + \sup_{\lambda > 0} \left\| \lambda (\overline{A} + \lambda I)^{-1} \right\|_{\mathcal{B}(L^2)} < \infty.$$

It follows that for all $r, s, p, q \in \mathbb{R}$ we have

$$\| S_\lambda u \|_{H^{r,s}} \lesssim \| u \|_{H^{p,q}} \quad \text{and} \quad \| S_\lambda u \|_{H^{r,s}} \lesssim \frac{1}{\lambda} \| u \|_{H^{p,q}},$$

and therefore

$$\| S_\lambda u \|_{H^{r,s}} \lesssim \min \left\{ 1, \frac{1}{\lambda} \right\} \| u \|_{H^{p,q}} \lesssim (1 + \lambda)^{-1} \| u \|_{H^{p,q}}, \qquad (4.3.30)$$

for all $u \in S(\mathbb{R}^d)$ and $\lambda > 0$.

These estimates imply that S_λ is a regularizing pseudo-differential operator, with Weyl symbol σ_λ satisfying for all $\alpha, \beta \in \mathbb{N}^d$ and $N \in \mathbb{N}$ the estimates

$$\left| D_\xi^\alpha D_x^\beta \sigma_\lambda(x, \xi) \right| \lesssim \frac{(1 + |x| + |\xi|)^{-N}}{1 + \lambda}, \quad \text{for all } (x, \xi) \in \mathbb{R}^{2d} \text{ and } \lambda > 0.$$

From (4.3.28) and (4.3.29) we obtain that $a^{\text{w}} (a^{\text{w}} + \lambda)^{-1}$ is a pseudo-differential operator with Weyl symbol

$$b_\lambda = a \# q_\lambda + q_\lambda \# a - q_\lambda \# a \# (a + \lambda) \# q_\lambda + \sigma_\lambda.$$

Then the result follows from Theorem 4.3.12, (4.3.19), (4.3.20), Lemmata 4.3.10 and 4.3.11. $\qquad \qquad \square$

Corollary 4.3.14. *For all $\lambda > 0$ we have that $(a^{\text{w}} + \lambda I)^{-1}$ is a pseudo-differential operator with hypoelliptic Weyl symbol*

$$\tilde{a}_\lambda = \frac{1}{\lambda} (1 - b_\lambda). \qquad (4.3.31)$$

Proof. We have

$$(a^{\mathrm{w}} + \lambda I)^{-1} = \frac{1}{\lambda}\left(I - a^{\mathrm{w}}\,(a^{\mathrm{w}} + \lambda I)^{-1}\right) = \frac{1}{\lambda}\left(I - b_\lambda^{\mathrm{w}}\right). \qquad \square$$

Remark 4.3.15. Since $a\#\tilde{a}_\lambda = b_\lambda$, for all $k \in \mathbb{N}$ we have the estimates

$$\left|a\#\tilde{a}_\lambda - \frac{a_0}{a_0 + \lambda}\right|_k \lesssim \left|\frac{a_0 h}{a_0 + \lambda}\right|, \tag{4.3.32}$$

for $\lambda > 0$.

4.3.2 Proof of Theorem 4.3.6

In all of this section we assume that the hypotheses of Theorem 4.3.6 are satisfied. Let $\tilde{a}_\lambda$ be the symbol in (4.3.31). We set

$$\tilde{a}_\lambda^{\#0} = 1 \quad \text{and} \quad \tilde{a}_\lambda^{\#k} = \underbrace{\tilde{a}_\lambda\#\cdots\#\tilde{a}_\lambda}_{k\text{-times}}, \qquad k \geq 1.$$

Lemma 4.3.16. *Given any symbol* $q \in S(M; \Phi, \Psi)$, *for all* $(x, \xi) \in \mathbb{R}^{2d}$ *and all* $k \in \mathbb{N}$, *the function*

$$\lambda \in \mathbb{R}_+ \mapsto \left(q\#\tilde{a}_\lambda^{\#k}\right)(x, \xi) \tag{4.3.33}$$

is smooth and

$$\frac{\partial}{\partial\lambda}\left(q\#\tilde{a}_\lambda^{\#k}\right)(x, \xi) = -k\left(q\#\tilde{a}_\lambda^{\#(k+1)}\right)(x, \xi). \tag{4.3.34}$$

Proof. Because $\tilde{a}_\lambda$ is the symbol of the resolvent $(a^{\mathrm{w}} + \lambda I)^{-1}$, it must satisfy the resolvent identity:

$$\tilde{a}_\lambda - \tilde{a}_{\lambda_0} = -(\lambda - \lambda_0)\tilde{a}_\lambda\#\tilde{a}_{\lambda_0},$$

from which, thanks to Theorem 4.3.13, Corollary 4.3.14 and (4.3.8), we obtain for all $k \in \mathbb{N}$ the estimate

$$|\tilde{a}_\lambda - \tilde{a}_{\lambda_0}|_k\,(x, \xi) \lesssim \frac{|(\lambda - \lambda_0)|}{\lambda\lambda_0}, \qquad (x, \xi) \in \mathbb{R}^{2d},\ \lambda > 0,$$

which implies that $\lambda \mapsto (\tilde{a}_\lambda)$ is continuous as a map valued in $S(1; \Phi, \Psi)$.
It follows that

$$\frac{(q\#\tilde{a}_\lambda)(x, \xi) - (q\#\tilde{a}_{\lambda_0})(x, \xi)}{\lambda - \lambda_0} = -(q\#\tilde{a}_\lambda\#\tilde{a}_{\lambda_0})(x, \xi)$$

converges to $-(q\#\tilde{a}_{\lambda_0}\#\tilde{a}_{\lambda_0})(x, \xi)$, as $\lambda \to \lambda_0$. This proves (4.3.34) for $k = 1$. The case corresponding to $k > 1$ follows by induction. Identity (4.3.34) implies also that (4.3.33) is smooth. $\square$

Lemma 4.3.17. *Given $z \in \mathbb{C}_+$ and $(x, \xi) \in \mathbb{R}^{2d}$, we have that the integral*

$$p_{a,z,k}(x, \xi) = \gamma_k(z) \int_0^\infty \lambda^{z-1} (a \# \tilde{a}_\lambda)^{\#k} (x, \xi) \, d\lambda$$

is convergent for all integers $k > \operatorname{Re} z$, where $\gamma_k(z)$ is defined in (4.3.1).
  Moreover, we have

$$p_{a,z,k}(x, \xi) = p_{a,z,k+1}(x, \xi), \quad k > \operatorname{Re} z. \tag{4.3.35}$$

Proof. From (4.3.32) and Lemma 4.3.11 for all k, $l \in \mathbb{N}$ we obtain the estimate:

$$\left| (a \# \tilde{a}_\lambda)^{\#k} \right|_l (x, \xi) \lesssim \left| \frac{a_0(x, \xi)}{a_0(x, \xi) + \lambda} \right|^k, \tag{4.3.36}$$

for all $(x, \xi) \in \mathbb{R}^{2d}$ and $\lambda > 0$. This implies (pointwise) integrability. So we have only to prove (4.3.35).
  Because $\tilde{a}_\lambda$ is the symbol of the resolvent $(a^{\mathrm{w}} + \lambda I)^{-1}$, a^{w} and $\tilde{a}_\lambda^{\mathrm{w}}$ commute: $a \# \tilde{a}_\lambda = \tilde{a}_\lambda \# a$. Therefore, thanks to Lemma 4.3.16, an integration by parts gives

$$\int_0^\infty \lambda^{z-1} (a \# \tilde{a}_\lambda)^{\#k} \, d\lambda = \int_0^\infty \lambda^{z-1} \left(a^{\#k} \# \tilde{a}_\lambda^{\#k} \right) d\lambda$$

$$= \frac{1}{z} \left[\lambda^z \left(a^{\#k} \# \tilde{a}_\lambda^{\#k} \right) \right]_{\lambda=0}^{\lambda=\infty} + \frac{k}{z} \int_0^\infty \lambda^z \left(a^{\#k} \# \tilde{a}_\lambda^{\#(k+1)} \right) d\lambda$$

$$= \frac{k}{z} \int_0^\infty \lambda^{z-1} (a \# \tilde{a}_\lambda)^{\#k} \, d\lambda - \frac{k}{z} \int_0^\infty \lambda^{z-1} (a \# \tilde{a}_\lambda)^{\#(k+1)} \, d\lambda,$$

because

$$\lambda \left(a^{\#k} \# \tilde{a}_\lambda^{\#(k+1)} \right) + \left(a^{\#(k+1)} \# \tilde{a}_\lambda^{\#(k+1)} \right) = (\lambda + a) \# \tilde{a}_\lambda \# \left(a^{\#k} \# \tilde{a}_\lambda^{\#k} \right)$$

$$= (a \# \tilde{a}_\lambda)^{\#k} .$$

It follows that

$$\int_0^\infty \lambda^{z-1} (a \# \tilde{a}_\lambda)^{\#k} \, d\lambda = \frac{k}{k-z} \int_0^\infty \lambda^{z-1} (a \# \tilde{a}_\lambda)^{\#(k+1)} \, d\lambda,$$

which implies (4.3.35) because from (4.3.1) we have $\gamma_k(z)k/(k-z) = \gamma_{k+1}(z)$. $\square$

  Thanks to Lemma 4.3.17, for all $(x, \xi) \in \mathbb{R}^{2d}$ and $z \in \mathbb{C}_+$, we may define

$$a^{\#z}(x, \xi) = p_{a,z,k}(x, \xi),$$

where k is any integer greater than $\operatorname{Re} z$.
  Now we show that for all $l \in \mathbb{N}$ and $z \in \mathbb{C}_+$ we have

$$\left| a^{\#z} - a_0^z \right|_l \lesssim |a_0|^{\operatorname{Re} z} h. \tag{4.3.37}$$

Because $h(x, \xi) \to 0$ as $|x| + |\xi| \to +\infty$, (4.3.37), (4.3.10) and Lemma 1.3.5 imply in particular that

$$a^{\#z} \in S\left(M^{\mathrm{Re}\, z}, M_0^{\mathrm{Re}\, z}\right).$$

From (4.3.32), Lemma 4.3.11 and Lemma 4.3.8 for all k, $l \in \mathbb{N}$ we obtain the estimate

$$\left| (a \# \tilde{a}_\lambda)^{\# k} - \left(\frac{a_0}{a_0 + \lambda} \right)^k \right|_l \lesssim \left(\frac{|a_0|}{|a_0| + \lambda} \right)^k h, \qquad (4.3.38)$$

for $\lambda > 0$.

On the other side, by using the identity (see, e.g., Gradshteyn and Ryzhik [94, 3.194.3, page 285])

$$\int_0^\infty \frac{\lambda^{z-1} w^k}{(w + \lambda)^k}\, d\lambda = \frac{w^z}{\gamma_k(z)},$$

with $k \in \mathbb{N}$, $z \in \mathbb{C}_+$ and $w \in \mathbb{C} \setminus 0$ such that $\mathrm{Re}\, z < k$ and $|\arg w| < \pi$, we obtain

$$a^{\#z} - a_0^z = \gamma_k(z) \int_0^\infty \lambda^{z-1} \left[(a \# \tilde{a}_\lambda)^{\# k} - \left(\frac{a_0}{a_0 + \lambda} \right)^k \right] d\lambda.$$

So from (4.3.38) we get

$$\left| a^{\#z} - a_0^z \right|_l \lesssim |\gamma_k(z)| \int_0^\infty \lambda^{\mathrm{Re}\, z - 1} \left(\frac{|a_0|}{|a_0| + \lambda} \right)^k h\, d\lambda$$

$$= \frac{|\gamma_k(z)|}{\gamma_k(\mathrm{Re}\, z)} |a_0|^{\mathrm{Re}\, z}\, h,$$

which is (4.3.37).

In order to prove (4.3.6) it suffices to verify that

$$\overline{A}^z u = \left(a^{\#z} \right)^{\mathrm{w}} u, \qquad u \in \mathrm{Dom}\left(\overline{A}^{[\mathrm{Re}\, z]+1} \right), \quad z \in \mathbb{C}_+.$$

Indeed, this shows that $\overline{A}^z$ is a closed extension of $\left(a^{\#z} \right)^{\mathrm{w}}|_{\mathcal{S}(\mathbb{R}^d)}$, and also that $\overline{\left(a^{\#z} \right)^{\mathrm{w}}|_{\mathcal{S}(\mathbb{R}^d)}}$, which has domain $\{ u \in L^2(\mathbb{R}^d) : \left(a^{\#z} \right)^{\mathrm{w}} u \in L^2(\mathbb{R}^d) \}$ by Proposition 4.2.1, is a closed extension of $J_{\overline{A}}^z$.

Now, when $u \in \mathrm{Dom}\left(\overline{A}^{[\mathrm{Re}\, z]+1} \right)$ we have

$$\overline{A}^z u = J_{\overline{A}}^z u = \gamma_k(z) \int_0^\infty \lambda^{z-1} \left(\overline{A} \left(\overline{A} + \lambda I \right)^{-1} \right)^k u\, d\lambda,$$

for $k > \mathrm{Re}\, z > 0$, with convergence in $L^2(\mathbb{R}^d)$. On the other hand,

$$\left(\overline{A} \left(\overline{A} + \lambda I \right)^{-1} \right)^k = \left((a \# \tilde{a}_\lambda)^{\# k} \right)^{\mathrm{w}},$$

on $L^2(\mathbb{R}^d)$, so we need only show that

$$\left(a^{\#z}\right)^{\mathrm{w}} u = \gamma_k(z) \int_0^\infty \lambda^{z-1} \left((a\#\tilde{a}_\lambda)^{\#k}\right)^{\mathrm{w}} u\, d\lambda, \qquad (4.3.39)$$

for all $k > \mathrm{Re}\, z > 0$ and all $u \in \mathcal{D}om\left(\overline{A}^{[\mathrm{Re}\, z]+1}\right)$.

We need the following result.

Lemma 4.3.18. *Consider a family of symbols $\phi_\lambda \in S(M)$, $\lambda > 0$, such that the map*

(a) $\lambda \mapsto \phi_\lambda(x,\xi)$ *is continuous on $\mathbb{R}_+$, for all $(x,\xi) \in \mathbb{R}^{2d}$,*

(b) $\lambda \mapsto \|\phi_\lambda\|_{k,S(M)}$ *is integrable on $\mathbb{R}_+$, for all $k \in \mathbb{N}$.*

Then

$$\psi(x,\xi) = \int_0^\infty \phi_\lambda(x,\xi)\, d\lambda$$

exists and belongs to $S(M)$ and

$$\psi^{\mathrm{w}} u = \int_0^\infty \phi_\lambda^{\mathrm{w}} u\, d\lambda, \qquad (4.3.40)$$

for all $u \in L^2(\mathbb{R}^d)$ such that the integral on the right-hand side of (4.3.40) converges in $L^2(\mathbb{R}^d)$ as an improper Riemann integral.

Proof. If we let $k = 0$ in hypothesis (b), we have that $\lambda \mapsto \phi_\lambda(x,\xi)$ is integrable on $\mathbb{R}_+$. By hypothesis (a), $\psi(x,\xi)$ is the pointwise limit of a sequence of Riemann sums $\sum_{j=1}^J \phi_{\lambda_j}(x,\xi)\Delta\lambda_j$. By (b), these Riemann sums are bounded in the symbol space $S(M)$ and therefore it follows from Proposition 1.1.2 that they converge to ψ also in $\mathcal{S}'(\mathbb{R}^{2d})$ and that $\psi \in S(M)$.

Now, the Riemann sums $\sum_{j=1}^J \phi_{\lambda_j}^{\mathrm{w}} u(x)\Delta\lambda_j$ converge by hypothesis in $L^2(\mathbb{R}^d)$, and therefore in $\mathcal{S}'(\mathbb{R}^d)$, to $\int_0^\infty \phi_\lambda^{\mathrm{w}} u(x)\, d\lambda$. Hence it suffices to show that they also converge in $\mathcal{S}'(\mathbb{R}^d)$ to $\psi^{\mathrm{w}} u$. This is clear if $u \in \mathcal{S}(\mathbb{R}^d)$, because of Proposition 1.2.2. The same conclusion actually holds for all $u \in L^2(\mathbb{R}^d)$ by a "3ϵ-argument", since by Proposition 1.5.5 (a) and the Closed Graph Theorem we have, for some $l \in \mathbb{N}$, the estimate

$$\|\phi^{\mathrm{w}} u\|_{H(M^{-1})} \lesssim \|\phi\|_{l,S(M)}\|u\|_{L^2}, \quad \phi \in S(M),\ u \in L^2(\mathbb{R}^d),$$

and the Sobolev space $H(M^{-1})$ is continuously embedded in $\mathcal{S}'(\mathbb{R}^d)$ by Proposition 1.5.4 (a). $\qquad\square$

We now apply Lemma 4.3.18 to the family of symbols

$$\phi_\lambda(x,\xi) := \lambda^{z-1}\, (a\#\tilde{a}_\lambda)^{\#k}\, (x,\xi),$$

which verifies (a) in Lemma 4.3.18 by Lemma 4.3.16. It remains to show that there exists a weight M such that the seminorms

$$\left\| \lambda^{z-1} \left(a \# \tilde{a}_\lambda \right)^{\# k} \right\|_{l, S(M)}$$

are integrable with respect to λ over $\mathbb{R}_+$, for $k > \operatorname{Re} z$ and $l \in \mathbb{N}$. Now, by (4.3.36) and Lemma 4.3.10 we obtain

$$\sup_{\lambda > 0} \; \sup_{(x,\xi) \in \mathbb{R}^{2d}} \frac{\left| (1 + \lambda)^k \left(a \# \tilde{a}_\lambda \right)^{\# k} \right|_l (x, \xi)}{h(x, \xi)^{-(\nu + \nu_0)/\delta}} < \infty, \quad k, l \in \mathbb{N}.$$

Hence, for all $k, l \in \mathbb{N}$ we have

$$\left\| \lambda^{z-1} \left(a \# \tilde{a}_\lambda \right)^{\# k} \right\|_{l, S(h^{-(\nu+\nu_0)/\delta}, g)} \lesssim \frac{\lambda^{\operatorname{Re} z - 1}}{(1 + \lambda)^k}, \quad \lambda > 0,$$

which is integrable for $k > \operatorname{Re} z$.

This concludes the proof of Theorem 4.3.6.

4.4 Hilbert-Schmidt and Trace-Class Operators

This section is devoted to the abstract theory of Hilbert-Schmidt and trace-class operators in a Hilbert space H (of infinite dimension), and to some criteria for a pseudo-differential operator to belong to those classes.

Denote by $\mathcal{B}(H)$ the space of bounded linear operators on a complex Hilbert space H. $\mathcal{B}(H)$ is complete with respect to the operator norm

$$\|T\|_{\mathcal{B}(H)} = \sup_{\|u\|_H = 1} \|Tu\|_H.$$

Since we are dealing in this section with bounded operators, we mention that such an operator T is self-adjoint if and only if it is symmetric: $(Tu, v)_H = (u, Tv)_H$ for all $u, v \in H$.

We now prove that any self-adjoint operator $T \in \mathcal{B}(H)$ which is non-negative, in the sense that $(Tu, u)_H \geq 0$ for all $u \in H$ (cf. Remark 4.3.2), admits a unique square root, i.e., a self-adjoint non-negative operator $S \in \mathcal{B}(H)$ such that $S^2 = T$. To this end, we need the following lemma.

Set, as usual, $\binom{r}{k} = r(r-1)(r-2) \cdots (r-k+1)/k!$ for $r \in \mathbb{R}$, $k \in \mathbb{N}$, $k \neq 0$, and $\binom{r}{0} = 1$ for $r \in \mathbb{R}$ (observe that, in particular, $\binom{r}{k} = 0$ if $r \in \mathbb{N}$, $k > r$).

Lemma 4.4.1. *Given a bounded operator B with norm $\|B\|_{\mathcal{B}(H)} \leq 1$, then for each $r > 0$ the series*

$$B_r = \sum_{k=0}^{\infty} \binom{r}{k} B^k \tag{4.4.1}$$

converges absolutely in $\mathcal{B}(H)$ *and therefore* B_r *commutes with all the operators commuting with* B. *Moreover we have*

$$B_{r+s} = B_r B_s, \quad r, s > 0. \tag{4.4.2}$$

Finally, if B *is self-adjoint, then* B_r *is self-adjoint and non-negative.*

Proof. The absolute convergence of the series is a consequence of the convergence of the numerical series $\sum_{k=0}^{\infty} \left| \binom{r}{k} \right|$. To prove this, observe that $\sum_{k=0}^{\infty} \left| \binom{r}{k} \right| x^n$ coincides, except for the first $[r] + 2$ terms, with the Taylor series of the function $-(1-x)^r$ if $[r]$ is even, or $(1-x)^r$ if $[r]$ is odd ($[r]$ stands for the integer part of r). In any case, the sum of the series $\sum_{k=0}^{\infty} \left| \binom{r}{k} \right| x^k$, which is an increasing function for $0 \leq x < 1$, is therefore bounded there. Hence,

$$\sum_{k=0}^{N} \left| \binom{r}{k} \right| = \lim_{x \to 1^-} \sum_{k=0}^{N} \left| \binom{r}{k} \right| x^k \leq \lim_{x \to 1^-} \sum_{k=0}^{\infty} \left| \binom{r}{k} \right| x^k < \infty.$$

Taking the limit as $N \to \infty$ gives the desired conclusion.

Since the convergence is absolute, equality (4.4.2) follows from the identity

$$\binom{r+s}{n} = \sum_{k=0}^{n} \binom{r}{k} \binom{s}{n-k}, \quad \text{for } n \in \mathbb{N},$$

which in turn can be proved by expanding as binomial series both sides of the equation

$$(1+x)^{r+s} = (1+x)^r (1+x)^s.$$

The remaining part of the statement is clear, because the map $T \to T^*$ is an isometry in $\mathcal{B}(H)$ and $B_r = B_{r/2}^2$. □

Theorem 4.4.2. *Given a non-negative self-adjoint operator* $T \in \mathcal{B}(H)$ *there exists a unique non-negative self-adjoint operator* $S \in \mathcal{B}(H)$ *such that* $S^2 = T$. *Moreover* S *commutes with all operators commuting with* T.

S is called the square root *of* T *and is denoted by* $\sqrt{T}$.

Proof. Let us prove uniqueness. If S_0 is another non-negative self-adjoint operator in $\mathcal{B}(H)$ such that $S_0^2 = T$, then we claim that S and S_0 must commute. Indeed, the operator $i(S_0 S - S S_0)$ is self-adjoint, so it suffices to show that $((S_0 S - S S_0)u, u)_H = 0$ for all $u \in H$. Now since S and S_0 commute with $T = S_0^2 = S^2$, we have

$$((S_0 S - S S_0)u, u)_H = 2T^2 - (S_0 S)^2 - (S S_0)^2 = -((S_0 S - S S_0)u, u)_H,$$

which gives the claim. Hence,

$$(S - S_0)^2 S + (S - S_0)^2 S_0 = (S - S_0)(S^2 - S_0^2) = 0.$$

But $(S - S_0)^2 S$ and $(S - S_0)^2 S_0$ are non-negative. Thus they must vanish, as well as their difference, which is $(S - S_0)^3$. This implies

$$(S - S_0)^4 = (S - S_0)(S - S_0)^3 = 0.$$

But then $(S - S_0)^2 = 0$ and in turn $S - S_0 = 0$ because $S - S_0$ is self-adjoint; in fact,

$$0 = \left((S - S_0)^4 u, u\right)_H = \left((S - S_0)^2 u, (S - S_0)^2 u\right)_H, \quad \text{for } u \in H,$$
$$0 = \left((S - S_0)^2 u, u\right)_H = \left((S - S_0)u, (S - S_0)u\right)_H, \quad \text{for } u \in H.$$

Now we prove the existence of a square root. If $T = 0$ we let $\sqrt{T} = 0$. If $T \neq 0$, set

$$B = \|T\|_{\mathcal{B}(H)}^{-1} T - I.$$

Then, if $\|u\|_H = 1$,

$$(Bu, u)_H = \|T\|_{\mathcal{B}(H)}^{-1}(Tu, u)_H - \|u\|_H^2 \in [-1, 0],$$

which implies $\|B\|_{\mathcal{B}(H)} \leq 1$. Thus we can define

$$\sqrt{T} = \sqrt{\|T\|_{\mathcal{B}(H)}} B_{1/2}.$$

We have

$$(B_{1/2})^2 = B_1 = \sum_{k=0}^{\infty} \binom{1}{k} B^k = I + B,$$

because $\binom{1}{k} = 0$ for $k > 1$. It follows that

$$(\sqrt{T})^2 = \|T\|_{\mathcal{B}(H)}(I + B) = T.$$

The last part of the statement follows from Lemma 4.4.1. □

As a consequence of the existence of the square root of a non-negative self-adjoint operator we can prove the existence and uniqueness of the *polar decomposition* of a bounded linear operator. It relies on two ingredients we are going to introduce. First, given a linear operator $T \in \mathcal{B}(H)$ we define its absolute value as

$$|T| := \sqrt{T^* T}.$$

Notice that $|T|$ is a self-adjoint operator. It is also compact if T is compact. Moreover, for every $u \in H$,

$$\||T|u\|_H^2 = (|T|u, |T|u)_H = (|T|^2 u, u)_H = (T^* T u, u)_H = \|Tu\|_H^2. \tag{4.4.3}$$

The second ingredient is the notion of partial isometry. A *partial isometry* is a linear operator $U \in \mathcal{B}(H)$ such that U is an isometry $(\operatorname{Ker} U)^\perp \to H$, i.e.,

$\|Ux\|_H = \|x\|_H$ for all $u \in (\operatorname{Ker} U)^\perp$. Observe that, by the polarization identity, we have $(Ux, Uy) = (x, y)$ for all $x, y \in (\operatorname{Ker} U)^\perp$. Trivially the same equality holds for $x \in (\operatorname{Ker} U)^\perp$, $y \in \operatorname{Ker} U$, so that we get $U^*U = I$ on $(\operatorname{Ker} U)^\perp$. This implies that U^*U is the orthogonal projection on $(\operatorname{Ker} U)^\perp$ and also that U^* is an isometry $U(H) = (\operatorname{Ker} U^*)^\perp \to H$. Hence U^* is a partial isometry too. Then the result just obtained, applied to U^*, tells us that UU^* is the orthogonal projection on $U(H)$.

When, in addition, $\operatorname{Ker} U = \{0\}$ and U is onto, then we refer to U as a *unitary operator*.

Theorem 4.4.3. *Given an operator $T \in \mathcal{B}(H)$ there exists a partial isometry U and a non-negative self-adjoint operator $S \in \mathcal{B}(H)$ such that*

$$T = US \tag{4.4.4}$$

and

$$\operatorname{Ker} U = \operatorname{Ker} S = \operatorname{Ker} T. \tag{4.4.5}$$

Moreover U and S are uniquely determined by the condition (4.4.4) and $\operatorname{Ker} U = \operatorname{Ker} S$.

In fact we have $S = |T|$. Thus (4.4.4) becomes

$$T = U|T|, \tag{4.4.6}$$

which is called the polar decomposition of T.

Proof. Let $S = |T|$. Let $Uu = 0$, if $u \in \operatorname{Ker} S = S(H)^\perp$. If $u \in S(H)$, let $Uu = Tv$, where v is any solution of $Sv = u$. Because of (4.4.3) we have that $\operatorname{Ker} S = \operatorname{Ker} T$. This shows that if $Sw = Sv$, then $Tw = Tv$, and the definition of U is consistent. Moreover from (4.4.3) we obtain also that U is an isometry from $S(H)$ in H. We can extend U to $\overline{S(H)}$ by continuity. In this way we obtain a partial isometry such that (4.4.4) and (4.4.5) are satisfied.

Let us prove uniqueness. From (4.4.4) and the condition $\operatorname{Ker} U = \operatorname{Ker} S$, therefore $(\operatorname{Ker} U)^\perp = \overline{S(H)}$, we obtain

$$S^2 = SU^*US = (US)^*US = T^*T.$$

Then $S = \sqrt{T^*T} = |T|$ by uniqueness of the square root. Since, as we have already seen, $\operatorname{Ker} S = \operatorname{Ker} T$, we have that U is uniquely determined by (4.4.4) on $S(H)$. So U is uniquely determined on $\overline{S(H)} = (\operatorname{Ker} S)^\perp$ by continuity, and the proof is complete. $\qquad\square$

We also recall the definition of the singular values of a compact operator.

If T is a compact operator, the eigenvalues $\{\mu_j\}_{j=0}^N$, $\mu_j \neq 0$, of $|T|$ are called *singular values* of T. Here it is understood that $N \in \mathbb{N}$ or $N = \infty$. They are the square roots of the non-zero eigenvalues of T^*T. In the sequel we assume that they are arranged in decreasing order and repeated as many times as their multiplicity. An important property is Courant's minimax principle.

Proposition 4.4.4. *μ_j is the minumum of the norms of the restrictions of T to the orthogonal complement of a j-dimensional subspace of H,*

$$\mu_j = \min\{\|T|_{E^\perp}\|_H : \dim E = j\}, \tag{4.4.7}$$

and this minimum is attained by taking for E the eigenspace corresponding to the first j eigenvalues $\mu_0, \mu_1, \ldots, \mu_{j-1}$ of $|T|$.

Proof. The operator T^*T is compact and self-adjoint, so there exists an orthonormal basis of $\mathrm{Ker}(T^*T)^\perp$ given by eigenvectors $\{\psi_k\}_{k=1}^N$ ($N \in \mathbb{N}$ or $N = \infty$) of T^*T, say $T^*T\psi_k = \mu_k^2 \psi_k$. Then, for all $u \in H$,

$$\|Tu\|_H^2 = (T^*Tu, u)_H = \sum_k \mu_k^2 |(u, \psi_k)_H|^2.$$

It is then clear that $\|Tu\|_H \leq \mu_j \|u\|_H$ if u is orthogonal to $\psi_0, \ldots, \psi_{j-1}$, and the equality holds for $u = \psi_j$.

On the other hand, for any j-dimensional subspace E, $E^\perp$ contains a vector $v \in \mathrm{span}\{\psi_0, \ldots, \psi_j\}$, $v \neq 0$, so that $\|Tv\|_H \geq \mu_j \|v\|_H$. $\qquad\square$

Notice that, if T is self-adjoint, the minimax principle gives a variational characterization of the eigenvalues.

Now we study the properties of Hilbert-Schmidt operators.

Definition 4.4.5. A linear operator T on H is a *Hilbert-Schmidt* operator if there exists an orthonormal basis $\{\psi_j\}_{j \in \mathbb{N}}$ of H such that

$$\sum_{j=0}^\infty \|T\psi_j\|_H^2 < \infty. \tag{4.4.8}$$

The set of Hilbert-Schmidt operators is denoted by $\mathcal{B}_2(H)$.

The square root of 4.4.8 is called the *Hilbert-Schmidt* norm of T:

$$\|T\|_{\mathcal{B}_2(H)} = \left\{ \sum_{j=0}^\infty \|T\psi_j\|_H^2 \right\}^{1/2}.$$

This is justified by the following property.

Proposition 4.4.6. *The sum of the series in (4.4.8) is independent of the basis $\{\psi_j\}$.*

Proof. Let $\{\eta_j\}$ be a second orthonormal basis of H. We have

$$\sum_{j=0}^\infty \|T\psi_j\|_H^2 = \sum_{j=0}^\infty \sum_{k=0}^\infty |(T\psi_j, \eta_k)_H|^2 = \sum_{k=0}^\infty \sum_{j=0}^\infty |(\psi_j, T^*\eta_k)_H|^2$$

$$= \sum_{k=0}^\infty \|T^*\eta_k\|_H^2. \tag{4.4.9}$$

An application of this formula with $\psi_j = \eta_j$ gives $\sum_{j=0}^{\infty} \|T\eta_j\|_H^2 = \sum_{k=0}^{\infty} \|T^*\eta_k\|_H^2$ as well, which concludes the proof. $\square$

We now list some properties of Hilbert-Schmidt operators.

Proposition 4.4.7.   (a) $T \in \mathcal{B}_2(H) \iff |T| \in \mathcal{B}_2(H)$ and

$$\|T\|_{\mathcal{B}_2(H)} = \||T|\|_{\mathcal{B}_2(H)}; \tag{4.4.10}$$

(b) $T \in \mathcal{B}_2(H) \iff T^* \in \mathcal{B}_2(H)$ and

$$\|T\|_{\mathcal{B}_2(H)} = \|T^*\|_{\mathcal{B}_2(H)}; \tag{4.4.11}$$

(c) Every $T \in \mathcal{B}_2(H)$ is compact, in particular bounded and

$$\|T\|_{\mathcal{B}(H)} \leq \|T\|_{\mathcal{B}_2(H)}; \tag{4.4.12}$$

(d) For each $S \in \mathcal{B}(H)$ and $T \in \mathcal{B}_2(H)$ we have $ST \in \mathcal{B}_2(H)$, $TS \in \mathcal{B}_2(H)$ and

$$\|ST\|_{\mathcal{B}_2(H)} \leq \|S\|_{\mathcal{B}(H)}\|T\|_{\mathcal{B}_2(H)}, \tag{4.4.13}$$
$$\|TS\|_{\mathcal{B}_2(H)} \leq \|T\|_{\mathcal{B}_2(H)}\|S\|_{\mathcal{B}(H)}. \tag{4.4.14}$$

Proof. Let $\{\psi_j\}$ be an orthonormal basis of H.
   (a) By (4.4.3) we have

$$\|T\|_{\mathcal{B}_2(H)}^2 = \sum_{j=0}^{\infty} \|T\psi_j\|_H^2 = \sum_{j=0}^{\infty} \||T|\psi_j\|_H^2 = \||T|\|_{\mathcal{B}_2(H)}^2.$$

(b) It follows from (4.4.9) with $\psi_j = \eta_j$.
   (c) Given $T \in \mathcal{B}_2(H)$, we have

$$\|Tu\|_H^2 = \sum_{j=0}^{\infty} |(Tu, \psi_j)_H|^2 = \sum_{j=0}^{\infty} |(u, T^*\psi_j)_H|^2$$

$$\leq \|u\|_H^2 \sum_{j=0}^{\infty} \|T^*\psi_j\|_H^2 = \|u\|_H^2 \|T^*\|_{\mathcal{B}_2(H)}^2 = \|u\|_H^2 \|T\|_{\mathcal{B}_2(H)}^2.$$

This proves that $T \in \mathcal{B}(H)$ and (4.4.12). Now for each $k \in \mathbb{N}$, let

$$T_k u = \sum_{j=0}^{k} (u, \psi_j)_H T\psi_j;$$

then T_k is compact because has finite rank. Moreover

$$\|T - T_k\|_{\mathcal{B}_2(H)}^2 = \sum_{j=k+1}^{\infty} \|T\psi_j\|_H^2 \to 0, \quad \text{as } k \to \infty.$$

In particular $T_k \to T$ in $\mathcal{B}(H)$ by (4.4.12) and so T is compact because it is the limit of a sequence of compact operators.

(d) We have

$$\|ST\|_{\mathcal{B}(H)}^2 = \sum_{j=0}^{\infty} \|ST\psi_j\|_H^2 \le \|S\|_{\mathcal{B}(H)}^2 \sum_{j=0}^{\infty} \|T\psi_j\|_H^2 = \|S\|_{\mathcal{B}(H)}^2 \|T\|_{\mathcal{B}_2(H)}^2.$$

Moreover

$$\|TS\|_{\mathcal{B}_2(H)} = \|S^*T^*\|_{\mathcal{B}_2(H)} \le \|S^*\|_{\mathcal{B}(H)} \|T^*\|_{\mathcal{B}_2(H)} = \|S\|_{\mathcal{B}(H)} \|T\|_{\mathcal{B}_2(H)}. \quad \square$$

Proposition 4.4.8. $\mathcal{B}_2(H)$ *is a Hilbert space with respect to the inner product (independent of the basis):*

$$(S,T)_{\mathcal{B}_2(H)} = \sum_{j=0}^{\infty} (S\psi_j, T\psi_j)_H.$$

Proof. It follows from Cauchy-Schwarz' inequality and Definition 4.4.5 that the series which defines $(S,T)_{\mathcal{B}_2(H)}$ converges. The independence of its sum of the choice of the basis is a consequence of Proposition 4.4.6 and the polarization identity. The axioms of inner product are also clearly satisfied.

Let us now prove completeness. Let T_n be a Cauchy sequence in $\mathcal{B}_2(H)$. Then, by (4.4.12), T_n is a Cauchy sequence in $\mathcal{B}(H)$, and hence converges in $\mathcal{B}(H)$, say $T_n \to T$. We claim that $T \in \mathcal{B}_2(H)$ and $T_n \to T$ in $\mathcal{B}_2(H)$. Indeed, observe that for every $\epsilon > 0$ we have

$$\sum_{j=0}^{J} \|(T_n - T_k)\psi_j\|_H^2 \le \epsilon$$

for n, k large enough, and every $J \in \mathbb{N}$. Taking the limits as $k \to \infty$ and then as $J \to \infty$ yields $\|T_n - T\|_{\mathcal{B}_2(H)}^2 \le \epsilon$, which gives the claim. $\quad \square$

We now give a characterization of the Hilbert-Schmidt operators in terms of singular values.

Proposition 4.4.9. $T \in \mathcal{B}_2(H)$ *if and only if it is compact and*

$$\sum_j \mu_j^2 < \infty,$$

where μ_j is the sequence of the singular values of T. Moreover,

$$\|T\|_{\mathcal{B}_2(H)} = \left(\sum_j \mu_j^2 \right)^{1/2}. \tag{4.4.15}$$

Proof. If $T \in \mathcal{B}_2(H)$, then it is compact by Proposition 4.4.7. If T is compact and $\{\psi_j\}$ is an orthonormal basis of H given by the eigenvectors of T^*T, then

$$\|T\|^2_{\mathcal{B}_2(H)} = \sum_{j=0}^{\infty} \|T\psi_j\|^2_H = \sum_{j=0}^{\infty}(T^*T\psi_j, \psi_j)_H = \sum_j \mu_j^2.$$

This completes the proof. $\square$

Now we turn to trace-class operators.

Definition 4.4.10. A linear operator $T \in \mathcal{B}(H)$ is called a *trace-class operator* if $\sqrt{|T|}$ is a Hilbert-Schmidt operator. The set of the trace-class operators is denoted by $\mathcal{B}_1(H)$. For each $T \in \mathcal{B}_1(H)$ we define the *trace of T* as

$$\operatorname{Tr} T = \sum_{j=0}^{\infty}(T\psi_j, \psi_j)_H, \qquad (4.4.16)$$

for any orthonormal basis $\{\psi_j\}$ of H.

The convergence of the series in (4.4.16) and the independence of its sum of the basis follow from Proposition 4.4.8, since one has

$$\operatorname{Tr} T = (\sqrt{|T|}, \sqrt{|T|}U^*)_{\mathcal{B}_2(H)},$$

where $T = U|T|$ is the polar decomposition of T, cf. (4.4.6).

Proposition 4.4.11. $\mathcal{B}_1(H)$ *is a linear space and* $T \in \mathcal{B}_1(H)$ *if and only if T is compact and*

$$\sum_j \mu_j < \infty, \qquad (4.4.17)$$

where μ_j is the sequence of singular values of T. Moreover, if $T \in \mathcal{B}_1(H)$,

$$\operatorname{Tr}|T| = \sum_j \mu_j. \qquad (4.4.18)$$

Finally, if $T \in \mathcal{B}_1(H)$ is self-adjoint with eigenvalues λ_j, then

$$\operatorname{Tr} T = \sum_{j=0}^{\infty}\lambda_j. \qquad (4.4.19)$$

Proof. Let $T = U|T|$ be the polar decomposition of T. By Proposition 4.4.7, $\sqrt{|T|}$ is compact, hence $T = U(\sqrt{|T|})^2$ is compact. The condition $T \in \mathcal{B}_2(H)$ reads $\sum_{j=0}^{\infty} \|\sqrt{|T|}\psi_j\|^2_H < \infty$, which is revealed to be equivalent to (4.4.17) by choosing as ψ_j an orthonormal basis given by eigenfunctions of $|T|$.

The remaining part of the statement is also immediate from (4.4.16), except the fact that $\mathcal{B}_1(H)$ is a linear space. To see this, observe that if E_1 and E_2 are subpaces of dimensions j and k respectively, the space $E_1 + E_2$ is contained in a space of dimension $j + k$. Therefore by (4.4.7) we have

$$\mu_{j+k}(T_1 + T_2) \leq \mu_j(T_1) + \mu_k(T_2)$$

for any couple of compact operators T_1, T_2, and $j, k \in \mathbb{N}$, where $\mu_j(T)$ stands for the j-th singular value of T. Hence by the characterization (4.4.17) we see that if $T_1, T_2 \in \mathcal{B}_1(H)$, then $T_1 + T_2 \in \mathcal{B}_1(H)$. On the other hand, trivially $T \in \mathcal{B}_1(H)$ and $z \in \mathbb{C}$ imply $zT \in \mathcal{B}_1(H)$. $\qquad\square$

Formula (4.4.19) actually holds for all trace-class operators (Lidskii's Theorem), but the proof is trickier.

Proposition 4.4.12. (a) $T \in \mathcal{B}_1(H) \iff |T| \in \mathcal{B}_1(H)$;

(b) $T \in \mathcal{B}_1(H) \iff T^* \in \mathcal{B}_1(H)$ *and*

$$\operatorname{Tr} T^* = \overline{\operatorname{Tr} T}; \tag{4.4.20}$$

(c) *if* $S \in \mathcal{B}(H)$ *and* $T \in \mathcal{B}_1(H)$, *then* $ST, TS \in \mathcal{B}_1(H)$ *and*

$$\operatorname{Tr}(ST) = \operatorname{Tr}(TS). \tag{4.4.21}$$

Proof. (a) It follows immediately from the definition.

(b) Let $T = U|T|$ be the polar decomposition of T. Then

$$|T^*| = U|T|U^*, \tag{4.4.22}$$

because $T^* = |T|U^*$. By using the fact that $U^*U = I$ on the range of $\sqrt{|T|}$, this implies that

$$(U\sqrt{|T|}U^*)^2 = U\sqrt{|T|}U^*U\sqrt{|T|}U^* = U|T|U^*,$$

i.e.,

$$\sqrt{|T^*|} = U\sqrt{|T|}U^*. \tag{4.4.23}$$

Thus by Proposition 4.4.7, $\sqrt{|T^*|} \in \mathcal{B}_2(H)$, that is $T^* \in \mathcal{B}_1(H)$. Because $T^{**} = T$ this shows that $T \in \mathcal{B}_1(H) \iff T^* \in \mathcal{B}_1(H)$. Finally, from (4.4.16) we get at once (4.4.20).

(c) Consider first the case in which S is unitary. Then

$$|ST| = \sqrt{T^*S^*ST} = \sqrt{T^*T} = |T|.$$

Then $ST \in \mathcal{B}_1(H)$. Moreover from (c) we have $T^* \in \mathcal{B}_1(H)$ and $(ST)^* = T^*S^* \in \mathcal{B}_1(H)$. Thus $TS \in \mathcal{B}_1(H)$.

Now let $\{\psi_j\}$ be an orthonormal basis of H. As S is unitary, also $\{S\psi_j\}$ is an orthonormal basis of H. Thanks to (4.4.16) this implies that

$$\operatorname{Tr}(ST) = \sum_{j=0}^{\infty}(ST\psi_j, \psi_j)_H = \sum_{j=0}^{\infty}(STS\psi_j, S\psi_j)_H = \sum_{j=0}^{\infty}(TS\psi_j, \psi_j)_H = \operatorname{Tr}(TS).$$

If S is not unitary, the result follows from the following lemma. $\qquad\square$

Lemma 4.4.13. *Every bounded operator is a linear combination of four unitary operators.*

Proof. If $S \in \mathcal{B}(H)$, then $S = S_1 + iS_2$ with

$$S_1 = \frac{1}{2}(S + S^*), \qquad S_2 = \frac{1}{2i}(S - S^*),$$

and S_1 and S_2 are self-adjoint. On the other hand, if $S \neq 0$ is self-adjoint, let $T = \|S\|_{\mathcal{B}(H)}^{-1} S$. Then $\|T\|_{\mathcal{B}(H)} = 1$ and

$$T = \frac{1}{2}\left(T + i\sqrt{I - T^2}\right) + \frac{1}{2}\left(T - i\sqrt{I - T^2}\right).$$

Because

$$\left(T \pm i\sqrt{I - T^2}\right)^* = T \mp i\sqrt{I - T^2}$$

and

$$\left(T + i\sqrt{I - T^2}\right)\left(T - i\sqrt{I - T^2}\right) = I,$$

$$\left(T - i\sqrt{I - T^2}\right)\left(T + i\sqrt{I - T^2}\right) = I,$$

the proof is complete.   □

Proposition 4.4.14. *If $S, T \in \mathcal{B}_2(H)$, then $ST \in \mathcal{B}_1(H)$ and*

$$\mathrm{Tr}\,(ST) = (T, S^*)_{\mathcal{B}_2(H)}. \tag{4.4.24}$$

Proof. Let $\{\psi_j\}$ be an orthonormal basis of H and let

$$ST = U|ST|$$

be the polar decomposition of ST. Then

$$\left\|\sqrt{|ST|}\right\|_{\mathcal{B}_2(H)}^2 = \sum_{j=0}^{\infty}\left(\sqrt{|ST|}\psi_j, \sqrt{|ST|}\psi_j\right)_H = \sum_{j=0}^{\infty}(|ST|\psi_j, \psi_j)_H$$

$$= \sum_{j=0}^{\infty}(U^*ST\psi_j, \psi_j)_H = \sum_{j=0}^{\infty}(T\psi_j, S^*U\psi_j)_H = (T, S^*U)_{\mathcal{B}_2(H)},$$

because $U^*S \in \mathcal{B}_2(H)$ by Proposition 4.4.7. Thus $ST \in \mathcal{B}_1(H)$. Moreover

$$\mathrm{Tr}\,(ST) = \sum_{j=0}^{\infty}(ST\psi_j, \psi_j)_H = \sum_{j=0}^{\infty}(T\psi_j, S^*\psi_j)_H = (T, S^*)_{\mathcal{B}_2(H)}. \qquad \square$$

Definition 4.4.15. We define the *trace-class norm* of an operator $T \in \mathcal{B}_1(H)$ as

$$\|T\|_{\mathcal{B}_1(H)} = \mathrm{Tr}\,|T|.$$

In Proposition 4.4.17 below we shall prove that in fact the trace-class norm satisfies the norm-axioms.

Proposition 4.4.16. *If $T \in \mathcal{B}_1(H)$, we have $T \in \mathcal{B}_2(H)$ and*

$$\|T\|_{\mathcal{B}_2(H)} \leq \|T\|_{\mathcal{B}_1(H)}. \tag{4.4.25}$$

Moreover

$$|\operatorname{Tr} T| \leq \|T\|_{\mathcal{B}_1(H)}, \tag{4.4.26}$$
$$\|T^*\|_{\mathcal{B}_1(H)} = \|T\|_{\mathcal{B}_1(H)}, \tag{4.4.27}$$

and

$$\|ST\|_{\mathcal{B}_1(H)} \leq \|S\|_{\mathcal{B}(H)}\|T\|_{\mathcal{B}_1(H)}, \tag{4.4.28}$$
$$\|TS\|_{\mathcal{B}_1(H)} \leq \|T\|_{\mathcal{B}_1(H)}\|S\|_{\mathcal{B}(H)}, \tag{4.4.29}$$

for all $S \in \mathcal{B}(H)$.
Finally

$$\|T\|_{\mathcal{B}_1(H)} = \sup\{|\operatorname{Tr}(ST)| : \ S \in \mathcal{B}(H), \|S\|_{\mathcal{B}(H)} = 1\}. \tag{4.4.30}$$

Proof. Let $T = U|T|$ be the polar decomposition of T. As $T = U\left(\sqrt{|T|}\right)^2$ and $\sqrt{|T|} \in \mathcal{B}_2(H)$, by Propositions 4.4.7 and 4.4.14 we have that $T \in \mathcal{B}_2(H)$, $|T|^2 \in \mathcal{B}_1(H)$, and

$$\|T\|_{\mathcal{B}_2(H)}^2 = \left\||T|\right\|_{\mathcal{B}_2(H)}^2 = \operatorname{Tr}\left(|T|^2\right).$$

Let μ_j be the sequence of singular values of $|T|$. Then $|T|^2$ has singular values μ_j^2 and by (4.4.18)

$$\operatorname{Tr}\left(|T|^2\right) = \sum_j \mu_j^2 \leq \left(\sum_j \mu_j\right)^2 = (\operatorname{Tr}|T|)^2 = \|T\|_{\mathcal{B}_1(H)}^2.$$

This proves (4.4.25).

Let $\{\psi_j\}$ be an orthonormal basis of H made of eigenvectors of $|T|$. Then by (4.4.18) we have

$$|\operatorname{Tr} T| = \left|\sum_{j=0}^{\infty} (U|T|\psi_j, \psi_j)_H\right| \leq \sum_j \mu_j = \operatorname{Tr}|T| = \|T\|_{\mathcal{B}_1(H)},$$

because $|(U\psi_j, \psi_j)_H| \leq 1$, for all $j \in \mathbb{N}$.

Moreover, by (4.4.22) and (4.4.24), we get

$$\operatorname{Tr}|T^*| = \operatorname{Tr}(U|T|U^*) = \left(U\sqrt{|T|}, U\sqrt{|T|}\right)_{\mathcal{B}_2(H)} = \left(\sqrt{|T|}, \sqrt{|T|}\right)_{\mathcal{B}_2(H)} = \operatorname{Tr}|T|,$$

from which (4.4.27) follows.

Now, if $S \in B(H)$ and $ST = V|ST|$ is the polar decomposition of ST, we have

$$\text{Tr}\,|ST| = \sum_{j=0}^{\infty} (V^*ST\psi_j, \psi_j)_H = \sum_{j=0}^{\infty} (V^*SU^*|T|\psi_j, \psi_j)_H$$

$$\leq \sum_j \mu_j \|S\|_{B(H)} = \|S\|_{B(H)} \text{Tr}\,|T|.$$

This gives (4.4.28). Formula (4.4.29) follows from (4.4.28) and (4.4.27).
It remains to prove (4.4.30). But, if $\|S\|_{B(H)} = 1$, we have

$$|\text{Tr}\,ST| \leq \|ST\|_{B_1(H)} \leq \|T\|_{B_1(H)}.$$

On the other hand
$$\text{Tr}\,|T| = \text{Tr}\,(U^*T)$$

with $\|U^*\|_{B(H)} = 1$, which completes the proof of (4.4.30). $\square$

Proposition 4.4.17. $B_1(H)$ *is a Banach space with respect to the trace-norm.*

Proof. First we have to prove that the trace-norm satisfies the norm-axioms. Everything is straightforward but the triangular inequality. We have

$$\|T_1 + T_2\|_{B_1(H)} = \sup_{\|S\|_{B(H)}=1} |\text{Tr}\,S(T_1 + T_2)|$$

$$\leq \sup_{\|S\|_{B(H)}=1} \{|\text{Tr}\,(ST_1)| + |\text{Tr}\,(ST_2)|\}$$

$$\leq \sup_{\|S\|_{B(H)}=1} |\text{Tr}\,(ST_1)| + \sup_{\|S\|_{B(H)}=1} |\text{Tr}\,(ST_2)|$$

$$= \|T_1\|_{B_1(H)} + \|T_2\|_{B_1(H)}.$$

Now we prove completeness. Let $T_n \in B_1(H)$ be a Cauchy sequence. By (4.4.25), T_n is a Cauchy sequence in $B_2(H)$, which is complete. Thus T_n converges to T in $B_2(H)$. We have to prove that $T \in B_1(H)$ and that $T_n \to T$ in $B_1(H)$.

To this end, observe first that, if $S \in B_1(H)$ and ψ_j is an orthonormal basis of H,

$$\sum_{j=0}^{\infty} |(S\psi_j, \psi_j)_H| \leq \|S\|_{B_1(H)}. \tag{4.4.31}$$

Indeed, let θ_j be such that $(S\psi_j, \psi_j)_H = e^{i\theta_j}|(S\psi_j, \psi_j)_H|$ and define the unitary operator $V\psi_j = e^{-i\theta_j}\psi_j$. Then we have

$$(VS\psi_j, \psi_j)_H = (S\psi_j, V^*\psi_j)_H = (S\psi_j, e^{i\theta_j}\psi_j)_H$$
$$= e^{-i\theta_j}(S\psi_j, \psi_j)_H = |(S\psi_j, \psi_j)_H|.$$

Hence, by (4.4.26) and (4.4.28) we get

$$\sum_{j=0}^{\infty} |(S\psi_j, \psi_j)_H| = \sum_{j=0}^{\infty} (VS\psi_j, \psi_j)_H = \operatorname{Tr}(VS) \leq \|S\|_{\mathcal{B}_1(H)},$$

as desired.

Now, since T_n is a Cauchy sequence, by (4.4.28) and (4.4.31) for every $\epsilon > 0$ we have

$$\sum_{j=0}^{J} |S(T_n - T_k)\psi_j, \psi_j)_H| \leq \epsilon,$$

for n, k large enough and every $J \in \mathbb{N}$, $S \in \mathcal{B}(H)$, $\|S\|_{\mathcal{B}(H)} = 1$. Taking the limit as $k \to \infty$ and then as $J \to \infty$ yields

$$\sum_{j=0}^{\infty} |(S(T_n - T)\psi_j, \psi_j)_H| \leq \epsilon.$$

If we choose S as the partial isometry satisfying $S(T_n - T) = |T_n - T|$ we get $\|T_n - T\|_{\mathcal{B}_1(H)} \leq \epsilon$ provided n is large enough, and also that $T = (T - T_n) + T_n$ belongs to $\mathcal{B}_1(H)$. $\qquad \square$

Remark 4.4.18. There is a strong analogy between the theory of Hilbert-Schmidt and trace-class operators and that of the functions in Lebesgue spaces $L^2(\mathbb{R}^d)$ and $L^1(\mathbb{R}^d)$ respectively. The trace for the operators corresponds to the integral for functions. In fact one could define, using (4.4.16), the trace for any non-negative self-adjoint operator (allowing that series to diverge); then an operator $T \in \mathcal{B}(H)$ belongs to $\mathcal{B}_r(H)$, $r = 1, 2$ if and only if $\operatorname{Tr} |T|^r < \infty$. The Banach spaces $\mathcal{B}_r(H)$, defined in this way for all $1 \leq r < \infty$, are two sided ideals in $\mathcal{B}(H)$ and consist of compact operators. They are called *Schatten-von Neumann classes*. This analogy is one of the main sources of inspiration in the development of Non-Commutative Geometry.

Now we characterize the Hilbert-Schmidt operators on $L^2(\mathbb{R}^d)$.

Proposition 4.4.19. $T \in \mathcal{B}_2(L^2(\mathbb{R}^d))$ *if and only if there exists (a unique)* $K_T \in L^2(\mathbb{R}^{2d})$ *such that*

$$(Tu)(x) = \int K_T(x, y) u(y) \, dy, \quad \text{for } u \in L^2(\mathbb{R}^d). \tag{4.4.32}$$

Moreover we have

$$\|T\|_{\mathcal{B}_2(L^2(\mathbb{R}^d))} = \|K_T\|_{L^2(\mathbb{R}^{2d})}. \tag{4.4.33}$$

Proof. Let $\{\psi_j\}_{j \in \mathbb{N}}$ be an orthonormal basis of $L^2(\mathbb{R}^d)$. Then

$$\{\psi_j(x)\overline{\psi_k(y)}\}_{(j,k) \in \mathbb{N} \times \mathbb{N}}$$

is an orthonormal basis of $L^2(\mathbb{R}^{2d})$. If $T \in \mathcal{B}_2(L^2(\mathbb{R}^d))$ we let

$$K_T(x, y) = \sum_{j,\,k=0}^{\infty} (T\psi_j, \psi_k)_{L^2} \psi_k(x)\overline{\psi_j(y)}. \qquad (4.4.34)$$

Then

$$\|K_T\|_{L^2}^2 = \sum_{j,k=0}^{\infty} |(T\psi_j, \psi_k)_{L^2}|^2 = \sum_{j=0}^{\infty} \|T\psi_j\|_{L^2}^2 = \|T\|_{\mathcal{B}_2(L^2)},$$

and

$$\int K_T(x, y)u(y)\,dy = \sum_{j,\,k=0}^{\infty} (T\psi_j, \psi_k)_{L^2}(u, \psi_j)_{L^2}\psi_k(x)$$

$$= \sum_{k=0}^{\infty} (Tu, \psi_k)_{L^2}\psi_k(x) = Tu(x).$$

Vice-versa, if $K_T \in L^2(\mathbb{R}^{2d})$, (4.4.32) defines an operator $T \in \mathcal{B}_2(L^2(\mathbb{R}^d))$ satisfying (4.4.34) and so (4.4.33). This completes the proof. $\qquad \square$

It is also easy to characterize the pseudo-differential operators which are Hilbert-Schmidt.

Proposition 4.4.20. *A pseudo-differential operator in $\mathbb{R}^d$ extends to a Hilbert-Schmidt operator on $L^2(\mathbb{R}^d)$ if and only if its Weyl symbol a belongs to $L^2(\mathbb{R}^{2d})$, and*

$$\|a^w\|_{\mathcal{B}_2(L^2(\mathbb{R}^d))} = (2\pi)^{-d/2}\|a\|_{L^2(\mathbb{R}^{2d})}.$$

Proof. The desired result is a consequence of Proposition 1.2.1 with $\tau = \frac{1}{2}$, in particular (1.2.4), and Proposition 4.4.19. $\qquad \square$

Now we examine under which conditions a pseudo-differential operator in $\mathbb{R}^d$ extends to a trace-class operator on $L^2(\mathbb{R}^d)$.

Theorem 4.4.21. *Every regularizing operator in $\mathbb{R}^d$ (i.e., having symbol in $\mathcal{S}(\mathbb{R}^{2d})$) extends to a trace-class operator on $L^2(\mathbb{R}^d)$. More generally, there exists $N \in \mathbb{N}$ depending only on d such that, for every symbol a having distribution partial derivatives $\partial^\alpha a \in L^1(\mathbb{R}^{2d})$, $|\alpha| \leq N$, $A = a^w$ extends to a trace-class operator on $L^2(\mathbb{R}^d)$, with the uniform estimate*

$$\|A\|_{\mathcal{B}_1(L^2)} \lesssim \sum_{|\alpha|\leq N} \int_{\mathbb{R}^{2d}} |\partial^\alpha a(z)|\,dz. \qquad (4.4.35)$$

For such operators we further have

$$\mathrm{Tr}\,A = (2\pi)^{-d} \int_{\mathbb{R}^{2d}} a(z)\,dz. \qquad (4.4.36)$$

Proof. Let A be a regularizing operator with Weyl symbol $a \in \mathcal{S}(\mathbb{R}^{2d})$, and let $P = -\Delta + |x|^2$ be the harmonic oscillator. We know from Proposition 4.4.11 and Theorem 2.2.3 that P^{-k} is trace-class if the integer k is large enough. Moreover, P^k belongs to $\mathrm{OP\Gamma}^{2k}(\mathbb{R}^d)$, see Definition 2.1.2. We can write $A = AP^k P^{-k}$, so that by Proposition 4.4.16,

$$\|A\|_{\mathcal{B}_1(L^2)} \leq \|AP^k\|_{\mathcal{B}(L^2)} \|P^{-k}\|_{\mathcal{B}_1(L^2)}$$

$$\lesssim \sup_{|\alpha| \leq N_0} \sup_{z \in \mathbb{R}^{2d}} (1 + |z|)^{N_0} |\partial^\alpha a(z)| \qquad (4.4.37)$$

for some $N_0 \in \mathbb{N}$, where we also used Theorems 1.4.1 and 1.2.16 (the left symbol is obtained by applying to a the operator $e^{\frac{i}{2} D_x \cdot D_\xi}$, which is continuous on $\mathcal{S}(\mathbb{R}^{2d})$, cf. (1.2.12)).

Now, let a satisfy

$$\operatorname{supp} a \subset B := \{z \in \mathbb{R}^{2d} : |z| \leq 1\}.$$

By (4.4.37) we have

$$\|A\|_{\mathcal{B}_1(L^2)} \lesssim \sup_{|\alpha| \leq N_0} \sup_{z \in \mathbb{R}^{2d}} |\partial^\alpha a(z)|.$$

Because a has compact support, we have

$$\partial^\alpha a(z) = \int_{-\infty}^{z_1} \cdots \int_{-\infty}^{z_{2d}} \partial_{w_1} \cdots \partial_{w_{2d}} \partial_w^\alpha a(w) \, dw_{2d} \cdots dw_1,$$

hence

$$\sup_{|\alpha| \leq N_0} \sup_{z \in \mathbb{R}^{2d}} |\partial^\alpha a(z)| \leq \sup_{|\alpha| \leq N_0 + 2d} \int |\partial^\alpha a(w)| \, dw.$$

This implies that (4.4.35) holds for smooth a supported in B, with $N = N_0 + 2d$.

Let us now observe that if A_0 is the pseudo-differential operator with Weyl symbol

$$a_0(z) = a(z - z_0), \quad z_0 = (x_0, \xi_0),$$

we have

$$A_0 u(x) = \int e^{i(x-y)\xi} a\left(\frac{x+y}{2} - x_0, \xi - \xi_0\right) u(y) \, dy \, d\xi$$

$$= \int e^{i(x-x_0-y')\xi'} a\left(\frac{x - x_0 + y'}{2}, \xi'\right) e^{i(x-x_0-y')\xi_0} u(x_0 + y') \, dy' \, d\xi'$$

$$= \left(U^{-1} A U u\right)(x),$$

where

$$U u(x) = e^{-ix\xi_0} u(x_0 + x).$$

Because U is unitary, we have

$$\|A_0\|_{\mathcal{B}_1(L^2)} = \|U^{-1}AU\|_{\mathcal{B}_1(L^2)} = \|A\|_{\mathcal{B}_1(L^2)}.$$

On the other hand, also

$$\int |\partial^\alpha a(z)|\, dz = \int |\partial^\alpha a(z - z_0)|\, dz = \int |\partial^\alpha a_0(z)|\, dz$$

is invariant, thus we obtain that (4.4.35) holds for every smooth function a which is supported in a ball of radius 1. Now we get rid of this last condition. Let $\{\theta_j(z)\}_{j\in\mathbb{N}}$ be a partition of unity of functions supported in balls of radius 1, and such that each supp θ_j intersects at most a fixed number of supp θ_k, $k \in \mathbb{N}$. We can easily construct $\{\theta_j\}$ in such a way that for each $\alpha \in \mathbb{N}^{2d}$ there exists C_α such that

$$\sup_{z\in\mathbb{R}^{2d}} |\partial^\alpha \theta_j(z)| \le C_\alpha, \quad \text{for } j \in \mathbb{N}. \tag{4.4.38}$$

Let A_j be the operator with Weyl symbol $\theta_j a$. By what we just proved we have

$$\|A\|_{\mathcal{B}_1(L^2)} \le \sum_{j=0}^\infty \|A_j\|_{\mathcal{B}_1(L^2)} \lesssim \sum_{j=0}^\infty \sup_{|\alpha|\le N} \int |\partial^\alpha(\theta_j a)(z)|\, dz.$$

Now, by (4.4.38) and Leibniz' formula, we have

$$|\partial^\alpha(\theta_j a)(z)| \le \sum_{\beta\le\alpha} \binom{\alpha}{\beta} C_\beta |\partial^{\alpha-\beta} a(z)|.$$

Thus

$$\sup_{|\alpha|\le N} \int |\partial^\alpha(\theta_j a)(z)|\, dz \lesssim \sup_{|\alpha|\le N} \int_{\mathrm{supp}\,\theta_j} |\partial^\alpha a(z)|\, dz.$$

Hence

$$\|A\|_{\mathcal{B}_1(H)} \lesssim \sup_{|\alpha|\le N} \int |\partial^\alpha a(z)|\, dz,$$

which proves (4.4.35) for Schwartz symbols. The same estimate for a general symbol a as in the statement follows at once from a limiting argument, which uses (4.4.35) for Schwartz symbols, the completeness of $\mathcal{B}(L^2)$ and Proposition 1.2.2.

Let us now consider the estimate (4.4.36). Again, by using (4.4.26) and a limiting argument, we may prove it just for Schwartz symbols. In terms of the corresponding integral kernel $K(x, y)$ the equality (4.4.36) then reads

$$\mathrm{Tr}\, A = \int_{\mathbb{R}^d} K(x, x)\, dx, \tag{4.4.39}$$

cf. (1.2.2). We may also prove this equality only for finite linear combinations of kernels of the type $\varphi_1(x)\varphi_2(y)$, $\varphi_1, \varphi_2 \in \mathcal{S}(\mathbb{R}^d)$. By linearity, we can further consider just a kernel $K(x, y) = \varphi_1(x)\varphi_2(y)$, $\|\varphi_1\|_{L^2} = 1$, for which it is clear that (4.4.39) holds: take, in the definition of Tr A, an orthonormal basis which contains φ_1 as an element.   □

4.5 Heat Kernel

Let A be a pseudo-differential operator with real-valued, positive, elliptic Weyl symbol $a \in S(M; \Phi, \Psi)$, with $M(x, \xi) \gtrsim (1 + |x| + |\xi|)^{\delta}$, for some $\delta > 0$, cf. Definition 1.3.1. Then, we know from Theorem 4.2.9 that its spectrum consists of a sequence of eigenvalues λ_j, $j \in \mathbb{N}$, diverging to $+\infty$ as $j \to \infty$, with the corresponding eigenfunctions ψ_j defining an orthonormal basis of $L^2(\mathbb{R}^d)$. We can then define, for $t \geq 0$, the operator

$$e^{-tA} f = \sum_{j=0}^{\infty} e^{-t\lambda_j} (f, \psi_j)_{L^2(\mathbb{R}^d)} \psi_j, \quad f \in L^2(\mathbb{R}^d),$$

with unconditional convergence in $L^2(\mathbb{R}^d)$. Hence e^{-tA} turns out to be a bounded operator on $L^2(\mathbb{R}^d)$.

The aim of the section is to show that this operator is in fact a regularizing operator for $t > 0$. The main result is the following one.

Theorem 4.5.1. *Let $a \in S(M; \Phi, \Psi)$ be real-valued, with $a \sim \sum_{j=0}^{\infty} a_j$, where $a_j \in S(Mh^j; \Phi, \Psi)$ are real-valued, and $a_0(x, \xi) \gtrsim M(x, \xi) \gtrsim (1 + |x| + |\xi|)^{\delta}$, for some $\delta > 0$. Let $A = a^{\mathrm{w}}$. Then the operator e^{-tA} is a pseudo-differential operator with Weyl symbol $u(t, x, \xi)$ satisfying, for every $k, l, N \in \mathbb{N}$, $J \geq 1$, $T > 0$, the estimates*

$$\left| t^N \partial_t^l \left(u(t, \cdot) - \sum_{j=0}^{J-1} u_j(t, \cdot) \right) \right|_k \lesssim M^{l-N} h^J, \quad t \in [0, T], \qquad (4.5.1)$$

where $u_0(t, x, \xi) = e^{-ta_0(x, \xi)}$ and, for $j \geq 1$,

$$u_j(t, x, \xi) = e^{-ta_0(x, \xi)} \sum_{l=1}^{2j} t^l u_{l,j}(x, \xi), \quad u_{l,j} \in S(M^l h^j; \Phi, \Psi), \qquad (4.5.2)$$

(see (4.2.4) for the notation $|\cdot|_k$).

Proof. We first look for a symbol $v(t, x, \xi)$ such that

$$\begin{cases} (\partial_t + a^{\mathrm{w}}) v^{\mathrm{w}} = K(t), \\ v^{\mathrm{w}}|_{t=0} = I, \end{cases} \qquad (4.5.3)$$

where $K(t)$ is an operator with smooth kernel $K_t(x, y)$ satisfying $\partial_t^l K_t \in \mathcal{S}(\mathbb{R}^{2d})$ uniformly for t in compact sets of $[0, +\infty)$. Precisely, we look for v in the form $v \sim \sum_{j=0}^{\infty} v_j$, where v_j have the same form as u_j in the statement, i.e., $v_0(t, x, \xi) = e^{-ta_0(x, \xi)}$ and, for $j \geq 1$,

$$v_j(t, x, \xi) = e^{-ta_0(x, \xi)} \sum_{l=1}^{2j} t^l v_{l,j}(x, \xi), \quad v_{l,j} \in S(M^l h^j; \Phi, \Psi). \qquad (4.5.4)$$

The formulas (4.5.3) and (4.2.1) yield the following *transport equations*:

$$\begin{cases} \partial_t v_j + \sum_{s+k+l=j} \frac{\{a_s, v_k\}_l}{(2i)^l l!} = 0, \\ v_0(0, x, \xi) = 1, \\ v_j(0, x, \xi) = 0, \quad j \geq 1. \end{cases}$$

For $j = 0$ we obtain $v_0(t, x, \xi) = e^{-ta_0(x,\xi)}$. It is easy to show by induction that the solutions v_j, $j \geq 1$, have the required form, as well. By using the assumption $a_0 \gtrsim M$, one also verifies from (4.5.4) that

$$|t^N \partial_t^l v_j(t, \cdot)|_k \lesssim M^{l-N} h^j,$$

uniformly for $t \in [0, +\infty)$. Moreover, by arguing as in the proof of Proposition 1.1.6 we can construct $v \sim \sum_{j=0}^{\infty} v_j$ which satisfies, for every $T > 0$,

$$\left| t^N \partial_t^l \left(v(t, \cdot) - \sum_{j=0}^{J-1} v_j(t, \cdot) \right) \right|_k \lesssim M^{l-N} h^J$$

uniformly for $t \in [0, T]$. This construction shows that v^w satisfies (4.5.3), for a suitable operator $K(t)$, regularizing for every $t \geq 0$, with a kernel satisfying the desired estimates.

We now look at the operator e^{-tA}. Since it solves the homogeneous version of (4.5.3), we have

$$v^w f - e^{-tA} f = \int_0^t e^{-(t-s)A} K(s) f \, ds, \quad f \in \mathcal{S}(\mathbb{R}^d).$$

It is therefore sufficient to prove that the operator $v^w - e^{-tA}$ has a symbol in $\mathcal{S}(\mathbb{R}^{2d})$, together with its derivatives with respect to t of any order $N \in \mathbb{N}$, uniformly for t in compact subsets of $[0, +\infty)$. To this end, observe that, since $M(x, \xi) \gtrsim (1 + |x| + |\xi|)^\delta$, we have $\mathcal{S}(\mathbb{R}^d) = \cap_{j \in \mathbb{N}} H(M^j)$, and the norm $\|f\|_{H(M^j)}$ in the Sobolev space $H(M^j)$ can be defined as $\|(A + cI)^j f\|_{L^2}$, c being a large constant so that $A + cI$ is invertible (see Section 1.5). Moreover $t \mapsto e^{-tA}$ is a strongly continuous map of bounded operators on $L^2(\mathbb{R}^d)$, and e^{-tA} commutes with $(A + cI)^j$. Taking into account the properties of the kernel of $K(s)$, this gives the desired estimate for $N = 0$. The estimates for the derivatives follow similarly, for $\frac{d^n}{dt^n} e^{-tA} f = (-A)^n e^{-tA} f$ for every $f \in \mathcal{S}(\mathbb{R}^d)$, $n \in \mathbb{N}$. $\qquad\square$

It follows from Theorem 4.5.1 that the operator e^{-tA} is regularizing for $t > 0$, hence trace-class by Theorem 4.4.21. Its trace is given by the sum of the eigenvalues, cf. (4.4.19), i.e.,

$$\operatorname{Tr} e^{-tA} = \sum_{j=0}^{\infty} e^{-t\lambda_j}.$$

Proposition 4.5.2. *Assume the hypotheses in Theorem 4.5.1, and let* $u(t, x, \xi)$ *be the symbol of* e^{-tA}. *Then, for* $t > 0$,

$$\sum_{j=0}^{\infty} e^{-t\lambda_j} = \int u(t, x, \xi) \, dx \, d\xi. \tag{4.5.5}$$

Proof. The desired result follows at once from Theorem 4.4.21 because $u(t, x, \xi) \in \mathcal{S}(\mathbb{R}^{2d})$ for every fixed $t > 0$. An alternative and direct proof goes as follows. Denote by $K_t(x, y)$ the distribution kernel of e^{-tA}. We have

$$K_t(x, y) = \sum_{j=0}^{\infty} e^{-t\lambda_j} \psi_j(x) \overline{\psi_j(y)},$$

with convergence in $L^2(\mathbb{R}^{2d})$. We have in fact $K_t \in \mathcal{S}(\mathbb{R}^{2d})$ for $t > 0$.
Since $e^{-tA} = e^{-\frac{t}{2}A} e^{-\frac{t}{2}A}$, we can write

$$K_t(x, y) = \int K_{t/2}(x, z) K_{t/2}(z, y) \, dz$$
$$= \int K_{t/2}(x, z) \overline{K_{t/2}(y, z)} \, dz,$$

where we also used that e^{-tA} is self-adjoint. Setting $x = y$ and integrating with respect to x yields

$$\int K_t(x, x) \, dx = \int |K_{t/2}(x, y)|^2 \, dx \, dy.$$

The right-hand side of this equality by Parseval's formula reads

$$\sum_{j=0}^{\infty} e^{-t\lambda_j} \|\psi_j\|_{L^2(\mathbb{R}^d)}^4 = \sum_{j=0}^{\infty} e^{-t\lambda_j}.$$

On the other hand,

$$\int K_t(x, x) \, dx = \int u(t, x, \xi) \, dx \, d\xi$$

by (1.2.2).
  This concludes the proof. $\qquad\square$

4.6   Weyl Asymptotics

We now study the asymptotic distribution of the eigenvalues of some special classes of elliptic self-adjoint operators A satisfying the assumptions of Theorem 4.5.1. Precisely, we consider the classes $\mathrm{OP}\Gamma_\rho^m(\mathbb{R}^d)$, $m > 0$, $0 < \rho \leq 1$, studied in

Chapter 2 (see Definition 2.1.2) and the classes $\mathrm{OPG}^{m,n}_{\mathrm{cl}(\xi,x)}(\mathbb{R}^d)$, $m > 0$, $n > 0$, of Chapter 3 (see Definition 3.2.5).

We use the notation in the previous section. In particular we denote by λ_j the eigenvalues (counted with multiplicity) of A. Moreover, we define the so-called *counting function*

$$N(\lambda) = \#\{j : \lambda_j \le \lambda\}.$$

Theorem 4.6.1. *Let $a \in \Gamma^m_\rho(\mathbb{R}^d)$, $m > 0$, $0 < \rho \le 1$ be real-valued, $a(x,\xi) = a_m(x,\xi) + a_{m-\rho}(x,\xi)$ for $|x| + |\xi|$ large, where $a_m(x,\xi)$ is real-valued and satisfies $0 < a_m(tx,t\xi) = t^m a_m(x,\xi)$, for $t > 0$, $(x,\xi) \in \mathbb{R}^d$, and $a_{m-\rho} \in \Gamma^{m-\rho}_\rho(\mathbb{R}^d)$. Then the trace of the heat kernel of $A = a^w$ has the asymptotic behaviour*

$$\sum_{j=0}^{\infty} e^{-t\lambda_j} \sim C\Gamma\left(1 + \frac{2d}{m}\right) t^{-\frac{2d}{m}}, \quad \text{as } t \to 0^+, \tag{4.6.1}$$

where

$$C = \frac{(2\pi)^{-d}}{2d} \int_{\mathbb{S}^{2d-1}} a_m(\Theta)^{-\frac{2d}{m}} d\Theta. \tag{4.6.2}$$

Proof. Let $\tilde{a}_m(x,\xi)$ be a smooth positive function which coincides with a_m for $|x| + |\xi|$ large. Hence $\tilde{a}_m \in \Gamma^m_1(\mathbb{R}^d)$. Let u and u_j be as in the statement of Theorem 4.5.1 (now $\Phi(x,\xi) = \Psi(x,\xi) = (1 + |x|^2 + |\xi|^2)^{\rho/2}$, $M(x,\xi) = (1 + |x|^2 + |\xi|^2)^{m/2}$, $h(x,\xi) = (1 + |x|^2 + |\xi|^2)^{-\rho}$). By (4.5.5) and Theorem 4.5.1 we have

$$\sum_{j=0}^{\infty} e^{-t\lambda_j} = \int_{\mathbb{R}^{2d}} e^{-t\tilde{a}_m(x,\xi)} \, dx \, d\xi + \sum_{j=1}^{J-1} \int_{\mathbb{R}^{2d}} u_j(t,x,\xi) \, dx \, d\xi$$

$$+ \int_{\mathbb{R}^{2d}} \left(u(t,x,\xi) - \sum_{j=0}^{J} u_j(t,x,\xi) \right) dx \, d\xi. \tag{4.6.3}$$

Using (4.5.1) with $N = l = 0$ and J large enough, the last integral is easily seen to be $O(1)$ as $t \to 0^+$. We now study the first integral. Since a and a_m coincide away from a compact set, we have

$$\int_{\mathbb{R}^{2d}} e^{-t\tilde{a}_m(x,\xi)} \, dx \, d\xi = \int_{\mathbb{R}^{2d}} e^{-t a_m(x,\xi)} \, dx \, d\xi + O(1), \quad \text{as } t \to 0^+.$$

On the other hand, using polar coordinates r, Θ in $\mathbb{R}^{2d}$ and the further change of variable $r \mapsto \sigma = tr^m a_m(\Theta)$ give

$$\int_{\mathbb{R}^{2d}} e^{-t a_m(x,\xi)} \, dx \, d\xi = \frac{(2\pi)^{-d}}{m} \Gamma\left(\frac{2d}{m}\right) t^{-\frac{2d}{m}} \int_{\mathbb{S}^{2d-1}} a_m(\Theta)^{-\frac{2d}{m}} d\Theta$$

$$= \frac{(2\pi)^{-d}}{2d} \Gamma\left(1 + \frac{2d}{m}\right) t^{-\frac{2d}{m}} \int_{\mathbb{S}^{2d-1}} a_m(\Theta)^{-\frac{2d}{m}} d\Theta.$$

Similarly one treats the integrals of u_j, $j = 1, \ldots, J - 1$. Namely, using the expressions (4.5.2), since $\tilde{a}_m(x, \xi) \gtrsim (1 + |x|^2 + |\xi|^2)^{m/2} = M(x, \xi)$, we have

$$\int_{\mathbb{R}^{2d}} |u_j(t, x, \xi)| \, dx \, d\xi \lesssim \sum_{l=1}^{2j} \int e^{-t\tilde{a}_m(x,\xi)} t^l \tilde{a}_m(x, \xi)^l (1 + |x|^2 + |\xi|^2)^{-\rho j} \, dx \, d\xi.$$

We can replace in this formula $\tilde{a}_m$ by a_m, with an error which is $O(1)$. The same changes of variables as above and the Dominated Convergence Theorem then show that the last integral is $o(t^{-\frac{2d}{m}})$, as $t \to 0^+$. This concludes the proof. □

We need the following Tauberian theorem (see, e.g., Taylor [187, Vol. II, Proposition 3.2, page 89] for the proof).

Theorem 4.6.2. *(Karamata's Tauberian Theorem) Let σ be a positive Borel measure on the real semi-axis $(0, \infty)$. Suppose that, for some $\alpha > 0$,*

$$\int_0^{+\infty} e^{-\lambda t} \, d\sigma(\lambda) \sim C t^{-\alpha}, \quad \text{as } t \to 0^+.$$

Then

$$\sigma(\lambda) \sim \frac{C}{\Gamma(\alpha + 1)} \lambda^\alpha, \quad \text{as } \lambda \to +\infty.$$

Theorem 4.6.3. *Under the same hypotheses as in Theorem 4.6.1, the counting function $N(\lambda)$ of the operator $A = a^w$ has the asymptotic behaviour*

$$N(\lambda) \sim C \lambda^{\frac{2d}{m}}, \quad \text{as } \lambda \to +\infty, \tag{4.6.4}$$

where C is given in (4.6.2).

Proof. The desired result follows at once from Theorem 4.6.1 and Theorem 4.6.2 applied to the measure $\sigma(\lambda) = \sigma([0, \lambda]) = N(\lambda)$. Indeed,

$$\sum_{j=0}^{\infty} e^{-t\lambda_j} = \int e^{-t\lambda} \, dN(\lambda). \qquad □$$

Notice that, by using the homogeneity of a_m we can re-write the formula (4.6.4) as

$$N(\lambda) \sim \int_{a_m(x,\xi) \leq \lambda} dx \, d\xi, \quad \text{as } \lambda \to +\infty$$

which is, up to the factor $(2\pi)^{-d}$, the volume of the set $\{(x, \xi) \in \mathbb{R}^{2d} : a_m(x, \xi) \leq \lambda\}$ (one can easily see that the same formula also holds with the whole symbol a in place of a_m). Heuristically this formula states that, on average, every eigenfunction occupies in phase space a set of volume comparable with 1, which agrees with the uncertainty principle in Harmonic Analysis.

We can also deduce the asymptotic behaviour of the eigenvalues themselves.

Proposition 4.6.4. *Under the hypotheses of Theorem* 4.6.3 *we have*

$$\lambda_j \sim C^{-\frac{m}{2d}} j^{\frac{m}{2d}}, \quad \text{as } j \to \infty,$$

where the constant C is given in (4.6.2).

Proof. It follows from (4.6.4) that for every $\epsilon > 0$ there exists λ_0 such that

$$1 - \epsilon \le N(\lambda) C^{-1} \lambda^{-\frac{2d}{m}} \le 1 + \epsilon, \tag{4.6.5}$$

for $\lambda > \lambda_0$. We now choose an integer $j_0 > 0$ such that $\lambda_{j_0} > \lambda_0$ and $\lambda_{j_0+1} > \lambda_{j_0}$. We claim that

$$1 - \epsilon \le j C^{-1} \lambda_j^{-\frac{2d}{m}} \le 1 + \epsilon, \tag{4.6.6}$$

for $j > j_0$. Indeed, for every $j > j_0$ there exist integers j_1 and j_2 such that $j_0 \le j_1 < j \le j_2$ and $\lambda_{j_1} < \lambda_{j_1+1} = \lambda_{j_2} < \lambda_{j_2+1}$. In particular, we have $N(\lambda_{j_1}) = j_1$ and $N(\lambda_{j_2}) = j_2$, so that from (4.6.5) we have

$$1 - \epsilon \le j_1 C^{-1} \lambda_{j_1}^{-\frac{2d}{m}} \le 1 + \epsilon, \tag{4.6.7}$$

and

$$1 - \epsilon \le j_2 C^{-1} \lambda_{j_2}^{-\frac{2d}{m}} \le 1 + \epsilon. \tag{4.6.8}$$

Moreover, $N(\lambda) = j_1$ for $\lambda_{j_1} \le \lambda < \lambda_{j_2}$, so that for these values of λ it turns out that

$$1 - \epsilon \le j_1 C^{-1} \lambda^{-\frac{2d}{m}} \le 1 + \epsilon,$$

and, by continuity,

$$1 - \epsilon \le j_1 C^{-1} \lambda_{j_2}^{-\frac{2d}{m}} \le 1 + \epsilon. \tag{4.6.9}$$

It follows from (4.6.8) and (4.6.9) that

$$1 - \epsilon \le j C^{-1} \lambda_{j_2}^{-\frac{2d}{m}} \le 1 + \epsilon,$$

and it suffices to observe that $\lambda_j = \lambda_{j_2}$. Hence (4.6.6) is proved. As a consequence of that formula we have

$$(1 + \epsilon)^{-\frac{m}{2d}} C^{-\frac{m}{2d}} j^{\frac{m}{2d}} \le \lambda_j \le (1 - \epsilon)^{-\frac{m}{2d}} C^{-\frac{m}{2d}} j^{\frac{m}{2d}},$$

which implies the desired result. $\square$

Example 4.6.5. For the eigenvalues of the harmonic oscillator, with Weyl symbol $a_2(x, \xi) = |\xi|^2 + |x|^2$, we obtain from (4.6.4)

$$N(\lambda) \sim C \lambda^d, \quad \text{as } \lambda \to +\infty,$$

with C given by (4.6.2)

$$C = \frac{(2\pi)^{-d}}{2d} \int_{\mathbb{S}^{2d-1}} d\Theta = \frac{1}{2^d d!},$$

corresponding to (2.2.32) directly computed on the eigenvalues in Chapter 2. Moreover, Proposition 4.6.4 gives for C as before

$$\lambda_j \sim C^{-\frac{1}{d}} j^{\frac{1}{d}}, \quad \text{as } j \to \infty.$$

Finally, we consider operators in the classes $\mathrm{OPG}^{m,n}_{\mathrm{cl}(\xi,x)}(\mathbb{R}^d)$, $m > 0$, $n > 0$. The main result is the following one.

Theorem 4.6.6. *Consider a real-valued symbol $a \in G^{m,n}_{\mathrm{cl}(\xi,x)}(\mathbb{R}^d)$, $m, n > 0$, with $\sigma^m_\psi(a)(x,\xi) > 0$ for all $x \in \mathbb{R}^d$ and $\xi \in \mathbb{R}^d \setminus \{0\}$, $\sigma^n_e(a)(x,\xi) > 0$ for all $x \in \mathbb{R}^d \setminus \{0\}$ and $\xi \in \mathbb{R}^d$, $\sigma^{m,n}_{\psi,e}(a)(x,\xi) > 0$ for all $x \in \mathbb{R}^d \setminus \{0\}$ and $\xi \in \mathbb{R}^d \setminus \{0\}$, cf. (3.2.2), (3.2.4) and (3.2.7). Then the counting function $N(\lambda)$ of the operator $A = a^{\mathrm{w}}$ has the asymptotic behaviour*

$$N(\lambda) \sim \begin{cases} C_m \lambda^{\frac{d}{m}} \log \lambda, & \text{for } m = n, \\ C'_m \lambda^{\frac{d}{m}}, & \text{for } m < n, \\ C''_n \lambda^{\frac{d}{n}}, & \text{for } m > n, \end{cases}$$

where

$$C_m = \frac{(2\pi)^{-d}}{dm} \int_{\mathbb{S}^{d-1}} \int_{\mathbb{S}^{d-1}} \sigma^{m,m}_{\psi,e}(a)^{-\frac{d}{m}} d\theta \, d\theta',$$

$$C'_m = \frac{(2\pi)^{-d}}{d} \int_{\mathbb{R}^d_x} \int_{\mathbb{S}^{d-1}} \sigma^m_\psi(a)^{-\frac{d}{m}} d\theta \, dx,$$

$$C''_n = \frac{(2\pi)^{-d}}{d} \int_{\mathbb{R}^d_\xi} \int_{\mathbb{S}^{d-1}} \sigma^n_e(a)^{-\frac{d}{n}} d\theta \, d\xi.$$

The proof follows the same pattern as that of Theorem 4.6.3. We omit it and refer the interested reader to Maniccia and Panarese [138].

Notes

The results collected in Section 4.1 are standard and can be found in most books of Functional Analysis, see, e.g., Reed and Simon [168, Chapter VIII] for a detailed account.

The characterization in Theorem 4.2.4 of the domain of $\overline{A}$, for a pseudo-differential operator A with hypoelliptic symbol, is a slight generalization of an argument of Hörmander [117] and is extracted from Buzano and Nicola [29]. The proof of Theorem 4.2.9 is inspired by the pattern in Boggiatto, Buzano and Rodino [19].

The brackets $\{\cdot, \cdot\}_j$ in (4.2.2) can be regarded as a type of higher order Poisson brackets, and in fact reduce to them for $j = 1$; they enjoy an invariance property with respect to linear symplectic changes of variables (see, e.g., Buzano and Nicola [30]). Their invariance, up to a certain extent, with respect to non-linear symplectomorphisms is discussed in Mughetti and Nicola [150].

Complex powers of pseudo-differential operators have been studied by several authors, starting from the work of Seeley [179], [181], [180], where the ζ-function $\zeta(z) = \text{Tr}\, A^z$ for boundary value problems was introduced. Generalizations have been then considered, among others, by Kumano-go and Tsutsumi [130], Kumano-go [129], Beals [9], [10], Robert [169], Helffer [109]. Indeed, the study of poles of the zeta function has important applications to index theory, as shown in the celebrated paper by Atiyah, Bott and Patodi [7], and Weyl asymptotics, for which we refer to Duistermaat and Guillemin [75] and also to Shubin [183]. Among other applications, we point out that the study of bounded imaginary powers of pseudo-differential operators also gives information on the maximal regularity for evolution equations, in view of the theorem of Dore and Venni [74]; see for example Coriasco, Schrohe and Seiler [66] and the references therein. Recently attention has been mostly fixed on complex powers of pseudo-differential operators on manifolds with boundary with a given boundary fibration structure; the relevant operators are then elliptic in a calculus which is not, in general, temperate (such as, e.g., Melrose's b-calculus [143]). In this context one is interested in the relationships with the geometric properties of the underlying manifold; we refer, for example, to the contributions by Schrohe [174], [176], Loya [135], [136], Melrose and Nistor [146], and Lauter and Moroianu [131]. In the last two papers the complex powers were also used to define various Wodzicki-type residues as generators of the Hochschild cohomology in dimension 0, cf. the next chapter.

The presentation in Section 4.3 follows the approach in [30], where more general pseudo-differential operators in $\mathbb{R}^d$, with Weyl symbol in Hörmander's classes $S(m,g)$, were considered. In this context complex powers were already treated by Robert [169] for (globally) elliptic symbols diverging at infinity, and consequently with compact resolvent. As already mentioned, here we allow the symbol to tend to zero in some direction and, also, the spectrum is allowed to have zero as an accumulation point. Applications to Schatten von-Neumann properties of pseudo-differential operators were given in [29], [30]; see also Buzano and Toft [32], where more general trace-class pseudo-differential operators are studied.

The material on Hilbert-Schmidt and trace-class operators in Section 4.4 is standard. Classical references are the books of Gohberg and Krein [93] and Schatten [173]. See also Connes [50] for applications in Non-Commutative Geometry and [19] for other applications to the Spectral Theory of pseudo-differential operators.

The construction of the heat semigroup e^{-tA} is classical too, see, e.g., Treves [191] and Taylor [187], but our presentation is also inspired by Maniccia [137], Maniccia and Panarese [138]. See also Buzano [26] for analytic semigroups generated by globally regular operators. Applications of the heat method to index theory can be found in [7]. The remainder term in the Weyl formula (4.6.4) can be estimated by other methods, see Tulovskiĭ and Shubin [193], [183], Hörmander [118], [19], the optimal formula being obtained by an analysis of the wave group

$e^{itA^{2/m}}$; see [109] and also the preceding work by Hörmander [116] for operators on compact manifolds. The results for G-operators in Theorem 4.6.6 appeared in [137], [138]. Estimates for the remainder term were given in Nicola [152]; further improvements were announced by Coriasco and Maniccia (personal communication). For a comprehensive study of Weyl asymptotics in several other contexts we refer the reader to Ivrii [122] and the references therein. The case of the multi-quasi-elliptic operators considered in Chapter 2 is treated in great detail in Boggiatto and Buzano [18] and in [19].

Finally we refer the reader interested in spectral asymptotics for *systems* of operators of Γ-type, generalizing in particular the scalar harmonic oscillator in Section 2.2, to the detailed study by Parmeggiani and Wakayama [160] and Parmeggiani [159].

Chapter 5

Non-Commutative Residue and Dixmier Trace

Summary

Let us recall the following definition, valid for any algebra $\mathcal{A}$ over $\mathbb{C}$.

Definition 5.0.1. The linear map $\tau : \mathcal{A} \to \mathbb{C}$ is called a *trace* if it vanishes on commutators, i.e., if

$$\tau([P,Q]) = \tau(PQ - QP) = 0 \quad \text{for all } P, Q \in \mathcal{A}. \tag{5.0.1}$$

If τ is a trace, then $\lambda\tau$ is a trace for all $\lambda \in \mathbb{C}$. As a basic example, take the algebra $\mathcal{A}$ of $r \times r$ matrices $A = (A_{jk})$ over $\mathbb{C}$; there is a unique non-trivial trace, up to a multiplicative constant, given by

$$\operatorname{Tr} A = \sum_{j=1}^{r} A_{jj}. \tag{5.0.2}$$

We want to study traces on algebras of pseudo-differential operators in $\mathbb{R}^d$. Take first the algebra $\mathcal{R}(\mathbb{R}^d)$ of all regularizing operators, i.e., operators K with kernel in $\mathcal{S}(\mathbb{R}^{2d})$; there is a unique trace functional:

$$\operatorname{Tr} K = \int K(x,x) \, dx, \tag{5.0.3}$$

where we write $K(x,y)$ for the kernel of K. After choosing a basis for $L^2(\mathbb{R}^d)$ given by functions in $\mathcal{S}(\mathbb{R}^d)$, we may regard K as an infinite matrix, and we see that (5.0.3) extends to all the trace-class pseudo-differential operators considered in Theorem 4.4.21; in fact, in terms of the Weyl symbol $a(x,\xi)$ of $A = a^w$ we have

$$\operatorname{Tr} A = \iint a(x,\xi) \, dx \, d\xi. \tag{5.0.4}$$

To fix ideas, consider $a \in \Gamma_{cl}^m(\mathbb{R}^d)$, that is $a(z) \sim \sum_{k=0}^\infty a_{m-k}(z)$, with $z = (x, \xi)$ and $a_{m-k}(z)$ positively homogeneous in z of degree $m - k$, cf. Definition 2.1.3. We assume $m \in \mathbb{Z}$. Then for $m < -2d$ the operator a^w is trace-class, and (5.0.4) makes sense, whereas for $m \geq -2d$ the integral is divergent in general. We would like to define a trace, according to Definition 5.0.1, on the whole algebra

$$\mathrm{OP}\Gamma_{cl}^\infty(\mathbb{R}^d) = \bigcup_{m \in \mathbb{Z}} \mathrm{OP}\Gamma_{cl}^m(\mathbb{R}^d). \qquad (5.0.5)$$

To this end we set for $A \in \mathrm{OP}\Gamma_{cl}^\infty(\mathbb{R}^d)$ with arbitrary order $m \in \mathbb{Z}$:

$$\mathrm{Res}\, A = \int_{\mathbb{S}^{2d-1}} a_{-2d}(\Theta)\, d\Theta, \qquad (5.0.6)$$

where $d\Theta$ is the usual surface measure on $\mathbb{S}^{2d-1}$ and $a_{-2d}(z)$ is the term homogeneous of degree $-2d$ in the asymptotic expansion of the symbol $a(z)$. Following the terminology of Wodzicki, who gave a similar definition for classical pseudo-differential operators on compact manifolds, we shall call $\mathrm{Res}\, A$ in (5.0.6) the *non-commutative residue* of the operator A. In Section 5.1 we shall prove that Res is a trace on the algebra $\mathrm{OP}\Gamma_{cl}^\infty(\mathbb{R}^d)$. This trace vanishes if the order of the operator is less than $-2d$. For one thing this shows that the non-commutative residue is not an extension of the usual trace functional; it also implies that it is zero on the ideal of regularizing operators $\mathcal{R}(\mathbb{R}^d)$ and therefore yields a trace on the quotient algebra $\mathcal{A} = \mathrm{OP}\Gamma_{cl}^\infty(\mathbb{R}^d)/\mathcal{R}(\mathbb{R}^d)$. In fact we shall prove that Res turns out to be the unique trace on $\mathcal{A}$ up to a multiplicative constant.

An interesting property of the non-commutative residue is that it coincides with Dixmier's trace on operators A in $\mathrm{OP}\Gamma_{cl}^{-2d}(\mathbb{R}^d)$. For classical pseudo-differential operators on compact manifolds, this was observed by Connes, in his work on non-commutative geometry.

Let us recall the definition of Dixmier trace on an (infinite dimensional) Hilbert space H. Let T belong to $\mathcal{K}(H)$, the ideal of the compact operators in $\mathcal{B}(H)$. Consider $|T| = (T^*T)^{1/2}$, cf. Theorem 4.4.2, and let $\mu_1(T) \geq \mu_2(T) \geq \cdots$ be the sequence of the eigenvalues of $|T|$, repeated according to their multiplicity. Denoted by

$$\sigma_N(T) = \sum_{n=1}^N \mu_n(T), \qquad (5.0.7)$$

we define

$$\mathcal{L}^{(1,\infty)}(H) = \{T \in \mathcal{K}(H) : \sigma_N(T) = O(\log N)\}, \qquad (5.0.8)$$

endowed with the norm $\|T\|_{1,\infty} = \sup_{N \geq 2} \sigma_N(T)/\log N$. The space $\mathcal{L}^{(1,\infty)}(H)$ is an ideal of $\mathcal{B}(H)$. The following preliminary definition will be sufficient for our practical use.

Definition 5.0.2. Let $T \in \mathcal{L}^{(1,\infty)}(H)$ be a non-negative operator. We define Dixmier's trace as

$$\text{Dixmier} - \text{Tr}\,(T) = \lim_{N \to \infty} \frac{1}{\log N} \sum_{n=1}^{N} \mu_n(T) \qquad (5.0.9)$$

provided the limit exists in $\mathbb{R}$.

To deal with the case when the limit in (5.0.9) does not exist, we consider a linear form ω on $C_b(1,\infty)$, with $\omega \geq 0$, $\omega(1) = 1$ and $\omega(f) = 0$ if $\lim_{x \to +\infty} f(x) = 0$. Given a bounded sequence $a = (a_N)_{N \geq 1}$, we construct the function $f_a = \sum_{N \geq 1} a_N \chi_{[N-1,N)} \in L^\infty(\mathbb{R}_+)$ and define the ω-limit $\lim_{\omega} a_N = \omega(M f_a)$, where, for $g \in L^\infty(\mathbb{R}_+)$, $Mg(t) = \frac{1}{\log t} \int_1^t \frac{g(s)}{s}\,ds$ is the Cesàro mean of g.

In the case of convergent sequences the ω-limit coincides with the usual limit. Hence in general one substitutes (5.0.9) with

$$\text{Tr}_\omega(T) = \lim_{\omega} \frac{1}{\log N} \sum_{n=1}^{N} \mu_n(T), \qquad (5.0.10)$$

depending on ω. We have that Tr_ω is additive on positive operators, so it can be extended to a linear map on $\mathcal{L}^{(1,\infty)}(H)$. Finally, observe that Tr_ω vanishes on trace-class operators and commutators, hence it is a trace in the sense of Definition 5.0.1.

Let us return to pseudo-differential operators. In Section 5.1 we shall prove that any $A \in \text{OP}\Gamma_{\text{cl}}^{-2d}(\mathbb{R}^d)$ belongs to $\mathcal{L}^{(1,\infty)}(L^2(\mathbb{R}^d))$ and

$$\text{Res}\,A = 2d(2\pi)^d \text{Tr}_\omega(A). \qquad (5.0.11)$$

The limit (5.0.9) exists in the present case, hence (5.0.11) is independent of ω.

As an example consider the inverse of the d-th power of the harmonic oscillator

$$H^{-d} = (-\Delta + |x|^2)^{-d}. \qquad (5.0.12)$$

We have $H^{-d} \in \text{OP}\Gamma_{\text{cl}}^{-2d}(\mathbb{R}^d)$ with principal symbol $a_{-2d}(z) = |z|^{-2d}$, hence

$$\text{Res}\,H^{-d} = \int_{\mathbb{S}^{2d-1}} d\Theta = \Omega_{2d},$$

where Ω_{2d} denotes the measure of $\mathbb{S}^{2d-1}$. On the other hand from Weyl's formula, and precisely Proposition 4.6.4, we have for the eigenvalues λ_j of $H^d = (-\Delta + |x|^2)^d$,

$$\lambda_j \sim \frac{j}{C}, \quad j \to \infty,$$

with

$$C = (2\pi)^{-d}(2d)^{-1}\text{Res}\,H^{-d}.$$

Passing then to the eigenvalues $\mu_j = \lambda_j^{-1}$ of H^{-d}, we deduce

$$\mu_j \sim \frac{\operatorname{Res} H^{-d}}{2d(2\pi)^d j}, \quad j \to \infty.$$

Since $\sum_{j=1}^{N} \frac{1}{j} \sim \log N$, from (5.0.10) we actually obtain

$$\operatorname{Res} H^{-d} = \Omega_{2d} = 2d(2\pi)^d \operatorname{Tr}_\omega H^{-d}.$$

Section 5.2 is devoted to similar results for the G-classes considered in Chapter 3. In short: for $A \in \operatorname{OPG}_{\operatorname{cl}(\xi,x)}^{m,n}(\mathbb{R}^d)$, $m \in \mathbb{Z}$, $n \in \mathbb{Z}$, we define

$$\operatorname{Tr}_{\psi,e} A = \int_{\mathbb{S}^{d-1}} \int_{\mathbb{S}^{d-1}} \sigma_{\psi,e}^{-d,-d}(a)(x,\xi) \, d\theta \, d\theta' \qquad (5.0.13)$$

where $\sigma_{\psi,e}^{-d,-d}(a)(x,\xi)$ is the term bi-homogeneous of orders $-d, -d$ in the double asymptotic expansion of the symbol $a(x,\xi)$ of A, cf. (3.2.7). We shall obtain (5.0.13) as coefficient in the Laurent expansion of a bi-holomorphic family of G-operators; the proceeding will provide in a natural way two other trace functionals on sub-algebras of the G-operators. We shall also obtain coincidence with Dixmier's trace, after a slight change in the definition (5.0.10), for $A \in \operatorname{OPG}_{\operatorname{cl}(\xi,x)}^{-d,-d}(\mathbb{R}^d)$. Finally in Section 5.3 we shall discuss Dixmier's traceability of the operators in the general classes of Chapter 1.

5.1   Non-Commutative Residue for Γ-Operators

We write in this section $\mathcal{H}^m(\mathbb{R}^{2d} \setminus \{0\})$, $m \in \mathbb{R}$, for the class of functions $a(z) \in C^\infty(\mathbb{R}^{2d} \setminus \{0\})$, $z = (x,\xi)$, which are positively homogeneous of degree m in $\mathbb{R}^{2d}$, i.e.,

$$a(tz) = t^m a(z), \quad \text{for } t > 0, \ z \in \mathbb{R}^{2d}, \ z \neq 0. \qquad (5.1.1)$$

If $a \in \mathcal{H}^m(\mathbb{R}^{2d} \setminus \{0\})$, then the Euler identity holds:

$$\sum_{j=1}^{2d} z_j \partial_{z_j} a = ma. \qquad (5.1.2)$$

Then recall that the symbol $a(z)$ belongs to $\Gamma_{\operatorname{cl}}^m(\mathbb{R}^d)$, the subspace of the classical symbols in $\Gamma^m(\mathbb{R}^d)$, if it admits an asymptotic expansion

$$a(z) \sim \sum_{k=0}^{\infty} a_{m-k}(z) \qquad (5.1.3)$$

where $a_{m-k} \in \mathcal{H}^{m-k}(\mathbb{R}^{2d} \setminus \{0\})$. In the following we shall assume $m \in \mathbb{Z}$. We recall that we may define elliptic symbols by assuming

$$a_m(z) \neq 0 \quad \text{for } z \in \mathbb{R}^{2d}, \ z \neq 0. \qquad (5.1.4)$$

In the first part of this section we shall consider pseudo-differential operators $A = \mathrm{Op}_0(a) \in \mathrm{OP\Gamma}_{\mathrm{cl}}^m(\mathbb{R}^d)$ in the standard quantization form:

$$Au(x) = \int e^{ix\xi} a(x,\xi) \hat{u}(\xi)\, d\xi \qquad (5.1.5)$$

(the results will be actually independent of the quantization, cf. Remark 5.1.5 below). We write here

$$\Gamma_{\mathrm{cl}}^\infty(\mathbb{R}^d) = \bigcup_{m\in\mathbb{Z}} \Gamma_{\mathrm{cl}}^m(\mathbb{R}^d), \quad \mathrm{OP\Gamma}_{\mathrm{cl}}^\infty(\mathbb{R}^d) = \bigcup_{m\in\mathbb{Z}} \mathrm{OP\Gamma}_{\mathrm{cl}}^m(\mathbb{R}^d). \qquad (5.1.6)$$

Note that $\Gamma_{\mathrm{cl}}^{-\infty}(\mathbb{R}^d) = \cap_{m\in\mathbb{Z}}\Gamma_{\mathrm{cl}}^m(\mathbb{R}^d)$ coincides with $\mathcal{S}(\mathbb{R}^{2d})$ and $\mathrm{OP\Gamma}_{\mathrm{cl}}^{-\infty}(\mathbb{R}^d) = \cap_{m\in\mathbb{Z}}\mathrm{OP\Gamma}_{\mathrm{cl}}^m(\mathbb{R}^d)$ coincides with the class $\mathcal{R}(\mathbb{R}^d)$ of the regularizing operators. Also note that every classical symbol determines its asymptotic expansion in a unique way. Indeed, if $a \sim \sum_{k=0}^\infty a_{m-k} \in \Gamma_{\mathrm{cl}}^m(\mathbb{R}^d)$, we can recover $a_m(z)$ from $\lim_{\lambda\to+\infty} \lambda^{-m} a(\lambda z)$. Take then an excision function $\chi \in C^\infty(\mathbb{R}^{2d})$, with $\chi(z) = 0$ for $|z| \le 1$, $\chi(z) = 1$ for $|z| \ge 2$. By applying the same procedure to $a(z) - \chi(z)a_m(z) \in \Gamma_{\mathrm{cl}}^{m-1}(\mathbb{R}^d)$ we obtain a_{m-1} and so on. Therefore it easily follows that there is an isomorphism of algebras

$$\mathcal{A} = \mathrm{OP\Gamma}_{\mathrm{cl}}^\infty(\mathbb{R}^d)/\mathcal{R}(\mathbb{R}^d) \simeq \bigcup_{m\in\mathbb{Z}} \bigoplus_{k\le m} \mathcal{H}^k(\mathbb{R}^{2d}\setminus\{0\}) \qquad (5.1.7)$$

where on the right the product is induced by the symbol product

$$a \circ b \sim \sum_\alpha (\alpha!)^{-1} \partial_\xi^\alpha a D_x^\beta b. \qquad (5.1.8)$$

The first result of this section is the existence and the uniqueness of a trace on the algebra $\mathcal{A}$ defined in (5.1.7), according to Definition 5.0.1.

Consider the $(2d-1)$-form on $\mathbb{R}^{2d}$ given by

$$\sigma(z) = \sum_{j=1}^{2d} (-1)^{j+1} z_j\, dz_1 \wedge \ldots \wedge \widehat{dz_j} \wedge \ldots \wedge dz_{2n}, \qquad (5.1.9)$$

where $\widehat{dz_j}$ means that dz_j is omitted.

Definition 5.1.1. Let $A = \mathrm{Op}_0(a) \in \mathrm{OP\Gamma}_{\mathrm{cl}}^m(\mathbb{R}^d)$ with $a \sim \sum_{k\ge0} a_{m-k}$, $m \in \mathbb{Z}$, and let $j : \mathbb{S}^{2d-1} \hookrightarrow \mathbb{R}^{2d}$ be the canonical injection. We define the non-commutative residue of A as

$$\mathrm{Res}\, A = \int_{\mathbb{S}^{2d-1}} a_{-2d}(z)\, j^*\sigma, \qquad (5.1.10)$$

where σ is defined in (5.1.9).

Note that $j^*\sigma$, restriction of σ in (5.1.9) to the unit sphere $\mathbb{S}^{2d-1}$ of $\mathbb{R}^{2d}$, coincides with the usual surface measure $d\Theta$, so we recapture the definition of Res A in (5.0.6). However the present form (5.1.10) will be more suitable for the computations which follow.

Clearly Res vanishes for operators of order $m < -2d$, in particular for regularizing operators. This, together with the linearity, implies that it is well defined on $\mathcal{A}$ in (5.1.7).

If $p \in \mathcal{H}^{-2d}(\mathbb{R}^{2d} \setminus \{0\})$, then Euler's identity (5.1.2) implies that the form $p\sigma$ on $\mathbb{R}^{2d} \setminus \{0\}$ is closed. Indeed,

$$d(p\sigma) = (dp) \wedge \sigma + p\,d\sigma = -2dp\,dz_1 \wedge \ldots \wedge dz_{2d} + 2dp\,dz_1 \wedge \ldots \wedge dz_{2d} = 0.$$

Hence in (5.1.10) we may replace $\mathbb{S}^{2d-1}$ with any $(2d-1)$-cycle homologous to $\mathbb{S}^{2d-1}$ in $\mathbb{R}^{2d} \setminus \{0\}$.

Now we establish the main result.

Theorem 5.1.2. *The non-commutative residue* Res *defined in* (5.1.10) *is a trace on the algebra* $\mathcal{A}$ *in* (5.1.7) *and any other trace is a multiple of* Res.

We will need the following lemmata.

Lemma 5.1.3. *Let* $g \in \mathcal{H}^{-2d+1}(\mathbb{R}^{2d} \setminus \{0\})$. *Then* $\int_{\mathbb{S}^{2d-1}} \partial_{z_k} g(z)\, j^*\sigma = 0$, $k = 1, \ldots, 2d$.

Proof. The statement follows by observing that the form $(\partial_{z_k} g)\sigma$ is exact. In fact we shall prove that

$$(\partial_{z_k} g)\sigma = d(g\sigma_k), \tag{5.1.11}$$

where σ_k is given by the contraction between $\frac{\partial}{\partial z_k}$ and σ, i.e.,

$$\sigma_k = \frac{\partial}{\partial z_k} \lrcorner \sigma = \sum_{i=1}^{k-1} (-1)^{i+k+1} z_i\, dz_1 \wedge \ldots \wedge \widehat{dz_i} \wedge \ldots \wedge \widehat{dz_k} \wedge \ldots \wedge dz_{2d}$$

$$+ \sum_{i=k+1}^{2d} (-1)^{i+k} z_i\, dz_1 \wedge \ldots \wedge \widehat{dz_k} \wedge \ldots \wedge \widehat{dz_i} \wedge \ldots \wedge dz_{2d}. \tag{5.1.12}$$

We have

$$d(g\sigma_k) = dg \wedge \sigma_k + g\,d\sigma_k, \tag{5.1.13}$$

and we easily compute

$$d\sigma_k = (-1)^k (2d-1)\, dz_1 \wedge \ldots \wedge \widehat{dz_k} \wedge \ldots \wedge dz_{2d} \tag{5.1.14}$$

and

$$dg \wedge \sigma_k = dg \wedge \left(\frac{\partial}{\partial z_k} \lrcorner \sigma \right)$$

$$= \left(\frac{\partial}{\partial z_k} \lrcorner dg \right) \sigma - \frac{\partial}{\partial z_k} \lrcorner (dg \wedge \sigma)$$

$$= (\partial_{z_k} g)\,\sigma - \frac{\partial}{\partial z_k} \lrcorner \, ((-2d+1)g\,dz_1 \wedge \ldots \wedge dz_{2d})$$

$$= (\partial_{z_k} g)\,\sigma - (-1)^k (2d-1)g\,dz_1 \wedge \ldots \wedge \widehat{dz_k} \wedge \ldots \wedge dz_{2d}. \qquad (5.1.15)$$

Substituting the expressions obtained in (5.1.14) and (5.1.15) in (5.1.13) we get (5.1.11). $\qquad\square$

Lemma 5.1.4. *Let* $f \in \mathcal{H}^m(\mathbb{R}^{2d} \setminus \{0\})$ *and suppose one of the following conditions is satisfied:*

(i) $m \neq -2d$;

(ii) $m = -2d$ *and* $\int_{\mathbb{S}^{2d-1}} f(z)\,j^*\sigma = 0$.

Then there exist functions $h_k \in \mathcal{H}^{m+1}(\mathbb{R}^{2d} \setminus \{0\})$, $k = 1, \ldots, 2d$, *such that* $f(z) = \sum_{k=1}^{2d} \partial_{z_k} h_k(z)$.

Proof. (i) If $m \neq -2d$, then

$$\sum_{k=1}^{2d} \partial_{z_k}(z_k f) = \sum_{k=1}^{2d} z_k \partial_{z_k} f + 2df = (m+2d)f,$$

by Euler's identity (5.1.2).

(ii) By hypothesis, the form $f\sigma$ on $\mathbb{R}^{2d} \setminus \{0\}$ is exact. So also $j^*(f\sigma)$ on $\mathbb{S}^{2d-1}$ is exact. On the other hand it is easy to verify that for every $x \in \mathbb{S}^{2d-1}$ the $(2d-2)$-linear forms $\{j^*(\sigma_k)(x)\}_{k=1}^{2d}$ (where σ_k is defined in (5.1.12)) span the space $\wedge^{2d-2} T_x^* \mathbb{S}^{2d-1}$ (which has dimension $2d-1$); hence we can write

$$j^*(f\sigma) = d\left(\sum_{k=1}^{2d} g_k j^*(\sigma_k) \right) \qquad (5.1.16)$$

for suitable functions $g_k \in C^\infty(\mathbb{S}^{2d-1})$.

Now, switching to polar coordinates (ρ, Θ), $\rho > 0$, $\Theta \in \mathbb{S}^{2d-1}$, (5.1.16) becomes

$$f(z(1,\Theta))j^*(\sigma) = d\left(\sum_{k=1}^{2d} g_k(\Theta)j^*(\sigma_k) \right)$$

or also

$$\left(\rho^{-2d} f(z(1,\Theta)) \right) \rho^{2d} j^*(\sigma) = d\left(\sum_{k=1}^{2d} \left(\rho^{-2d+1} g_k(\Theta) \right) \rho^{2d-1} j^*(\sigma_k) \right),$$

as an equality between forms on $\mathbb{R}^{2d} \setminus \{0\}$.

Since σ and therefore σ_k in polar coordinates do not contain $d\rho$, by homogeneity we have $\rho^{2d} j^*(\sigma) = \sigma$ and $\rho^{2d-1} j^*(\sigma_k) = \sigma_k$. Then, setting

$$h_k(z) := |z|^{-2d+1} g_k(\Theta(z)) \in \mathcal{H}^{-2d+1}(\mathbb{R}^{2d} \setminus \{0\}),$$

we obtain

$$f\sigma = \mathrm{d}\left(\sum_{k=1}^{2d} h_k \sigma_k\right).$$

In view of (5.1.11) we get $f = \sum_{k=1}^{2d} \partial_{z_k} h_k$. $\qquad\qquad\square$

Proof of Theorem 5.1.2. The first assertion is proved if we show that $\mathrm{Res}([A,B])=0$ for $A, B \in \mathcal{A}$. Let $a \in \Gamma_{\mathrm{cl}}^m(\mathbb{R}^d)$, $b \in \Gamma_{\mathrm{cl}}^{m'}(\mathbb{R}^d)$ be the classical symbols of A and B respectively; then the symbol of the commutator has asymptotic expansion, cf. (5.1.8),

$$\sum_\alpha (\alpha!)^{-1}\left(\partial_\xi^\alpha a D_x^\alpha b - \partial_\xi^\alpha b D_x^\alpha a\right), \quad (x,\xi) = z \in \mathbb{R}^{2d}.$$

We may rewrite this expression as $\sum_{j=1}^d \partial_{\xi_j} A_j + \partial_{x_j} B_j$ for suitable asymptotic expansions A_j, B_j. Then the integrals over $\mathbb{S}^{2d-1}$ of $(\partial_{\xi_j} A_j)_{-2d}$ and $(\partial_{x_j} B_j)_{-2d}$ are zero by Lemma 5.1.3.

We now prove uniqueness. Let us suppose τ is another trace on $\mathcal{A}$. Consider $A = \mathrm{Op}_0(a) \in \mathrm{OP}\Gamma_{\mathrm{cl}}^m(\mathbb{R}^d)$, $m \in \mathbb{Z}$, $a \sim \sum_{j\geq0} a_{m-j}$; then the functions $D_{z_k} a$ for $1 \leq k \leq d$ and $-D_{z_k} a$ for $d+1 \leq k \leq 2d$ are the symbols of the operators $[A, \mathrm{Op}_0(z_k)]$, but τ vanishes on commutators, so $\tau\left(\mathrm{Op}_0(\partial_z^\alpha a)\right) = 0$ for all multi-indices $\alpha \neq 0$.

Applying Lemma 5.1.4 (i) to the function $a_{m-j}(z)$ for $m - j \neq -2d$, we can write $a_{m-j}(z) = \sum_{k=1}^{2d} \partial_{z_k} b_{k,m-j}(z)$ for suitable functions $b_{k,m-j}(z) \in \mathcal{H}^{m-j+1}$ $(\mathbb{R}^{2d} \setminus \{0\})$. If we set $b_k(z) \sim \sum_{j\geq0, j\neq m+2d} b_{k,m-j}(z)$, cf. Proposition 1.1.6, then we have

$$a(z) \sim \sum_{k=1}^{2d} \partial_{z_k} b_k(z) + \chi(z) a_{-2d}(z), \qquad (5.1.17)$$

where $\chi \in C_0^\infty(\mathbb{R}^{2d})$, $\chi(z) = 0$ for $|z| \leq 1$, $\chi(z) = 1$ for $|z| \geq 2$.

Set

$$r = \Omega_{2d}^{-1}\int_{\mathbb{S}^{2d-1}} a_{-2d}(z)\, j^*\sigma,$$

where Ω_{2d} is the measure of $\mathbb{S}^{2d-1}$. We may rewrite (5.1.17) as

$$a(z) \sim r\chi(z)|z|^{-2d} + \sum_{k=1}^{2d} \partial_{z_k} b_k(z) + \chi(z)\left(a_{-2d}(z) - r|z|^{-2d}\right). \qquad (5.1.18)$$

Since $a_{-2d}(z) - r|z|^{-2d}$ is homogeneous of degree $-2d$ and

$$\int_{\mathbb{S}^{2d-1}}\left(a_{-2d}(z) - r|z|^{-2d}\right) j^*\sigma = 0,$$

by Lemma 5.1.4 (ii) it is a finite sum of derivatives of homogeneous functions. From (5.1.18) we therefore obtain

$$\tau\left(\mathrm{Op}_0(a)\right) = \tau\left(\mathrm{Op}_0(r\chi(z)|z|^{-2d})\right)$$
$$= \frac{\tau\left(\mathrm{Op}_0(\chi(z)|z|^{-2d})\right)}{\Omega_{2d}} \mathrm{Res}\, A,$$

which proves the theorem. □

Remark 5.1.5. (a) From symbolic calculus, cf. Remark 1.2.6, it turns out that, if $A = \mathrm{Op}_0(a)$ has τ-symbol b_τ, the difference $a - b_\tau$ is a formal series of derivatives and therefore, by Lemma 5.1.3, $\int_{\mathbb{S}^{2d-1}} (a - b_\tau)_{-2d}\, j^*\sigma = 0$. Hence in Definition 5.1.1 we may replace a and its asymptotic expansion by b_τ with the corresponding expansion. In particular, we may use the Weyl symbol.

(b) Theorem 5.1.2 remains valid for pseudo-differential operators with symbols in the classes $\Gamma_{\mathrm{cl}}^m(\mathbb{R}^d, \mathbb{C}^M, \mathbb{C}^M) := \Gamma_{\mathrm{cl}}^m(\mathbb{R}^d) \otimes \mathbb{C}^M \otimes \mathbb{C}^M$. In that case the definition of the non-commutative residue will read

$$\mathrm{Res}\, A = \int_{\mathbb{S}^{2d-1}} \mathrm{Tr}\, a_{-2d}(z)\, j^*\sigma,$$

where Tr is the matrix trace.

(c) Theorem 5.1.2 tells us that Res spans the vector space $(\mathcal{A}/[\mathcal{A}, \mathcal{A}])' \simeq \mathbb{C}$ of all traces on $\mathcal{A}$. More precisely, introduced in $\mathcal{H}^{-2d}(\mathbb{R}^{2d} \setminus \{0\})$ the equivalence relation $\sim$ defined by $p \sim q$ if $(p - q)\sigma$ is exact, we have the following isomorphisms of vector spaces:

$$(\mathcal{A}/[\mathcal{A}, \mathcal{A}])' \to \left(\mathcal{H}^{-2d}(\mathbb{R}^{2d} \setminus \{0\})/\sim\right)' \to \left(H^{2d-1}(\mathbb{R}^{2d} \setminus \{0\})\right)' \simeq \mathbb{C}$$

given by $\lambda \mathrm{Res} \mapsto \lambda \int_{\mathbb{S}^{2d-1}} (\cdot) j^*\sigma \mapsto \lambda \int_{\mathbb{S}^{2d-1}} (\cdot)$.

Let us now show that the non-commutative residue coincides with Dixmier's trace for operators $A \in \mathrm{OP}\Gamma_{\mathrm{cl}}^{-2d}(\mathbb{R}^d)$. Observe that such operators are compact on $L^2(\mathbb{R}^d)$ by Theorem 1.4.2.

Let $\mathcal{L}^{(1,\infty)}(L^2(\mathbb{R}^d))$ be defined according to (5.0.8) and for $T \in \mathcal{L}^{(1,\infty)}$ $(L^2(\mathbb{R}^d))$ define $\mathrm{Tr}_\omega(T)$ as in (5.0.10). We also need to recall some basic facts on the spectrum of operators of positive order (see Chapter 4). Namely, let $A = a^w \in \mathrm{OP}\Gamma_{\mathrm{cl}}^m(\mathbb{R}^d)$, $m > 0$, be elliptic with real Weyl symbol $a \sim \sum_{j \geq 0} a_{m-j}$. Then A has a self-adjoint realization in $L^2(\mathbb{R}^d)$ and its spectrum is an unbounded sequence of real isolated eigenvalues of finite multiplicity. This sequence diverges to $+\infty$ or to $-\infty$. Let $(\lambda_k)_{k \in \mathbb{N}}$ be the sequence of the eigenvalues repeated according to their multiplicity. Modulo a change of sign we can suppose here that it diverges to $+\infty$. By Theorem 4.6.3 we have the following asymptotic formula for the counting function $N(\lambda)$:

$$N(\lambda) \sim C\lambda^{\frac{2d}{m}} \quad \text{as } \lambda \to +\infty, \tag{5.1.19}$$

where

$$C = \frac{(2\pi)^{-d}}{2d} \int_{\mathbb{S}^{2d-1}} a_m(z)^{-\frac{2d}{m}} j^* \sigma, \qquad (5.1.20)$$

whereas for the eigenvalues we have

$$\lambda_k \sim \left(\frac{k}{C}\right)^{\frac{m}{2d}} \qquad \text{as } k \to \infty, \qquad (5.1.21)$$

cf. Proposition 4.6.4.

Theorem 5.1.6. *Let* $A \in \mathrm{OP\Gamma}_{\mathrm{cl}}^{-2d}(\mathbb{R}^d)$, *regarded as a compact operator on* $L^2(\mathbb{R}^d)$. *Then* $A \in \mathcal{L}^{(1,\infty)}(L^2(\mathbb{R}^d))$ *and*

$$\mathrm{Res}\, A = 2d(2\pi)^d \mathrm{Tr}_\omega(A), \qquad (5.1.22)$$

independently of ω.

Proof. We first verify the statement of the theorem when $A \in \mathrm{OP\Gamma}_{\mathrm{cl}}^{-2d}(\mathbb{R}^d)$ is non-negative as an operator on $L^2(\mathbb{R}^d)$, injective on $\mathcal{S}(\mathbb{R}^d)$ and has a real elliptic Weyl symbol. Under these assumptions it follows that $A = B^{-1}$ for an operator $B = b^{\mathrm{w}} \in \mathrm{OP\Gamma}_{\mathrm{cl}}^{2d}(\mathbb{R}^d)$ with real Weyl symbol $b \sim \sum_{j \geq 0} b_{2d-j}$, semibounded from below, cf. Remark 1.7.13 and Theorem 4.2.9. Note that $b_{2d}(z) = a_{-2d}(z)^{-1}$. From (5.1.21), (5.1.20) we have the following asymptotic behaviour for the eigenvalues λ_k of B:

$$\lambda_k \sim \frac{k}{C} = \frac{2d(2\pi)^d k}{\int_{\mathbb{S}^{2d-1}} b_{2d}(z)^{-1} j^* \sigma} = 2d(2\pi)^d (\mathrm{Res}\, A)^{-1} k \quad \text{as } k \to \infty.$$

This gives for the eigenvalues of A, that are λ_k^{-1}, the formula

$$\lambda_k^{-1} \sim \frac{\mathrm{Res}\, A}{2d(2\pi)^d} k^{-1} \quad \text{as } k \to \infty.$$

Then

$$\frac{\sigma_N(A)}{\log N} \sim \frac{\mathrm{Res}\, A}{2d(2\pi)^d} \quad \text{as } N \to \infty,$$

and (5.1.22) holds for A.

Now, if we fix an operator B as above, given any $A \in \mathrm{OP\Gamma}_{\mathrm{cl}}^{-2d}(\mathbb{R}^d)$ we can write $A = (AB)B^{-1}$, with $AB \in \mathrm{OP\Gamma}_{\mathrm{cl}}^0(\mathbb{R}^d) \subset \mathcal{B}(L^2(\mathbb{R}^d))$, cf. Theorem 1.4.1, and therefore $A \in \mathcal{L}^{(1,\infty)}(L^2(\mathbb{R}^d))$, since $\mathcal{L}^{(1,\infty)}(L^2(\mathbb{R}^d))$ is an ideal of $\mathcal{B}(L^2(\mathbb{R}^d))$.

To prove (5.1.22) in the general case, observe that by linearity it suffices to prove the result for operators with real Weyl symbol. Now if $A = a^{\mathrm{w}}$ is such an operator we may write $a = (a + Cq) - Cq$ where $q(x, \xi) = (1 + |x|^2 + |\xi|^2)^{-d}$ and $C = -\inf a/q + 1$. So we can assume that A has an elliptic real Weyl symbol. By arguing as in the proof of Lemma 4.2.8 we can further suppose that A is non-negative, modulo operators in $\mathrm{OP\Gamma}_{\mathrm{cl}}^{-2d-1}(\mathbb{R}^d)$, which are trace-class by Theorem

4.4.21 and on which both the traces Res and Tr_ω vanish. Finally, as by Fredholm theory and global regularity $V = \operatorname{Ker} A$ is a finite dimensional subspace of $\mathcal{S}(\mathbb{R}^d)$, cf. Theorem 2.1.14, the orthogonal projection P_V on V is regularizing. Now $A = (A + P_V) - P_V$ and $A + P_V$ falls in the class of operators considered at the beginning of the present proof.

Hence Theorem 5.1.6 is proved. $\qquad\square$

5.2 Trace Functionals for G-Operators

In this section we consider classical pseudo-differential operators of type G, namely $A \in OPG^{m,n}_{\mathrm{cl}(\xi,x)}(\mathbb{R}^d)$, $m \in \mathbb{Z}$, $n \in \mathbb{Z}$. We recall from Section 3.2 that the corresponding class of symbols $G^{m,n}_{\mathrm{cl}(\xi,x)}(\mathbb{R}^d)$ consists of all $a(x,\xi)$ which belong to $G^{m,n}(\mathbb{R}^d)$, i.e., satisfying for every $\alpha \in \mathbb{N}^d$, $\beta \in \mathbb{N}^d$ the estimates

$$|\partial_\xi^\alpha \partial_x^\beta a(x,\xi)| \lesssim \langle \xi \rangle^{m-|\alpha|} \langle x \rangle^{n-|\beta|} \quad \text{for all } x \in \mathbb{R}^d,\ \xi \in \mathbb{R}^d, \tag{5.2.1}$$

and admit a double asymptotic expansion

$$a(x,\xi) \sim \sum_{j=0}^{\infty} \sigma_\psi^{m-j}(a)(x,\xi), \tag{5.2.2}$$

$$a(x,\xi) \sim \sum_{k=0}^{\infty} \sigma_e^{n-k}(a)(x,\xi), \tag{5.2.3}$$

where $\sigma_\psi^{m-j}(a)(x,\xi)$ is homogeneous of degree $m-j$ with respect to $\xi \in \mathbb{R}^d \setminus \{0\}$ and $\sigma_e^{n-k}(a)(x,\xi)$ is homogeneous of degree $n-k$ with respect to $x \in \mathbb{R}^d \setminus \{0\}$; moreover:

$$\sigma_\psi^{m-j}(a)(x,\xi) \sim \sum_{k=0}^{\infty} \sigma_{\psi,e}^{m-j,n-k}(a)(x,\xi), \tag{5.2.4}$$

$$\sigma_e^{n-k}(a)(x,\xi) \sim \sum_{j=0}^{\infty} \sigma_{\psi,e}^{m-j,n-k}(a)(x,\xi), \tag{5.2.5}$$

where $\sigma_{\psi,e}^{m-j,n-k}(a)(x,\xi)$ are the same in (5.2.4), (5.2.5), homogeneous separately with respect to ξ of degree $m-j$, with respect to x of degree $n-k$.

We finally recall that elliptic symbols $a(x,\xi)$ can be characterized by imposing simultaneously $\sigma_\psi^m(a)(x,\xi) \neq 0$ for all $x \in \mathbb{R}^d$, $\xi \in \mathbb{R}^d \setminus \{0\}$; $\sigma_e^n(a)(x,\xi)$ for all $x \in \mathbb{R}^d \setminus \{0\}$, $\xi \in \mathbb{R}^d$; $\sigma_{\psi,e}^{m,n}(a)(x,\xi) \neq 0$ for all $x \in \mathbb{R}^d \setminus \{0\}$, $\xi \in \mathbb{R}^d \setminus \{0\}$. For other details, in particular concerning the symbolic calculus, we refer to Chapter 3. In the following we shall adopt Weyl quantization, denoting $A = a^w$ the Weyl operator with symbol a. We shall write for short $G^{m,n}_{\text{cl}(\xi,x)}$, $\text{OPG}^{m,n}_{\text{cl}(\xi,x)}$ instead of $G^{m,n}_{\text{cl}(\xi,x)}(\mathbb{R}^d)$ and $\text{OPG}^{m,n}_{\text{cl}(\xi,x)}(\mathbb{R}^d)$. Observe also that

$$\text{OPG}^{-\infty,-\infty}_{\text{cl}(\xi,x)} := \bigcap_{m\in\mathbb{Z}} \bigcap_{n\in\mathbb{Z}} \text{OPG}^{m,n}_{\text{cl}(\xi,x)}$$

coincides with the class $\mathcal{R}(\mathbb{R}^d)$ of the regularizing operators.

Let us now define the following operator algebras.

Definition 5.2.1. Let $\mathcal{A} = \cup_{n\in\mathbb{Z}}\cup_{m\in\mathbb{Z}} \text{OPG}^{m,n}_{\text{cl}(\xi,x)}/\mathcal{R}$. We define the two-sided ideals of $\mathcal{A}$

$$\mathcal{I}_\psi = \bigcup_{m\in\mathbb{Z}} \bigcap_{n\in\mathbb{Z}} \text{OPG}^{m,n}_{\text{cl}(\xi,x)}/\mathcal{R}, \qquad \mathcal{I}_e = \bigcup_{n\in\mathbb{Z}} \bigcap_{m\in\mathbb{Z}} \text{OPG}^{m,n}_{\text{cl}(\xi,x)}/\mathcal{R},$$

and the quotient algebras

$$\mathcal{A}_\psi = \mathcal{A}/\mathcal{I}_e, \qquad \mathcal{A}_e = \mathcal{A}/\mathcal{I}_\psi, \qquad \mathcal{A}_{\psi,e} = \mathcal{A}/(\mathcal{I}_\psi + \mathcal{I}_e).$$

The first result of this section is the explicit construction of trace functionals for each of the algebras in Definition 5.2.1. These traces come from residues of the trace of holomorphic operator families. As we observed in the summary, on the ideal of regularizing operators every trace is a multiple of the functional

$$\text{Tr}(a^w) = \iint a(x,\xi)\, dx\, d\xi, \tag{5.2.6}$$

i.e., the usual operator trace. That formula still holds for any $a \in G^{m,n}_{\text{cl}(\xi,x)}$ provided $m < -d, n < -d$, see Theorem 4.4.21. In order to extend it further, we need to regularize the resultant divergent integral, and we do this by means of holomorphic families of symbols. Namely, let $a(x,\xi) \in G^{m,n}_{\text{cl}(\xi,x)}$, and consider the family of symbols

$$\tilde{a}(\tau,z) = \tilde{a}(\tau,z;x,\xi) = [x]^\tau[\xi]^z a(x,\xi) \in G^{m+\text{Re}\,z,\,n+\text{Re}\,\tau}_{\text{cl}(\xi,x)}, \quad \tau, z \in \mathbb{C}, \tag{5.2.7}$$

where $[\cdot]$ denotes an arbitrary strictly positive C^∞ function on $\mathbb{R}^d$ with $[y] = |y|$ for $|y| \geq 1$. Notice that $\tilde{a}(0,0) = a$.

Lemma 5.2.2. Let $a \in G^{m,n}_{\text{cl}(\xi,x)}$ and let $\tilde{a}$ be as in (5.2.7). Then the function $t(\tau,z) := \text{Tr}(\tilde{a}(\tau,z)^w)$ is defined and holomorphic for $\text{Re}\,z < -m - d$, $\text{Re}\,\tau < -n - d$, and extends to a meromorphic function of τ, z with at most simple poles on the lines $z = -m - d + j$, $\tau = -n - d + k$, $j, k \in \mathbb{N}$.

Proof. By Theorem 4.4.21 we have

$$t(\tau,z) = \iint \tilde{a}(\tau,z;x,\xi)\, dx\, d\xi = \iint [x]^\tau[\xi]^z a(x,\xi)\, dx\, d\xi,$$

provided $\mathrm{Re}\, z < -m - d$, $\mathrm{Re}\, \tau < -n - d$, and the function $t(\tau, z)$ is then holomorphic there. In order to show the desired meromorphic extension, write $t(\tau, z) = t_1(\tau, z) + t_2(\tau, z) + t_3(\tau, z) + t_4(\tau, z)$ where t_1, t_2, t_3, t_4 are the integrals respectively on $A_1 = \{|x| \le \epsilon, |\xi| \le 1\}$, $A_2 = \{|x| \le \epsilon, |\xi| \ge 1\}$, $A_3 = \{|x| \ge \epsilon, |\xi| \le 1\}$, $A_4 = \{|x| \ge \epsilon, |\xi| \ge 1\}$. In fact, it would suffice to set $\epsilon = 1$, but in view of future developments it is useful to work with an arbitrary $\epsilon \ge 1$.

Clearly $t_1(\tau, z)$ is an entire function. As t_2 is concerned, we note that for $|\xi| \ge 1$ and every $p \in \mathbb{N}$, $p \ge 1$, we have

$$\tilde{a}(\tau, z; x, \xi) = \sum_{j=0}^{p-1} \sigma_\psi^{m-j}(a)(x, \xi/|\xi|)|\xi|^{z+m-j}[x]^\tau + r_p(x, \xi)[x]^\tau |\xi|^z,$$

with a remainder $r_p \in G_{\mathrm{cl}(\xi,x)}^{m-p,n}$. Substituting this expression for $\tilde{a}(\tau, z)$ in the integral

$$t_2(\tau, z) = \int_{|x| \le \epsilon} \int_{|\xi| \ge 1} \tilde{a}(\tau, z; x, \xi)\, dx\, d\xi$$

and introducing polar coordinates for the integration in the variables ξ yield

$$t_2(\tau, z) = -(2\pi)^{-d} \sum_{j=0}^{p-1} \frac{1}{z + m + d - j} \int_{|x| \le \epsilon} \int_{\mathbb{S}^{d-1}} [x]^\tau \sigma_\psi^{m-j}(a)\, d\theta\, dx$$

$$+ R_{p,\epsilon}(\tau, z), \quad (5.2.8)$$

where $R_{p,\epsilon}(\tau, z)$ is holomorphic for $\mathrm{Re}\, z < -m - d + p$ and all $\tau \in \mathbb{C}$.

Interchanging the roles of the variables x, ξ we obtain

$$t_3(\tau, z) = -(2\pi)^{-d} \sum_{k=0}^{q-1} \frac{\epsilon^{\tau+n+d-k}}{\tau + n + d - k} \int_{|\xi| \le 1} \int_{\mathbb{S}^{d-1}} [\xi]^z \sigma_e^{n-k}(a)\, d\theta\, d\xi$$

$$+ R'_{q,\epsilon}(\tau, z), \quad (5.2.9)$$

where $R'_{q,\epsilon}(\tau, z)$ is holomorphic for $\mathrm{Re}\, \tau < -n - d + q$ and all $z \in \mathbb{C}$.

Finally, repeating the same argument twice, we get

$$t_4(\tau, z) = (2\pi)^{-d} \sum_{j=0}^{p-1} \sum_{k=0}^{q-1} \frac{1}{z + m + d - j} \frac{\epsilon^{\tau+n+d-k}}{\tau + n + d - k}$$

$$\times \int_{\mathbb{S}^{d-1}} \int_{\mathbb{S}^{d-1}} \sigma_{\psi,e}^{m-j,n-k}(a)\, d\theta\, d\theta' + \sum_{j=0}^{p-1} \frac{1}{z + m + d - j} R''_{q,j,\epsilon}(\tau)$$

$$+ \sum_{k=0}^{q-1} \frac{\epsilon^{\tau+n+d-k}}{\tau + n + d - k} R'''_{p,k}(z) + R''''_{p,q,\epsilon}(\tau, z). \quad (5.2.10)$$

Here we set

$$R''_{q,j,\epsilon}(\tau) = \int_{|x| \geq \epsilon} \int_{\mathbb{S}^{d-1}} [x]^\tau r_{q,j}(x, \theta) \, d\theta \, dx,$$

with $r_{q,j} \in G^{m-j,n-q}_{\mathrm{cl}(\xi,x)}$. Hence $R''_{q,j,\epsilon}(\tau)$ is holomorphic for $\mathrm{Re}\,\tau < -n - d + q$. Notice, for future reference, that $R''_{q,j,\epsilon}(\tau) \to 0$ as $\epsilon \to +\infty$ uniformly for τ in compact subsets of $\{\tau \in \mathbb{C} : \mathrm{Re}\,\tau < -n - d + q\}$.

Similarly, $R'''_{p,k}(z)$ is holomorphic for $\mathrm{Re}\,z < -m - d + p$, whereas $R''''_{p,q,\epsilon}(\tau, z)$ is holomorphic for $\mathrm{Re}\,\tau < -n - d + q$, $\mathrm{Re}\,z < -m - d + p$.

So, we have verified that $t(\tau, z)$ extends to a meromorphic function on $\mathrm{Re}\,\tau < -n - d + q$, $\mathrm{Re}\,z < -m - d + p$. As p and q are arbitrary, this concludes the proof. $\square$

For $a \in G^{m,n}_{\mathrm{cl}(\xi,x)}$, $m \in \mathbb{Z}$, $n \in \mathbb{Z}$, and $\tilde{a}(\tau, z)$ as in (5.2.7), we now consider the functionals defined by

$$\tau z \mathrm{Tr}(\tilde{a}(\tau, z)^{\mathrm{w}}) = \mathrm{Tr}_{\psi,e}(a^{\mathrm{w}}) - \tau \widehat{\mathrm{Tr}}_\psi(a^{\mathrm{w}}) - z \widehat{\mathrm{Tr}}_e(a^{\mathrm{w}}) + \tau^2 V + \tau z V' + z^2 V'', \quad (5.2.11)$$

where V, V', V'' are holomorphic near $(0,0)$.

Proposition 5.2.3. *The functionals* $\mathrm{Tr}_{\psi,e}, \widehat{\mathrm{Tr}}_\psi, \widehat{\mathrm{Tr}}_e$ *defined in* (5.2.11) *have the following explicit expressions:*

$$\mathrm{Tr}_{\psi,e}(a^{\mathrm{w}}) = (2\pi)^{-d} \int_{\mathbb{S}^{d-1}} \int_{\mathbb{S}^{d-1}} \sigma^{-d,-d}_{\psi,e}(a) \, d\theta \, d\theta', \quad (5.2.12)$$

$$\widehat{\mathrm{Tr}}_\psi(a^{\mathrm{w}}) = (2\pi)^{-d} \lim_{\epsilon \to +\infty} \left(\int_{|x| \leq \epsilon} \int_{\mathbb{S}^{d-1}} \sigma^{-d}_\psi(a) \, d\theta \, dx - \log \epsilon \, \mathrm{Tr}_{\psi,e}(a^{\mathrm{w}}) \right.$$
$$\left. - \sum_{i=1}^{n+d} \frac{\epsilon^i}{i} \int_{\mathbb{S}^{d-1}} \int_{\mathbb{S}^{d-1}} \sigma^{-d,i-d}_{\psi,e}(a) \, d\theta \, d\theta' \right), \quad (5.2.13)$$

$$\widehat{\mathrm{Tr}}_e(a^{\mathrm{w}}) = (2\pi)^{-d} \lim_{\epsilon \to +\infty} \left(\int_{|\xi| \leq \epsilon} \int_{\mathbb{S}^{d-1}} \sigma^{-d}_e(a) \, d\theta \, d\xi - \log \epsilon \, \mathrm{Tr}_{\psi,e}(a^{\mathrm{w}}) \right.$$
$$\left. - \sum_{i=1}^{m+d} \frac{\epsilon^i}{i} \int_{\mathbb{S}^{d-1}} \int_{\mathbb{S}^{d-1}} \sigma^{i-d,-d}_{\psi,e}(a) \, d\theta \, d\theta' \right). \quad (5.2.14)$$

Proof. We refer to the proof of Lemma 5.2.2, where now we take $p > m + d$, $q > n + d$.

Formula (5.2.12) follows at once as the limit $\lim_{(\tau,z) \to (0,0)} \tau z \mathrm{Tr}(a(\tau, z)^{\mathrm{w}})$ using the expressions (5.2.8), (5.2.9), (5.2.10).

To prove (5.2.13), we observe that we can obtain $\widehat{\mathrm{Tr}}_\psi$ as

$$\widehat{\mathrm{Tr}}_\psi(a^{\mathrm{w}}) = -\lim_{\tau \to 0} \tau^{-1} \lim_{z \to 0} (\tau z \mathrm{Tr}(\tilde{a}(\tau, z)^{\mathrm{w}}) - \mathrm{Tr}_{\psi,e}(a^{\mathrm{w}})). \quad (5.2.15)$$

We use the decomposition $\tau z \mathrm{Tr}(\tilde{a}(\tau, z)^{\mathrm{w}}) = \sum_{i=1}^{4} \tau z t_i(\tau, z)$ as in the proof of Lemma 5.2.2, and we perform the limit as $z \to 0$: the expressions $\tau z t_1(\tau, z)$ and $\tau z t_3(\tau, z)$ tend to 0, as well as all the terms in (5.2.8) and (5.2.10), except possibly those corresponding to $j = m+d$. What remains is, on the whole, independent of ϵ but, on the other hand, as $\epsilon \to +\infty$ the expression $R''_{q,m+d,\epsilon}(\tau)$ and the terms of the first sum in (5.2.10) with $k > n+d$ tend to zero uniformly for small τ. Then we have

$$\lim_{z \to 0} (\tau z \mathrm{Tr}(\tilde{a}(\tau, z)^{\mathrm{w}}) - \mathrm{Tr}_{\psi,e}(a^{\mathrm{w}}))$$

$$= (2\pi)^{-d} \tau \left(-\int_{|x| \le \epsilon} \int_{\mathbb{S}^{d-1}} [x]^{\tau} \sigma_{\psi}^{-d}(a) \, d\theta \, dx + \frac{\epsilon^{\tau}}{\tau} \int_{\mathbb{S}^{d-1}} \int_{\mathbb{S}^{d-1}} \sigma_{\psi,e}^{-d,-d}(a) \, d\theta \, d\theta' \right.$$

$$\left. + \sum_{k=0}^{n+d-1} \frac{\epsilon^{\tau+n+d-k}}{\tau+n+d-k} \int_{\mathbb{S}^{d-1}} \int_{\mathbb{S}^{d-1}} \sigma_{\psi,e}^{-d,n-k}(a) \, d\theta \, d\theta' - \tau^{-1} \mathrm{Tr}_{\psi,e}(a^{\mathrm{w}}) \right) + o(1),$$

where $o(1)$ stands for a function of τ which tends to 0 as $\epsilon \to +\infty$ uniformly for small τ. Hence (5.2.13) follows from (5.2.15).

In the same way one proves (5.2.14). □

Remark 5.2.4. Let us note that the restrictions Tr_{ψ} and Tr_e of $\widehat{\mathrm{Tr}}_{\psi}$ and $\widehat{\mathrm{Tr}}_e$ to $\bigcup_{m \in \mathbb{Z}} \mathrm{OPG}_{\mathrm{cl}(\xi,x)}^{m,-d-1}$ and $\bigcup_{n \in \mathbb{Z}} \mathrm{OPG}_{\mathrm{cl}(\xi,x)}^{-d-1,n}$ are given by

$$\mathrm{Tr}_{\psi}(a^{\mathrm{w}}) = (2\pi)^{-d} \int_{\mathbb{R}_x^d} \int_{\mathbb{S}^{d-1}} \sigma_{\psi}^{-d}(a) \, d\theta \, dx, \qquad a \in \bigcup_{m \in \mathbb{Z}} G_{\mathrm{cl}(\xi,x)}^{m,-d-1}, \qquad (5.2.16)$$

$$\mathrm{Tr}_e(a^{\mathrm{w}}) = (2\pi)^{-d} \int_{\mathbb{R}_\xi^d} \int_{\mathbb{S}^{d-1}} \sigma_o^{-d}(a) \, d\theta \, d\xi, \qquad a \in \bigcup_{n \in \mathbb{Z}} G_{\mathrm{cl}(\xi,x)}^{-d-1,n}, \qquad (5.2.17)$$

and $\widehat{\mathrm{Tr}}_{\psi}$ and $\widehat{\mathrm{Tr}}_e$ turn out just the finite parts of the integrals in (5.2.16) and (5.2.17) when $a \in \bigcup_{m \in \mathbb{Z}, n \in \mathbb{Z}} G_{\mathrm{cl}(\xi,x)}^{m,n}$. Furthermore, the functionals $\widehat{\mathrm{Tr}}_{\psi}$ and $\widehat{\mathrm{Tr}}_e$ vanish on $\mathcal{J}_e$ and $\mathcal{J}_{\psi}$ respectively, so that they are well defined on $\mathcal{A}_{\psi}$ and $\mathcal{A}_e$ as extensions of Tr_{ψ} and Tr_e.

Theorem 5.2.5. *The functional* $\mathrm{Tr}_{\psi,e}$ *defines a trace on the algebra* $\mathcal{A}$ *which vanishes on* $\mathcal{J}_{\psi}$ *and* $\mathcal{J}_e$ *and therefore it induces traces on* $\mathcal{A}_{\psi}, \mathcal{A}_e$ *and* $\mathcal{A}_{\psi,e}$. *On* $\mathcal{J}_{\psi}$ *and* $\mathcal{J}_e$ *trace functionals are given respectively by* Tr_{ψ} *and* Tr_e *defined in* (5.2.16) *and* (5.2.17).

When $n \ge 2$, *for all these algebras the above functionals are the unique traces up to multiplication by a constant.*

Proof. In all cases the desired conclusion easily follows by the same arguments as in the proof of Theorem 5.1.2. To avoid an overweight of this section, we prefer then to omit any detail. □

We pass now to consider Dixmier traces. With respect to the discussion in the Summary, cf. Definition 5.0.2 and (5.0.10), we use here a somewhat more general notion. We consider traces whose natural domain is contained in the ideal $\mathcal{K}(H)$ of compact operators on the Hilbert space H.

For $T \in \mathcal{K}(H)$, let $\mu_n(T)$, $n = 1, 2, \ldots$, be the sequence of the eigenvalues of $|T|$, counted with their multiplicity and labelled in decreasing order and let $\sigma_N(T) = \sum_{n=1}^{N} \mu_n(T)$, $N = 1, 2, \ldots$, as in (5.0.7). For a fixed sequence α of positive numbers α_N such that

(i) $\alpha_N \to +\infty$;

(ii) $\alpha_0 > \alpha_1 - \alpha_0$ and $\alpha_{N+1} - \alpha_N \geq \alpha_{N+2} - \alpha_{N+1}$ for $N \in \mathbb{N}$;

(iii) $\alpha_N^{-1} \alpha_{2N} \to 1$;

we define the ideal $I_\alpha(H) := \{T \in \mathcal{K}(H) : \alpha_N^{-1} \sigma_N(T) \in l^\infty(\mathbb{N})\}$. Repeating the arguments in the summary, we then consider a linear form ω on $C_b(1, \infty)$, the space of the continuous bounded functions on $[1, \infty]$, with $\omega \geq 0$, $\omega(1) = 1$ and $\omega(f) = 0$ if $\lim_{x \to +\infty} f(x) = 0$. Given a bounded sequence $a = (a_n)_{n \geq 1}$, we construct the function $f_a = \sum_{n \geq 1} a_n \chi_{[n-1,n)} \in L^\infty(\mathbb{R}_+)$ and define the ω-limit $\lim_\omega a_n = \omega(M f_a)$ where, for $g \in L^\infty(\mathbb{R}_+)$, $Mg(t) := \frac{1}{\log t} \int_1^t \frac{g(s)}{s} \, ds$ is the Cesàro mean of g. In the case of convergent sequences the ω-limit coincides with the usual limit.

Definition 5.2.6. Let $\alpha = (\alpha_N)$ be a sequence as above and $T \in I_\alpha(H)$, $T \geq 0$. We define the Dixmier trace of T as

$$\mathrm{Tr}_{\alpha, \omega}(T) = \lim_\omega \alpha_N^{-1} \sigma_N(T).$$

Dixmier's trace extends to a linear map on $I_\alpha(H)$. In the case of the sequence $\alpha_N = \log N$ we recapture the definition in (5.0.10); we shall continue to use the notation Tr_ω for the Dixmier trace associated with that sequence and to denote by $\mathcal{L}^{(1,\infty)}(H)$ its domain, cf. the following more general definition.

Definition 5.2.7. For $1 < p < \infty$ we define the subspace $\mathcal{L}^{(p,\infty)}(H) \subset \mathcal{K}(H)$ as the set of all compact operators T with $\sigma_N(T) = O(N^{1-1/p})$. Similarly we define $\mathcal{L}^{(1,\infty)}(H) \subset \mathcal{K}(H)$ by the condition $\sigma_N(T) = O(\log N)$.

For $1 < p < \infty$, we define the subspace $\mathcal{L}_{\log}^{(p,\infty)}(H) \subset \mathcal{K}(H)$ as the set of all compact operators T with $\sigma_N(T) = O(N^{1-1/p}(\log N)^{-1/p})$; $\mathcal{L}_{\log}^{(1,\infty)}(H) \subset \mathcal{K}(H)$ will be defined by the condition $\sigma_N(T) = O((\log N)^2)$.

All these spaces are normed ideals contained in $\mathcal{K}(H)$, containing the ideal $\mathcal{B}_1(H)$ of trace-class operators.

Remark 5.2.8. Let us observe that for $p = 1$ the ideal $\mathcal{L}_{\log}^{(1,\infty)}(H)$ is the natural domain of the Dixmier trace associated with the sequence $\alpha_N = (\log N)^2$. In short we shall denote it by Tr_ω'.

All the spaces $\mathrm{OPG}^{m,n}_{\mathrm{cl}(\xi,x)}$ with $m < 0$, $n < 0$ are contained in $\mathcal{K}(L^2(\mathbb{R}^d))$. In order to establish relations between these spaces and the ideals in Definition 5.2.1, we have to study the asymptotic behaviour of the spectrum of such operators.

We recall some basic facts from Chapter 4. Let $a \in G^{m,n}_{\mathrm{cl}(\xi,x)}$, $m > 0$, $n > 0$, be a real elliptic symbol, semibounded from below. Then the corresponding operator a^{w} has a self-adjoint realization $L^2(\mathbb{R}^d)$; it is bounded from below and has discrete spectrum $\{\lambda_k\}_{k\in\mathbb{N}}$ diverging to $+\infty$. Denote by $N(\lambda)$ the counting function associated with the operator a^{w}. Then

$$N(\lambda) \sim \begin{cases} C_m \lambda^{\frac{d}{m}} \log \lambda & \text{for } m = n, \\ C'_m \lambda^{\frac{d}{m}} & \text{for } m < n, \\ C''_n \lambda^{\frac{d}{n}} & \text{for } m > n, \end{cases} \tag{5.2.18}$$

where

$$C_m = \frac{(2\pi)^{-d}}{dm} \int_{\mathbb{S}^{d-1}} \int_{\mathbb{S}^{d-1}} \sigma^{m,m}_{\psi,e}(a)^{-\frac{d}{m}} \, d\theta \, d\theta', \tag{5.2.19}$$

$$C'_m = \frac{(2\pi)^{-d}}{d} \int_{\mathbb{R}^d_x} \int_{\mathbb{S}^{d-1}} \sigma^m_\psi(a)^{-\frac{d}{m}} \, d\theta \, dx, \tag{5.2.20}$$

$$C''_n = \frac{(2\pi)^{-d}}{d} \int_{\mathbb{R}^d_\xi} \int_{\mathbb{S}^{d-1}} \sigma^n_e(a)^{-\frac{d}{n}} \, d\theta \, d\xi, \tag{5.2.21}$$

cf. Theorem 4.6.6. We shall need the following simple lemma.

Lemma 5.2.9. *For $1 \le p < \infty$, let g_p be the inverse function of $f_p : (1,\infty) \to \mathbb{R}_+$, $f_p(x) = x^p \log x$. Then*

(a) *if (a_n) and (b_n) are positive sequences with $a_n \sim b_n$ we have $g_p(a_n) \sim g_p(b_n)$;*

(b) *for every positive sequence (k_n) diverging to $+\infty$ we have*

$$g_p(k_n) \sim (pk_n / \log k_n)^{1/p}.$$

Proof. (a) The statement follows by observing that, for $0 < x < x'$, we have

$$0 < \frac{\log g_p(x) - \log g_p(x')}{x - x'} \le \frac{1}{px},$$

as one verifies by Lagrange's formula.

(b) Note that $f_p\left((pk_n/\log k_n)^{1/p}\right) \sim k_n$ and then use (a). □

Theorem 5.2.10. *Let $m < 0$, $n < 0$, with $m \ge -d$ or $n \ge -d$, so that $\mathrm{OPG}^{m,n}_{\mathrm{cl}(\xi,x)} \subset \mathcal{K}(L^2(\mathbb{R}^d))$ but $\mathrm{OPG}^{m,n}_{\mathrm{cl}(\xi,x)} \not\subset \mathcal{B}_1(L^2(\mathbb{R}^d))$. Then the following inclusions hold:*

$$\mathrm{OPG}^{m,n}_{\mathrm{cl}(\xi,x)} \subset \begin{cases} \mathcal{L}^{(-d/m,\infty)}_{\log}(L^2(\mathbb{R}^d)) & \text{if } m = n, \\ \mathcal{L}^{(-d/m,\infty)}(L^2(\mathbb{R}^d)) & \text{if } m > n, \\ \mathcal{L}^{(-d/n,\infty)}(L^2(\mathbb{R}^d)) & \text{if } m < n. \end{cases} \tag{5.2.22}$$

Furthermore we have

$$\mathrm{Tr}_{\psi,e}(a^w) = 2d^2 \mathrm{Tr}'_\omega(a^w) \quad for\ a \in G^{-d,-d}_{\mathrm{cl}(\xi,x)}, \tag{5.2.23}$$

$$\mathrm{Tr}_\psi(a^w) = d\,\mathrm{Tr}_\omega(a^w) \quad for\ a \in G^{-d,n}_{\mathrm{cl}(\xi,x)}\ with\ n \in \mathbb{Z},\ n < -d, \tag{5.2.24}$$

$$\mathrm{Tr}_e(a^w) = d\,\mathrm{Tr}_\omega(a^w) \quad for\ a \in G^{m,-d}_{\mathrm{cl}(\xi,x)}\ with\ m \in \mathbb{Z},\ m < -d, \tag{5.2.25}$$

independently of ω.

Proof. We verify the first inclusion in (5.2.22). The other cases can be proved in the same way.

Consider first the case of an operator $A \in \mathrm{OPG}^{m,m}_{\mathrm{cl}(\xi,x)}$, non-negative as an operator on $L^2(\mathbb{R}^d)$, injective on $\mathcal{S}(\mathbb{R}^d)$ and with a real elliptic Weyl symbol. Then A is invertible on $\mathcal{S}(\mathbb{R}^d)$ and $\mathcal{S}'(\mathbb{R}^d)$ and $A^{-1} \in \mathrm{OPG}^{-m,-m}_{\mathrm{cl}(\xi,x)}$ has real elliptic Weyl symbol, semibounded from below, cf. Remark 1.7.13 and Theorem 4.2.9. Hence from (5.2.18) and (5.2.19) (with $-m$ in place of m) we have the formula

$$N_{A^{-1}}(\lambda) \sim \tilde{C}_m \lambda^{-\frac{d}{m}} \log \lambda, \tag{5.2.26}$$

with

$$\tilde{C}_m = -\frac{(2\pi)^{-d}}{dm} \int_{\mathbb{S}^{d-1}} \int_{\mathbb{S}^{d-1}} \sigma^{m,m}_{\psi,e}(a)^{-\frac{d}{m}}\, d\theta\, d\theta'.$$

On the other hand, (5.2.26) is equivalent to the following formula for the eigenvalues λ_k of A^{-1}:

$$\lambda_k^{-\frac{d}{m}} \log \lambda_k \sim \tilde{C}_m^{-1} k,$$

cf. the proof of Proposition 4.6.4, which by Lemma 5.2.9 implies

$$\lambda_k \sim g_{-\frac{d}{m}}(\tilde{C}_m^{-1} k) \sim (-dk/(m\tilde{C}_m \log k))^{-m/d}.$$

For the eigenvalues of A, that are λ_k^{-1}, we obtain the formula

$$\lambda_k^{-1} \sim (-dk/(m\tilde{C}_m \log k))^{m/d}. \tag{5.2.27}$$

From (5.2.27) it follows that

$$\sum_{k=1}^{N} \lambda_k^{-1} \sim \left(-\frac{d}{m}\tilde{C}_m^{-1}\right)^{\frac{m}{d}} \int_1^N \left(\frac{\log x}{x}\right)^{-\frac{m}{d}} dx$$

$$\sim \begin{cases} \frac{d}{d+m}\left(-\frac{d}{m}\tilde{C}_m^{-1}\right)^{\frac{m}{d}} N^{1+\frac{m}{d}}(\log N)^{-\frac{m}{d}} & for\ -d < m < 0, \\ \frac{1}{2}\tilde{C}_{-d}(\log N)^2 & for\ m = -d. \end{cases} \tag{5.2.28}$$

Hence $A \in \mathcal{L}^{(-d/m,\infty)}_{\log}(L^2(\mathbb{R}^d))$. As $\mathcal{L}^{(-d/m,\infty)}_{\log}(L^2(\mathbb{R}^d))$ is an ideal of $\mathcal{B}(L^2(\mathbb{R}^d))$ the first inclusion in (5.2.22) follows, since one can write $P \in \mathrm{OPG}^{m,m}_{\mathrm{cl}(\xi,x)}$ as $P = (PA^{-1})A$ where PA^{-1} is bounded in $L^2(\mathbb{R}^d)$.

Now we come to the relations (5.2.23), (5.2.24), (5.2.25) between the traces $\mathrm{Tr}_{\psi,e}, \mathrm{Tr}_\psi, \mathrm{Tr}_e$ and the Dixmier traces. We limit ourselves to consider (5.2.23).

It follows from (5.2.28) that (5.2.22) holds for an operator $A = a^{\mathrm{w}}$ as in the first part of the present proof (with $m = -d$). The extension to every symbol $a \in G^{-d,-d}_{\mathrm{cl}(\xi,x)}$ goes exactly as in the last part of the proof of Theorem 5.1.6. $\qquad\square$

5.3 Dixmier Traceability for General Pseudo-Differential Operators

A natural question is whether the results of Sections 5.1 and 5.2 can be extended to Weyl operators $A = a^{\mathrm{w}}$ with symbols a in the classes $S(M; \Phi, \Psi)$ considered in Chapter 1, i.e., satisfying the estimates

$$|\partial_\xi^\alpha \partial_x^\beta a(x,\xi)| \lesssim M(x,\xi) \Psi(x,\xi)^{-|\alpha|} \Phi(x,\xi)^{-|\beta|}, \quad x \in \mathbb{R}^d, \, \xi \in \mathbb{R}^d. \tag{5.3.1}$$

The definition of non-commutative residue cannot be reproduced in this general setting, because of the lack of homogeneity, however we may investigate Dixmier traceability, i.e., whether such general pseudo-differential operators belong to

$$\mathcal{L}^{(1,\infty)}(L^2(\mathbb{R}^d)) = \{ A \in \mathcal{K}(L^2(\mathbb{R}^d)) : \sigma_N(A) = O(\log N) \}, \tag{5.3.2}$$

with $\sigma_N(A) = \sum_{n=1}^N \mu_n(A)$, where $\mu_n(A)$, $n = 1, 2, \ldots$, is the sequence of the eigenvalues of $|A|$.

We will assume the strong uncertainty principle:

$$h(x,\xi) := \Phi(x,\xi)^{-1} \Psi(x,\xi)^{-1} \lesssim (1 + |x| + |\xi|)^{-\delta}, \quad x \in \mathbb{R}^d, \, \xi \in \mathbb{R}^d, \tag{5.3.3}$$

for some $\delta > 0$. We also suppose that M is a regular weight, cf. Definition 1.5.1, satisfying

$$M(x,\xi) \to 0 \quad \text{as } (x,\xi) \to \infty, \tag{5.3.4}$$

so that $a^{\mathrm{w}} \in \mathcal{K}(L^2(\mathbb{R}^d))$ by Theorem 1.4.2. We shall now express a sufficient condition on the weight M in (5.3.1) to have $a^{\mathrm{w}} \in \mathcal{L}^{(1,\infty)}(L^2(\mathbb{R}^d))$ for all $a \in S(M; \Phi, \Psi)$. Before stating the result, we recall the definition of $L^1_w(\mathbb{R}^n)$, the Lorentz-Marcinkiewicz space of the L^1-weak functions in $\mathbb{R}^n$. Namely,

$$L^1_w(\mathbb{R}^n) = \{ f : \mathbb{R}^n \to \mathbb{C} \text{ measurable: } \sup_{s>0} s \cdot \text{measure}(\{|f| > s\}) < \infty \}. \tag{5.3.5}$$

For example, the functions $|z|^{-n}$ and $(1 + |z|^2)^{-n/2}$, $z \in \mathbb{R}^n$, are in $L^1_w(\mathbb{R}^n)$.

Theorem 5.3.1. *Let the strong uncertainty principle (5.3.3) be satisfied and let M be a regular weight fulfilling (5.3.4). Then, if $M \in L^1_w(\mathbb{R}^{2d})$,*

$$a \in S(M; \Phi, \Psi) \Rightarrow a^{\mathrm{w}} \in \mathcal{L}^{(1,\infty)}(L^2(\mathbb{R}^d)). \tag{5.3.6}$$

As examples, we may recapture in part the results of Sections 5.1 and 5.2. In fact, for $\Phi(x,\xi) = \Psi(x,\xi) = (1 + |x|^2 + |\xi|^2)^{1/2}$, $M(x,\xi) = (1 + |x|^2 + |\xi|^2)^{-d}$ we have $M(x,\xi) \in L_w^1(\mathbb{R}^{2d})$ and we obtain $\mathrm{OP}\Gamma^{-2d}(\mathbb{R}^d) \subset \mathcal{L}^{(1,\infty)}(L^2(\mathbb{R}^d))$, cf. Theorem 5.1.6. In the case $\Phi(x,\xi) = \langle x \rangle$, $\Psi(x,\xi) = \langle \xi \rangle$, $M(x,\xi) = \langle x \rangle^{-d} \langle \xi \rangle^{-d-\epsilon}$, or $M(x,\xi) = \langle x \rangle^{-d-\epsilon} \langle \xi \rangle^{-d}$ for some $\epsilon > 0$, we have $M \in L_w^1(\mathbb{R}^{2d})$ and we obtain $\mathrm{OP}G^{-d-\epsilon,-d}(\mathbb{R}^d) \subset \mathcal{L}^{(1,\infty)}(L^2(\mathbb{R}^d))$, $\mathrm{OP}G^{-d,-d-\epsilon}(\mathbb{R}^d) \subset \mathcal{L}^{(1,\infty)}(L^2(\mathbb{R}^d))$, cf. Theorem 5.2.10.

Theorem 5.3.1 will be obtained as a consequence of some auxiliary propositions. First observe that, since M is a regular weight, $M \in S(M; \Phi, \Psi)$. Consider $(M^{-1})^w$, the operator with Weyl symbol M^{-1}. In view of (5.3.4), it is a self-adjoint operator in $L^2(\mathbb{R}^d)$ with a spectrum made of a sequence of eigenvalues bounded from below. It follows that $(M^{-1} + c)^w$ is bounded from below, say, by 1 if c is large enough, and the inverse A^{-1} on $\mathcal{S}(\mathbb{R}^d)$, $\mathcal{S}'(\mathbb{R}^d)$ is well defined as a pseudo-differential operator with Weyl symbol in $S(M; \Phi, \Psi)$, cf. Remark 1.7.13. Let us now write $a = M^{-1} + c$, so that $A = a^w$. Notice that we have as well

$$a^{-1} \in L_w^1(\mathbb{R}^{2d}). \tag{5.3.7}$$

We observe that Theorem 5.1.3 is proved if we verify that $A^{-1} \in \mathcal{L}^{(1,\infty)}(L^2(\mathbb{R}^d))$. Indeed, given any pseudo-differential operator P with Weyl symbol in $S(M; \Phi, \Psi)$ we can write $P = PAA^{-1}$. Since PA has a symbol in $S(1; \Phi, \Psi)$, it is bounded in $L^2(\mathbb{R}^n)$ by Theorem 1.4.1, and $\mathcal{L}^{(1,\infty)}$ is an ideal in the space of bounded operators. Hence we deduce that $P \in \mathcal{L}^{(1,\infty)}(L^2(\mathbb{R}^d))$. Thus we are reduced to prove that the eigenvalues λ_j of A satisfy

$$\lambda_j^{-1} = O(1/j) \quad \text{as } j \to \infty. \tag{5.3.8}$$

Theorem 5.3.2. *The operator e^{-tA}, $t \geq 0$ can be written as*

$$e^{-tA} = b_t^w + S(t),$$

where b_t is a bounded family of symbols in $S(1; \Phi, \Psi)$ for $t \geq 0$, satisfying for all $\alpha \in \mathbb{N}^d$, $\beta \in \mathbb{N}^d$,

$$|\partial_\xi^\alpha \partial_x^\beta b_t(x,\xi)| \lesssim e^{-ta(x,\xi)/2} \quad \text{for all } t \geq 0, \ (x,\xi) \in \mathbb{R}^{2d}, \tag{5.3.9}$$

and $S(t)$ is a trace-class operator with

$$\|S(t)\|_{\mathcal{B}_1(L^2)} \lesssim t \quad \text{for all } t \geq 0. \tag{5.3.10}$$

Proof. The first part of the proof of Theorem 4.5.1 shows that there exist symbols $v_j(t,x,\xi)$, $j = 0,1,\ldots$, of the form $v_0(t,x,\xi) = e^{-ta_0(x,\xi)}$ and, for $j \geq 1$,

$$v_j(t,x,\xi) = e^{-ta_0(x,\xi)} \sum_{l=1}^{2j} t^l v_{l,j}(x,\xi), \quad v_{l,j} \in S(M^l h^j; \Phi, \Psi), \tag{5.3.11}$$

such that $b_t(x, \xi) := \sum_{j=0}^{N} v_j(t, x, \xi)$ satisfies

$$\begin{cases} (\partial_t + a^w) b_t^w = K(t), \\ b_0^w = I, \end{cases} \tag{5.3.12}$$

for some operator $K(t)$ with Weyl symbol belonging to a bounded subset of $S(M^{-1} h^{N+1}, g)$ when $t \geq 0$. It follows therefore from Theorem 4.4.21 that, if N is chosen so that $M^{-1} h^{N+1} \in L^1(\mathbb{R}^{2d})$ (which is possible in view of (5.3.3)), $K(t)$ turns out to be trace-class and $\|K(t)\|_{\mathcal{B}_1(L^2)} \leq C$ for every $t \geq 0$.

In order to verify (5.3.10) we observe that, since the operator e^{-tA} solves (5.3.12) with $K = 0$, we have

$$b_t^w - e^{-tA} = \int_0^t e^{-(t-s)A} K(s) \, ds. \tag{5.3.13}$$

Then, by (4.4.28),

$$\|b_t^w - e^{-tA}\|_{\mathcal{B}_1(L^2)} \leq \int_0^t \|e^{-(t-s)A} K(s)\|_{\mathcal{B}_1(L^2)} ds$$

$$\leq \int_0^t \|e^{-(t-s)A}\|_{\mathcal{B}(L^2)} \|K(s)\|_{\mathcal{B}_1(L^2)} ds \lesssim t.$$

This concludes the proof. $\qquad\square$

Proposition 5.3.3. *We have*

$$\sum_{j=1}^{\infty} e^{-t\lambda_j} = O(t^{-1}), \quad \text{as } t \searrow 0.$$

Proof. By Theorem 5.3.2 we have

$$\sum_{j=1}^{\infty} e^{-t\lambda_j} = \operatorname{Tr} e^{-tA} = \|e^{-tA}\|_{\mathcal{B}_1(L^2)} \leq \|b_t^w\|_{\mathcal{B}_1(L^2)} + \|S(t)\|_{\mathcal{B}_1(L^2)}, \tag{5.3.14}$$

with $\|S(t)\|_{\mathcal{B}_1} \lesssim t$, whereas it follows from Theorem 4.4.21 that, for N large enough,

$$\|b_t^w\|_{\mathcal{B}_1(L^2)} \lesssim \sum_{|\gamma| \leq N} \|\partial^\gamma b_t\|_{L^1}. \tag{5.3.15}$$

On the other hand, by (5.3.9),

$$\|\partial^\gamma b_t\|_{L^1} \lesssim \int e^{-ta(x,\xi)/2} dx \, d\xi = \int_0^{+\infty} e^{-ts} d\lambda(s), \tag{5.3.16}$$

where we set

$$\lambda(s) = \text{measure}(\{a/2 \leq s\}).$$

Now we have $\lambda(s) = 0$ for s in a right neighbourhood of 0 and $\lambda(s) \leq Cs$, since $a^{-1} \in L^1_w(\mathbb{R}^{2d})$, see (5.3.7). Thus, integrating by parts in (5.3.16) yields

$$\int_0^{+\infty} e^{-ts}d\lambda(s) = t\int_0^{+\infty} e^{-ts}\lambda(s)ds \leq Ct\int_0^{+\infty} e^{-ts}s\,ds = Ct^{-1}.$$

This shows that $\|b_t^w\|_{\mathcal{B}_1(L^2)} = O(t^{-1})$ as $t \searrow 0$, which together with (5.3.14) and (5.3.10) concludes the proof. $\square$

Let now $N(\lambda)$ be the counting function of A.

Proposition 5.3.4. *We have $N(x) = O(x)$ as $x \to +\infty$.*

Proof. Since

$$N(t^{-1}) = \int_0^{t^{-1}} dN(\lambda) = \int_0^{+\infty} \chi_{[0,t^{-1}]}(\lambda)dN(\lambda),$$

we get

$$N(t^{-1}) \leq e\int_0^{+\infty} e^{-t\lambda}dN(\lambda). \qquad (5.3.17)$$

On the other hand, the right-hand side of (5.3.17) is exactly $e\sum_{j=1}^{\infty} e^{-t\lambda_j}$, which is $O(t^{-1})$ as $t \searrow 0$ in view of Proposition 5.3.3. $\square$

We finally prove (5.3.8).

Proposition 5.3.5. *We have $\lambda_j^{-1} = O(1/j)$ as $j \to \infty$.*

Proof. We know from Proposition 5.3.4 that $N(x) \leq Cx$ for $x \geq 0$. Now, given any j, take $j_1 \geq j$ such that $\lambda_j = \lambda_{j_1} < \lambda_{j_1+1}$. Then $N(\lambda_{j_1}) = j_1$ so that

$$j \leq j_1 \leq C\lambda_{j_1} = C\lambda_j.$$

This concludes the proof. $\square$

Theorem 5.3.1 is therefore proved.

Notes

We first review related results for pseudo-differential operators on compact manifolds. The non-commutative residue had been used initially in the one-dimensional case by Manin [139] and Adler [2] in their work on algebraic aspects of the Korteweg-de Vries equation. In 1987 Wodzicki gave a more detailed account of the non-commutative residue and related topics, cf. [196] and also the survey by Kassel [126]. Guillemin discovered the non-commutative residue independently in the context of a new proof of Weyl's formula [106]. Connes obtained (5.0.11) on compact manifolds in his work on Non-Commutative Geometry. Different variants and generalizations of the non-commutative residue of Wodzicki are presented in Guillemin [107] concerning Fourier integral operators, Fedosov, Golse, Leichtman and Schrohe [76] concerning operators on manifolds with boundary, Schrohe [177] about manifolds with conical singularities, Nicola [151] about anisotropic operators on foliated manifolds. We refer also to Grubb [101], Melrose and Nistor [146], Paycha and Scott [155] for connections with the index problem and Laurent expansions of holomorphic families of pseudo-differential operators.

About pseudo-differential operators in $\mathbb{R}^d$, we mention first the contribution of Boggiatto and Nicola [20]; the contents in Section 5.1 correspond to a particular case of their results, concerning a more general class of anisotropic Γ-operators. The main reference for Section 5.2 is Nicola [152], which in turn follows the lines of [146]; see also Lauter and Moroianu [131]. For the non-normal traces used in Section 5.2, we refer to the original paper of Dixmier [73]. Finally, for Section 5.3 we refer to Nicola and Rodino [154], where a more general result was given in terms of the Weyl-Hörmander classes. Indeed the problem of the Dixmier traceability in $\mathbb{R}^d$, treated in Section 5.3, is richer than on compact manifolds, see Gayral, Gracia-Bondía, Iochum, Schücker and Vàrilly [89] for related arguments in non-compact Non-Commutative Geometry.

Chapter 6

Exponential Decay and Holomorphic Extension of Solutions

Summary

In the preceding chapters we proved, under different assumptions of global hypoellipticity on the symbol in $\mathbb{R}^d$, that the solutions in $S'(\mathbb{R}^d)$ of the homogeneous equation belong to $S(\mathbb{R}^d)$. In particular, for the self-adjoint operators discussed in Chapter 4, all the eigenfunctions belong to $S(\mathbb{R}^d)$. In the present chapter we show that this information can be strongly improved for G and Γ operators, namely we may give precise results of exponential decay and holomorphic extension of solutions.

The next Section 6.1 introduces the spaces $S_\nu^\mu(\mathbb{R}^d)$, $\mu > 0$, $\nu > 0$, which provide a precise language to describe the above mentioned properties.

In short: the index ν expresses an exponential decay of order $1/\nu$, i.e.,

$$|f(x)| \lesssim e^{-\epsilon|x|^{1/\nu}}, \quad x \in \mathbb{R}^d, \qquad (6.0.1)$$

for some $\epsilon > 0$; the index μ corresponds to exponential decay of order $1/\mu$ of the Fourier transform:

$$|\widehat{f}(\xi)| \lesssim e^{-\epsilon|\xi|^{1/\mu}}, \quad \xi \in \mathbb{R}^d. \qquad (6.0.2)$$

If $\mu < 1$, then (6.0.1), (6.0.2) imply extension of f to an entire function satisfying

$$|f(x + iy)| \lesssim e^{-\epsilon|x|^{1/\nu} + \delta|y|^{1/1-\mu}}, \quad x \in \mathbb{R}^d, \ y \in \mathbb{R}^d, \qquad (6.0.3)$$

for some $\epsilon > 0$, $\delta > 0$; if $\mu = 1$ the extension is limited to a strip of the complex domain.

Let us clarify our objectives on some model operators, on which we can test S_ν^μ-regularity of the solutions. Consider first the Schrödinger harmonic oscillator:

$$H = -\Delta + |x|^2. \tag{6.0.4}$$

Corresponding to the eigenvalue $\lambda = \lambda_k = \sum_{j=1}^d (2k_j + 1)$, $k = (k_1, \ldots, k_d) \in \mathbb{N}^d$, we have the eigenfunction, cf. Section 2.2:

$$u(x) = u_k(x) = \prod_{j=1}^d P_{k_j}(x_j) e^{-\frac{|x|^2}{2}} \tag{6.0.5}$$

where $P_r(t)$ stands for the r-th Hermite polynomial, cf. (2.2.26). Observe that the decay at infinity of $u(x)$ in (6.0.5) is super-exponential of order 2, i.e., for a certain constant $\epsilon > 0$:

$$|u(x)| \lesssim e^{-\epsilon|x|^2}, \quad x \in \mathbb{R}^d. \tag{6.0.6}$$

The second aspect is holomorphic extension, namely u extends to an entire function, satisfying in the complex domain the following improved version of (6.0.6), for suitable positive constants ϵ and δ:

$$|u(x + iy)| \lesssim e^{-\epsilon|x|^2 + \delta|y|^2}, \quad x \in \mathbb{R}^d, \ y \in \mathbb{R}^d. \tag{6.0.7}$$

Note also that the eigenfunctions (6.0.5) are invariant under Fourier transformation. In terms of the spaces $S_\nu^\mu(\mathbb{R}^d)$, we therefore have $\nu = 1/2$, cf. (6.0.1), and $\mu = 1/2$, cf. (6.0.2), (6.0.3).

Our aim in Section 6.2 will be to extend the properties (6.0.6), (6.0.7) to the eigenfunctions $u(x)$ of all higher order Γ-elliptic equations with polynomial coefficients, namely we shall prove $u \in S_{\frac{1}{2}}^{\frac{1}{2}}(\mathbb{R}^d)$. We shall also consider semilinear perturbations of the harmonic oscillator:

$$Hu - \lambda u = -\Delta u + |x|^2 u - \lambda u = u^k \tag{6.0.8}$$

where $\lambda \in \mathbb{R}$ and $k \geq 2$. For the eigenfunctions of (6.0.8), i.e., homoclinics, we still have super-exponential decay of order 2, cf. (6.0.6), however (6.0.7) fails in general, and the extension to the complex domain $u(x + iy)$ is limited to a strip $\{x + iy \in \mathbb{C}^d : |y| < T\}$ for some $T > 0$, as we shall show by a counterexample. In other terms, we have now $u \in S_{\frac{1}{2}}^1(\mathbb{R}^d)$.

The subsequent Sections 6.3, 6.4, 6.5 are devoted to similar results for G-elliptic equations. It is clear that in the G-case the optimal result is $u \in S_1^1(\mathbb{R}^d)$, that is the estimate for $\epsilon > 0$,

$$|u(x)| \lesssim e^{-\epsilon|x|}, \quad x \in \mathbb{R}^d, \tag{6.0.9}$$

combined with analytic extension in a strip. Consider for example the G-elliptic ordinary differential operator

$$Lu = -(1 + x^2)u'' - 2xu' + x^2 u, \quad x \in \mathbb{R}. \tag{6.0.10}$$

The operator L is self-adjoint with compact resolvent and then there exists a sequence $\lambda_j \in \mathbb{R}$, $j = 1, 2, \ldots$, such that $Lu_j = \lambda_j u_j$ for some non-trivial $u_j \in \mathcal{S}(\mathbb{R}^d)$. From the theory of the asymptotic integration, cf. Section 3.3, we have

$$u_j(x) = Cx^{-1}e^{-|x|} + O(x^{-2}e^{-|x|}) \quad \text{for } |x| \to +\infty,$$

and from Fuchs theory we may expect singularities at $x = \pm i$.

Initially in Section 6.3 we shall present a version of the G-pseudo-differential calculus for $S_\nu^\mu(\mathbb{R}^d)$ classes, $\mu \geq 1$, $\nu \geq 1$, arguing on symbols $p(x, \xi)$ satisfying, for some $C > 0$,

$$|\partial_\xi^\alpha \partial_x^\beta p(x, \xi)| \lesssim C^{|\alpha|+|\beta|} (\alpha!)^\mu (\beta!)^\nu \langle \xi \rangle^{m-|\alpha|} \langle x \rangle^{n-|\beta|}$$

$$\text{for all } x \in \mathbb{R}^d, \ \xi \in \mathbb{R}^d, \ \alpha \in \mathbb{N}^d, \ \beta \in \mathbb{N}^d. \quad (6.0.11)$$

In particular, we shall construct parametrices for the G-elliptic operators.

Because of the technical difficulties coming from the analytic case, our regularity results will be limited here to $u \in S_\theta^\theta(\mathbb{R}^d)$, $\theta > 1$, $\theta \geq \mu + \nu - 1$.

For symbols satisfying (6.0.11) with $\mu = \nu = 1$, we shall reach the expected optimal regularity $S_1^1(\mathbb{R}^d)$ in the conclusive Section 6.5 by a technique of a priori estimates and iterative methods, providing results also for semilinear operators. Applications mainly concern travelling wave equations, with G-elliptic linear part and semilinear perturbation. On the subject, we give in Section 6.4 a short survey for readers interested in the Mathematical Physics aspects. Relevant examples are:

$$u'' - Vu + u^2 = 0 \quad \text{in } \mathbb{R} \quad (6.0.12)$$

for the solitary waves $v(t, x) = u(x - Vt)$, $V > 0$, of the Korteweg-de Vries equation, and higher order generalizations; the d-dimensional extension

$$-\Delta u + u = u^k, \quad k \geq 2, \quad (6.0.13)$$

appearing in Plasma Physics and Nonlinear Optics; the intermediate-long-wave equation in $\mathbb{R}$

$$N(D)u + \gamma u = u^2, \quad \gamma > -1, \quad (6.0.14)$$

with $N(\xi) = \xi \mathrm{Ctgh}\, \xi$, which is a symbol of type (6.0.11) with $\mu = 1$. Note that the linear parts of (6.0.12), (6.0.13), (6.0.14) have no eigenfunctions in $\mathcal{S}(\mathbb{R}^d)$, whereas the semilinearity produces homoclinics $u \in S_1^1(\mathbb{R}^d)$.

6.1 The Function Spaces $S_\nu^\mu(\mathbb{R}^d)$

The asymptotic information given by the Schwartz space $\mathcal{S}(\mathbb{R}^d)$ is not so satisfactory in Applied Mathematics, in particular one would like to know more precisely how fast the decay of $f \in \mathcal{S}(\mathbb{R}^d)$ is at infinity. To this end, it is convenient to use the spaces $S_\nu^\mu(\mathbb{R}^d)$, subspaces of $\mathcal{S}(\mathbb{R}^d)$. We shall define them in terms of simultaneous estimates of exponential type for $f(x)$ and $\widehat{f}(\xi)$. This seems the appropriate approach for applications to the operators of the preceding chapters, because of the symmetrical role of the variables x and ξ.

Definition 6.1.1. The function $f(x)$ is in $S_\nu^\mu(\mathbb{R}^d)$, $\mu > 0$, $\nu > 0$, if $f(x) \in \mathcal{S}(\mathbb{R}^d)$ and there exists a constant $\epsilon > 0$ such that

$$|f(x)| \lesssim e^{-\epsilon|x|^{\frac{1}{\nu}}}, \quad x \in \mathbb{R}^d; \tag{6.1.1}$$

$$\left|\widehat{f}(\xi)\right| \lesssim e^{-\epsilon|\xi|^{\frac{1}{\mu}}}, \quad \xi \in \mathbb{R}^d. \tag{6.1.2}$$

We have the obvious inclusions $S_\nu^\mu(\mathbb{R}^d) \subset S_{\nu'}^{\mu'}(\mathbb{R}^d)$ for $\mu \leq \mu'$, $\nu \leq \nu'$. It is also evident that application of the Fourier transform interchanges the indices μ and ν in Definition 6.1.1, namely:

Theorem 6.1.2. *For $f \in \mathcal{S}(\mathbb{R}^d)$, we have $f \in S_\nu^\mu(\mathbb{R}^d)$ if and only if $\widehat{f} \in S_\mu^\nu(\mathbb{R}^d)$.*

In particular the spaces $S_\mu^\mu(\mathbb{R}^d)$, $\mu > 0$, are invariant under the action of the Fourier transform. It will be clear from the arguments in the sequel that Definition 6.1.1 does not change meaning, if referred to $f \in L^2(\mathbb{R}^d)$, or even $f \in \mathcal{S}'(\mathbb{R}^d)$, provided (6.1.1), (6.1.2) make sense. So we shall keep $\mathcal{S}(\mathbb{R}^d)$ as an universe-set, in the whole section.

Example 6.1.3. The Gaussian function $f(x) = e^{-\frac{|x|^2}{2}}$ belongs to $S_{\frac{1}{2}}^{\frac{1}{2}}(\mathbb{R}^d)$. In fact (6.1.1), (6.1.2) are satisfied with $\mu = \nu = \frac{1}{2}$, since $\widehat{f}(\xi) = e^{-\frac{|\xi|^2}{2}}$. More generally, all the functions in $\mathbb{R}^d$ of the type $P(x)e^{-a|x|^2}$, with $a > 0$ and $P(x)$ any polynomial, belong to $S_{\frac{1}{2}}^{\frac{1}{2}}(\mathbb{R}^d)$.

Example 6.1.4. Consider then in $\mathcal{S}(\mathbb{R})$, for a fixed integer $k > 0$, the function

$$g(t) = e^{-t^{2k}}, \quad t \in \mathbb{R}, \tag{6.1.3}$$

which satisfies (6.1.1) with $\nu = \frac{1}{2k}$. The Fourier transform

$$h(\tau) = \widehat{g}(\tau) = \int e^{-it\tau - t^{2k}} \, dt, \quad \tau \in \mathbb{R}, \tag{6.1.4}$$

cannot be computed in elementary terms. Note however that $g(t)$ solves

$$g' + 2kt^{2k-1}g = 0, \tag{6.1.5}$$

hence $h(\tau)$ is, modulo a multiplicative constant, the unique solution in $\mathcal{S}(\mathbb{R})$ of the transformed equation

$$h^{(2k-1)} + \frac{(-1)^{k+1}}{2k}\tau h = 0. \tag{6.1.6}$$

From the classical theory of asymptotic integration, we know that the bounded solutions of (6.1.6) must satisfy, for some $\epsilon > 0$,

$$|h(\tau)| \lesssim e^{-\epsilon|\tau|^{\frac{2k}{2k-1}}},$$

hence (6.1.2) holds with $\mu = 1 - \frac{1}{2k}$ and $g \in S_{\frac{1}{2k}}^{1-\frac{1}{2k}}(\mathbb{R})$. Symmetrically, the function $h(\tau)$ in (6.1.4) belongs to $S_{1-\frac{1}{2k}}^{\frac{1}{2k}}(\mathbb{R})$. $\square$

We would like now to pass from the estimates (6.1.1), (6.1.2) to estimates involving only $f(x)$. The first step is to convert exponential bounds into factorial bounds.

Proposition 6.1.5. *The following conditions are equivalent:*

(1) *the condition* (6.1.1) *holds, i.e., there exists a constant* $\epsilon > 0$ *such that*

$$|f(x)| \lesssim e^{-\epsilon|x|^{\frac{1}{\nu}}}, \quad x \in \mathbb{R}^d; \tag{6.1.7}$$

(2) *there exists a constant* $C > 0$ *such that*

$$|x^\alpha f(x)| \lesssim C^{|\alpha|}(\alpha!)^\nu, \quad x \in \mathbb{R}^d, \ \alpha \in \mathbb{N}^d. \tag{6.1.8}$$

Proof. It will be convenient to re-write (6.1.7) in the form

$$|f(x)|^{\frac{1}{\nu}} \lesssim e^{-\epsilon|x|^{\frac{1}{\nu}}}, \quad x \in \mathbb{R}^d, \tag{6.1.9}$$

for a new constant $\epsilon > 0$. In turn, (6.1.9) can be re-written as

$$\sup_{x \in \mathbb{R}^d} \sum_{n=0}^{\infty} \epsilon^n (n!)^{-1} |x|^{\frac{n}{\nu}} |f(x)|^{\frac{1}{\nu}} < \infty. \tag{6.1.10}$$

Hence the sequence of the terms of the series is uniformly bounded, as well as the sequence of the ν-th powers:

$$\epsilon^{\nu n} (n!)^{-\nu} |x|^n |f(x)|, \quad n = 0, 1, \ldots,$$

and we obtain

$$|x|^n |f(x)| \lesssim \epsilon^{-\nu n} (n!)^\nu, \quad x \in \mathbb{R}^d, \ n = 0, 1, \ldots.$$

Hence, writing $|\alpha| = n$ and applying (0.3.3):

$$|x^\alpha f(x)| \lesssim \epsilon^{-\nu n} (n!)^\nu \lesssim C^{|\alpha|}(\alpha!)^\nu, \quad x \in \mathbb{R}^d, \ \alpha \in \mathbb{N}^d,$$

for a suitable constant $C > 0$. Therefore (6.1.8) is proved.

In the opposite direction, let (6.1.8) be satisfied. Since $|x|^n \leq k^n \sum_{|\alpha|=n} |x^\alpha|$ for a constant k depending only on the dimension d, cf. (0.3.1) and (0.3.3), then

$$|x|^n |f(x)| \leq k^n \sum_{|\alpha|=n} |x^\alpha f(x)| \lesssim (kC)^n \sum_{|\alpha|=n} (\alpha!)^\nu, \quad x \in \mathbb{R}^d, \ n = 0, 1, \ldots.$$

Since $\alpha! \leq n!$ and the number of the multi-indices α in the sum does not exceed 2^{d+n-1}, cf. (0.3.9), (0.3.16), we may further estimate with a new constant $C > 0$

$$|x|^n |f(x)| \lesssim C^n (n!)^\nu, \quad x \in \mathbb{R}^d, \ n = 0, 1, \ldots.$$

Therefore the sequence

$$C^{-n} (n!)^{-\nu} |x|^n |f(x)|, \quad n = 0, 1, \ldots$$

is uniformly bounded for $x \in \mathbb{R}^d$, as well as the sequence

$$C^{-\frac{n}{\nu}} (n!)^{-1} |x|^{\frac{n}{\nu}} |f(x)|^{\frac{1}{\nu}}, \quad n = 0, 1, \ldots.$$

If we choose $\epsilon = \frac{C^{-\frac{1}{\nu}}}{2}$, we conclude that

$$e^{\epsilon |x|^{\frac{1}{\nu}}} |f(x)|^{\frac{1}{\nu}} = \sum_{n=0}^\infty \frac{C^{-\frac{n}{\nu}}}{2^n} (n!)^{-1} |x|^{\frac{n}{\nu}} |f(x)|^{\frac{1}{\nu}} \lesssim \sum_{n=0}^\infty \frac{1}{2^n}.$$

We obtain (6.1.9), hence (6.1.7). The proof of Proposition 6.1.5 is complete. □

The following result gives some equivalent definitions of the class $S_\nu^\mu(\mathbb{R}^d)$.

Theorem 6.1.6. *Assume $\mu > 0$, $\nu > 0$, $\mu + \nu \geq 1$. For $f \in \mathcal{S}(\mathbb{R}^d)$ the following conditions are equivalent:*

(i) $f \in S_\nu^\mu(\mathbb{R}^d)$.

(ii) *There exists $C > 0$ such that*

$$|x^\alpha f(x)| \lesssim C^{|\alpha|} (\alpha!)^\nu, \quad x \in \mathbb{R}^d, \ \alpha \in \mathbb{N}^d;$$
$$\left| \xi^\beta \widehat{f}(\xi) \right| \lesssim C^{|\beta|} (\beta!)^\mu, \quad \xi \in \mathbb{R}^d, \ \beta \in \mathbb{N}^d.$$

(iii) *There exists $C > 0$ such that*

$$\|x^\alpha f(x)\|_{L^2} \lesssim C^{|\alpha|} (\alpha!)^\nu, \quad \alpha \in \mathbb{N}^d;$$
$$\left\| \xi^\beta \widehat{f}(\xi) \right\|_{L^2} \lesssim C^{|\beta|} (\beta!)^\mu, \quad \beta \in \mathbb{N}^d.$$

(iv) *There exists $C > 0$ such that*

$$\|x^\alpha f(x)\|_{L^2} \lesssim C^{|\alpha|} (\alpha!)^\nu, \quad \alpha \in \mathbb{N}^d;$$
$$\|\partial^\beta f(x)\|_{L^2} \lesssim C^{|\beta|} (\beta!)^\mu, \quad \beta \in \mathbb{N}^d.$$

(v) *There exists $C > 0$ such that*

$$\left\| x^\alpha \partial^\beta f(x) \right\|_{L^2} \lesssim C^{|\alpha|+|\beta|} (\alpha!)^\nu (\beta!)^\mu, \quad \alpha \in \mathbb{N}^d, \ \beta \in \mathbb{N}^d.$$

(vi) *There exists $C > 0$ such that*

$$\left| x^\alpha \partial^\beta f(x) \right| \lesssim C^{|\alpha|+|\beta|} (\alpha!)^\nu (\beta!)^\mu, \quad x \in \mathbb{R}^d, \ \alpha \in \mathbb{N}^d, \ \beta \in \mathbb{N}^d.$$

Proof. The preceding Proposition 6.1.5 shows that (i) is equivalent to (ii). We shall prove (ii) $\Rightarrow$ (iii) $\Rightarrow$ (iv) $\Rightarrow$ (v) $\Rightarrow$ (vi) $\Rightarrow$ (ii). First, we assume (ii) and prove (iii). Fixing an integer $M > \frac{d}{4}$ so that $\left\| (1 + |x|^2)^{-M} \right\|_{L^2} < \infty$, we have

$$\left\| x^\alpha f(x) \right\|_{L^2} \lesssim \sup_{x \in \mathbb{R}^d} (1 + |x|^2)^M \left| x^\alpha f(x) \right|, \quad \alpha \in \mathbb{N}^d.$$

Write then

$$(1 + |x|^2)^M = \sum_{|\gamma| \leq M} c_\gamma x^{2\gamma}$$

where the c_γ are positive integers, which we can estimate in terms of the fixed integer M. Therefore:

$$\left\| x^\alpha f(x) \right\|_{L^2} \lesssim \sum_{|\gamma| \leq M} \sup_{x \in \mathbb{R}^d} \left| x^{\alpha+2\gamma} f(x) \right|, \quad \alpha \in \mathbb{N}^d.$$

At this moment we apply the first estimate in (ii) and we obtain

$$\left\| x^\alpha f(x) \right\|_{L^2} \lesssim \sum_{|\gamma| \leq M} C^{|\alpha+2\gamma|} ((\alpha + 2\gamma)!)^\nu, \quad \alpha \in \mathbb{N}^d.$$

In view of (0.3.6) we have $(\alpha + 2\gamma)! \leq 2^{|\alpha+2\gamma|} \alpha! (2\gamma)!$, where $|\gamma| \leq M$. Note also that the number of the terms in the sum in the right-hand side is estimate by 2^{M+d}, cf. (0.3.9), (0.3.15) hence the first inequality in (iii) follows, for a new constant $C > 0$. Arguing similarly in the ξ variables, we obtain the second inequality.

By Plancherel's formula, we have (iii) $\Leftrightarrow$ (iv). Let us prove (iv) $\Rightarrow$ (v). Integrating by parts and using Leibniz' formula we have:

$$\left\| x^\alpha \partial^\beta f(x) \right\|_{L^2}^2 = (\partial^\beta f, x^{2\alpha} \partial^\beta f)_{L^2} = \left| (f, \partial^\beta (x^{2\alpha} \partial^\beta f))_{L^2} \right|$$

$$\leq \sum_{\gamma \leq \beta, \gamma \leq 2\alpha} \binom{\beta}{\gamma} \binom{2\alpha}{\gamma} \gamma! \left| (x^{2\alpha-\gamma} f, \partial^{2\beta-\gamma} f)_{L^2} \right|.$$

Let us observe that $\binom{\beta}{\gamma}\binom{2\alpha}{\gamma} \leq 2^{|\beta|+2|\alpha|}$, in view of (0.3.9). Applying then the Cauchy-Schwarz inequality, we obtain

$$\left\| x^\alpha \partial^\beta f(x) \right\|_{L^2}^2 \leq 2^{|\beta|+2|\alpha|} \sum_{\gamma \leq \beta, \gamma \leq 2\alpha} \gamma! \left\| x^{2\alpha-\gamma} f(x) \right\|_{L^2} \left\| \partial^{2\beta-\gamma} f(x) \right\|_{L^2}. \quad (6.1.11)$$

Using the assumptions (iv), where we may assume $C \geq 1$, we have for $\gamma \leq \beta$, $\gamma \leq 2\alpha$:

$$\gamma! \left\| x^{2\alpha-\gamma} f(x) \right\|_{L^2} \left\| \partial^{2\beta-\gamma} f(x) \right\|_{L^2} \lesssim C^{2|\alpha|+2|\beta|} \gamma! (2\alpha-\gamma)!^{\nu} (2\beta-\gamma)!^{\mu}$$
$$\lesssim C^{2|\alpha|+2|\beta|} (2\alpha)!^{\nu} (2\beta)!^{\mu}, \quad \alpha \in \mathbb{N}^d, \ \beta \in \mathbb{N}^d. \qquad (6.1.12)$$

In fact, the last estimate follows by writing

$$\gamma! \leq (\gamma!)^{\mu} (\gamma!)^{\nu} \qquad (6.1.13)$$

in view of the hypothesis $\mu + \nu \geq 1$.

   We now return to (6.1.11), where we apply (6.1.12) and we observe that the number of the terms in the sum does not exceed $2^{2|\alpha|+|\beta|+2d}$, cf. (0.3.15). We then conclude, for a new constant $C > 0$,

$$\left\| x^{\alpha} \partial^{\beta} f(x) \right\|_{L^2}^{2} \lesssim C^{|\alpha|+|\beta|} (2\alpha)!^{\nu} (2\beta)!^{\mu}, \quad \alpha \in \mathbb{N}^d, \ \beta \in \mathbb{N}^d.$$

Since $(2\alpha)! \leq 2^{2|\alpha|} (\alpha!)^2$ and $(2\beta)! \leq 2^{2|\beta|} (\beta!)^2$ in view of (0.3.6), we obtain (v).

   To prove (v) $\Rightarrow$ (vi) we use the Sobolev embedding theorem (0.2.6). Namely, fixing an integer $s > \frac{d}{2}$, we have

$$\left| x^{\alpha} \partial^{\beta} f(x) \right| \lesssim \left\| x^{\alpha} \partial^{\beta} f(x) \right\|_{H^s}$$
$$\lesssim \sum_{|\gamma| \leq s} \left\| \partial^{\gamma} (x^{\alpha} \partial^{\beta} f(x)) \right\|_{L^2}, \quad x \in \mathbb{R}^d, \ \alpha \in \mathbb{N}^d, \ \beta \in \mathbb{N}^d. \qquad (6.1.14)$$

Applying Leibniz' rule we may further estimate the right-hand side of (6.1.14) by

$$\sum_{|\gamma| \leq s} \sum_{\substack{\delta \leq \gamma \\ \delta \leq \alpha}} \binom{\gamma}{\delta} \binom{\alpha}{\delta} \delta! \left\| x^{\alpha-\delta} \partial^{\beta+\gamma-\delta} f(x) \right\|_{L^2}. \qquad (6.1.15)$$

From (0.3.9) we have $\binom{\gamma}{\delta} \binom{\alpha}{\delta} \delta! \leq C_s 2^{|\alpha|}$, where the constant C_s does not depend on α. Moreover, the number of the terms of the sums in (6.1.15) can be estimated by an integer independent of α. On the other hand, using the assumption (v) and (0.3.6) we obtain

$$\left\| x^{\alpha-\delta} \partial^{\beta+\gamma-\delta} f(x) \right\|_{L^2} \lesssim C^{|\alpha|+|\beta|-2|\delta|+|\gamma|} (\alpha-\delta)!^{\nu} (\beta+\gamma-\delta)!^{\mu}$$
$$\lesssim C_s^{|\alpha|+|\beta|} (\alpha!)^{\nu} (\beta!)^{\mu}, \quad \alpha \in \mathbb{N}^d, \ \beta \in \mathbb{N}^d, \qquad (6.1.16)$$

for a constant C_s depending on s. Combining (6.1.14), (6.1.15) and (6.1.16) we obtain, for a new constant $C > 0$,

$$\left| x^{\alpha} \partial^{\beta} f(x) \right| \lesssim C^{|\alpha|+|\beta|} (\alpha!)^{\nu} (\beta!)^{\mu}, \quad x \in \mathbb{R}^d, \ \alpha \in \mathbb{N}^d, \ \beta \in \mathbb{N}^d,$$

that is (vi).

It remains to prove (vi) $\Rightarrow$ (ii). To this end, first observe that (vi) with $\beta = 0$ gives the first inequality of (ii). On the other hand

$$\left|\xi^\beta \widehat{f}(\xi)\right| = \left|\widehat{\partial^\beta f}(\xi)\right| \le (2\pi)^{-\frac{d}{2}} \left\|\partial^\beta f\right\|_{L^1}, \quad \xi \in \mathbb{R}^d. \tag{6.1.17}$$

Fixing an integer $M > \frac{d}{2}$, so that $\left\|(1+|x|^2)^{-M}\right\|_{L^1} < \infty$, we obtain

$$\left\|\partial^\beta f\right\|_{L^1} \lesssim \sup_{x \in \mathbb{R}^d} (1+|x|^2)^M \left|\partial^\beta f(x)\right|. \tag{6.1.18}$$

From (vi) we have

$$\sup_{x \in \mathbb{R}^d} (1+|x|^2)^M \left|\partial^\beta f(x)\right| \lesssim C^{|\beta|}(\beta!)^\mu, \quad \beta \in \mathbb{N}^d, \tag{6.1.19}$$

and combining (6.1.17), (6.1.18) and (6.1.19) we get the second inequality in (ii). The proof of Theorem 6.1.6 is therefore complete. $\qquad\square$

The assumption $\mu + \nu \ge 1$ in the statement of Theorem 6.1.6 is not restrictive. In fact we shall now prove that for $\mu + \nu < 1$ the spaces $S^\mu_\nu(\mathbb{R}^d)$ are trivial, i.e., they contain only the zero function. To this end, we first give the following propositions, deserving independent interest.

Proposition 6.1.7. *Assume $f \in \mathcal{S}(\mathbb{R}^d)$, $\mu > 0$, $\nu > 0$. Then the estimates (vi) in Theorem 6.1.6 are valid if and only if there exist positive constants C and ϵ such that*

$$\left|\partial^\beta f(x)\right| \lesssim C^{|\beta|}(\beta!)^\mu e^{-\epsilon|x|^{\frac{1}{\nu}}}, \quad x \in \mathbb{R}^d, \ \beta \in \mathbb{N}^d. \tag{6.1.20}$$

Hence, when $\mu + \nu \ge 1$ the estimates (6.1.20) give an equivalent definition of $S^\mu_\nu(\mathbb{R}^d)$.

Proof. Just apply Proposition 6.1.5 to the function $C^{-|\beta|}(\beta!)^{-\mu}\partial^\beta f(x)$. An inspection of the proof shows that the bounds are independent of β, hence (6.1.20) is equivalent for a new constant $C > 0$ to

$$C^{-|\beta|}(\beta!)^{-\mu}\left|x^\alpha \partial^\beta f(x)\right| \lesssim C^{|\alpha|}(\alpha!)^\nu,$$

which gives the estimates (vi) in Theorem 6.1.6. $\qquad\square$

Proposition 6.1.8. *Assume $f \in \mathcal{S}(\mathbb{R}^d)$, $0 < \mu < 1$, $\nu > 0$. Let (6.1.20) be satisfied for suitable constants $C > 0$, $\epsilon > 0$. Then f extends to an entire analytic function $f(x + iy)$ in $\mathbb{C}^d$, with*

$$|f(x+iy)| \lesssim e^{-\epsilon|x|^{\frac{1}{\nu}}+\delta|y|^{\frac{1}{1-\mu}}}, \quad x \in \mathbb{R}^d, \ y \in \mathbb{R}^d, \tag{6.1.21}$$

where δ is a suitable positive constant. In the case $\mu = 1$, $\nu > 0$, f extends to an analytic function $f(x + iy)$ in the strip $\{x + iy \in \mathbb{C}^d : |y| < T\}$ with

$$|f(x+iy)| \lesssim e^{-\epsilon|x|^{\frac{1}{\nu}}}, \quad x \in \mathbb{R}^d, \ |y| < T, \tag{6.1.22}$$

for a suitable $T > 0$.

Proof. For $\nu > 0$ and $0 < \mu \leq 1$ the estimates (6.1.20) imply

$$|\partial^\beta f(x)| \lesssim C^{|\beta|}\beta!, \quad x \in \mathbb{R}^d, \ \beta \in \mathbb{N}^d.$$

This is sufficient to conclude analyticity in $\mathbb{R}^d$ and analytic extension to the strip $\{x + iy \in \mathbb{C}^d : |y| < T\}$ with T given by $\frac{1}{C}$. If $\mu < 1$, then f extends further to the entire function defined by the Taylor expansion

$$f(z) = \sum_\beta \frac{\partial^\beta f(0)}{\beta!} z^\beta, \quad z = x + iy \in \mathbb{C}^d. \tag{6.1.23}$$

In fact $\frac{|\partial^\beta f(0)|}{\beta!} \lesssim C^{|\beta|}(\beta!)^{\mu-1}$ in view of (6.1.20) and the radius of convergence of the power series (6.1.23) is ∞. To prove (6.1.21), we now evaluate $f(x + iy)$ by Taylor expanding at the points $x \in \mathbb{R}^d$:

$$f(x + iy) = \sum_\beta \frac{\partial^\beta f(x)}{\beta!}(iy)^\beta.$$

Hence from (6.1.20) we have

$$|f(x + iy)| \lesssim e^{-\epsilon|x|^{\frac{1}{\nu}}} \sum_\beta C^{|\beta|}(\beta!)^{\mu-1} |y^\beta|. \tag{6.1.24}$$

Write the sum in the right-hand side of (6.1.24) as

$$\sum_\beta C^{|\beta|}(\beta!)^{\mu-1} |y^\beta| = \prod_{j=1}^d M(|y_j|) \tag{6.1.25}$$

with

$$M(t) = \sum_{n=0}^\infty C^n (n!)^{\mu-1} t^n, \quad t > 0. \tag{6.1.26}$$

To estimate $M(t)$, set

$$\lambda = (2Ct)^{\frac{1}{1-\mu}}, \quad t > 0, \tag{6.1.27}$$

and re-write

$$M(t) = \sum_{n=0}^\infty \frac{1}{2^n} \left(\frac{\lambda^n}{n!}\right)^{1-\mu}. \tag{6.1.28}$$

In view of (0.3.13) we have

$$\left(\frac{\lambda^n}{n!}\right)^{1-\mu} \leq e^{(1-\mu)\lambda}, \tag{6.1.29}$$

and applying (6.1.27), (6.1.29) in (6.1.28) we conclude, for some $\delta > 0$,

$$M(t) \lesssim e^{\delta t^{\frac{1}{1-\mu}}}, \quad t > 0.$$

Returning to (6.1.25) and then to (6.1.24), we obtain (6.1.21). The same arguments give (6.1.22). Proposition 6.1.8 is therefore proved. □

Hence, in the case $\mu + \nu \geq 1$, every $f \in S^\mu_\nu(\mathbb{R}^d)$ extends to the complex domain, as an entire function satisfying (6.1.21) if $\mu < 1$, as a holomorphic function in a strip satisfying (6.1.22) if $\mu = 1$. On the other hand, in the case $\mu + \nu < 1$ we may apply the following Liouville-type proposition.

Proposition 6.1.9. *Assume* $0 < \lambda < \theta$. *Let* $f(x + iy)$ *be an entire function in* $\mathbb{C}^d$ *satisfying for positive constants* ϵ *and* δ:

$$|f(x+iy)| \lesssim e^{-\epsilon|x|^\theta + \delta|y|^\lambda}, \quad x \in \mathbb{R}^d, \ y \in \mathbb{R}^d. \tag{6.1.30}$$

Then $f \equiv 0$.

Proof. Consider the entire function $f(iz) = f(ix - y)$. For the product $f(z)f(iz)$ we obtain, from (6.1.30),

$$|f(z)f(iz)| \lesssim e^{-\epsilon|x|^\theta + \delta|y|^\lambda} e^{-\epsilon|y|^\theta + \delta|x|^\lambda}, \quad x \in \mathbb{R}^d, \ y \in \mathbb{R}^d.$$

Since we assume $\lambda < \theta$, the right-hand side tends to zero for $x + iy \to \infty$. Hence, according to the Liouville theorem, the entire function $f(z)f(iz)$ is identically zero. But then $f \equiv 0$, and Proposition 6.1.9 is proved. □

We may now return to the question of the triviality of the classes $S^\mu_\nu(\mathbb{R}^d)$, $\mu + \nu < 1$. The result can be read as a version of the uncertainty principle of Heisenberg, generically expressed by the statement that a function $f(x)$ and its Fourier transform $\widehat{f}(\xi)$ cannot both be small at infinity.

Theorem 6.1.10. *Let* $f \in S(\mathbb{R}^d)$ *satisfy*

$$|f(x)| \lesssim e^{-\epsilon|x|^{\frac{1}{\nu}}} \text{ for } x \in \mathbb{R}^d, \quad |\widehat{f}(\xi)| \lesssim e^{-\epsilon|\xi|^{\frac{1}{\mu}}} \text{ for } \xi \in \mathbb{R}^d, \tag{6.1.31}$$

for some $\epsilon > 0$, $\mu > 0$, $\nu > 0$ *and* $\mu + \nu < 1$. *Then* $f \equiv 0$. *In other words, according to Definition 6.1.1, the classes* $S^\mu_\nu(\mathbb{R}^d)$ *are trivial if* $\mu + \nu < 1$.

Moreover, if $\mu + \nu < 1$, *each of the conditions* (ii), (iii), (iv), (v), (vi) *in Theorem 6.1.6, as well as* (6.1.20), *implies* $f \equiv 0$.

Proof. First, let us go back to the proof of Theorem 6.1.6 and observe that the assumption $\mu + \nu \geq 1$ was only used to prove (iv) $\Rightarrow$ (v), namely to obtain the key estimate (6.1.13). In the case $\mu + \nu < 1$ we may replace (6.1.13) by the obvious identity

$$\gamma! = (\gamma!)^\mu (\gamma!)^\nu (\gamma!)^{1-(\mu+\nu)}. \tag{6.1.32}$$

Since $\gamma \leq \beta$ in (6.1.12), we can estimate in (6.1.32):

$$(\gamma!)^{1-(\mu+\nu)} \leq (\beta!)^{1-(\mu+\nu)} \leq (2\beta!)^{\frac{1-(\mu+\nu)}{2}},$$

hence

$$\gamma! \leq (\gamma!)^{\mu}(\gamma!)^{\nu}(2\beta!)^{\frac{1-(\mu+\nu)}{2}}. \tag{6.1.33}$$

Using (6.1.33) instead of (6.1.13), we obtain a weaker version of (6.1.12), namely

$$\gamma! \left\| x^{2\alpha-\gamma}f(x) \right\|_{L^2} \left\| \partial^{2\beta-\gamma}f(x) \right\|_{L^2}$$
$$\lesssim C^{2|\alpha|+2|\beta|}(2\alpha)!^{\nu}(2\beta)!^{\mu^*}, \quad \alpha \in \mathbb{N}^d, \ \beta \in \mathbb{N}^d, \tag{6.1.34}$$

with

$$\mu^* = \mu + \frac{1-(\mu+\nu)}{2}. \tag{6.1.35}$$

Continuing to argue as in the proof of Theorem 6.1.6, for $\mu+\nu < 1$ we then obtain (v) for a new couple of indices μ^*, ν, with

$$\mu + \nu < \mu^* + \nu < 1. \tag{6.1.36}$$

This is sufficient for our purposes. Namely, starting from (6.1.31), that is $f \in S_\nu^\mu(\mathbb{R}^d)$ with $\mu+\nu < 1$, we obtain (ii), (iii), (iv) in Theorem 6.1.6 with same μ, ν, and (v), (vi), hence (6.1.20), for the new couple μ^*, ν, in view of Proposition 6.1.7. From Proposition 6.1.8 we obtain analytic extension to an entire function with bounds

$$|f(x+iy)| \lesssim e^{-\epsilon|x|^{\frac{1}{\nu}}+\delta|y|^{\frac{1}{1-\mu^*}}}, \quad x \in \mathbb{R}^d, \ y \in \mathbb{R}^d.$$

Finally we apply Proposition 6.1.9 with $\lambda = \frac{1}{1-\mu^*}$, $\theta = \frac{1}{\nu}$. In fact then, from (6.1.36), or directly from (6.1.35), we have $\lambda < \theta$ in (6.1.30). Hence $f \equiv 0$. Theorem 6.1.10 is proved. $\qquad \square$

Finally, we give other equivalent definitions of the classes $S_\nu^\mu(\mathbb{R}^d)$, $\mu+\nu \geq 1$, which we shall use in the applications of the next sections. Let us introduce the notation

$$\|f\|_{s,\nu;\epsilon} = \sum_{\alpha \in \mathbb{N}^d} \frac{\epsilon^{|\alpha|}}{(\alpha!)^\nu} \left\| x^\alpha f(x) \right\|_{H^s}, \tag{6.1.37}$$

$$\|f\|_{\{s,\mu;\delta\}} = \sum_{\beta \in \mathbb{N}^d} \frac{\delta^{|\beta|}}{(\beta!)^\mu} \left\| \partial^\beta f(x) \right\|_{H^s}, \tag{6.1.38}$$

$$\|f\|_{s,\mu,\nu;\delta,\epsilon} = \sum_{\alpha,\beta \in \mathbb{N}^d} \frac{\epsilon^{|\alpha|}\delta^{|\beta|}}{(\alpha!)^\nu(\beta!)^\mu} \left\| x^\alpha \partial^\beta f(x) \right\|_{H^s}, \tag{6.1.39}$$

where $\mu > 0$, $\nu > 0$, $\epsilon > 0$, $\delta > 0$ and we consider Sobolev norms with $s \geq 0$.

Proposition 6.1.11. *Let $\mu > 0$, $\nu > 0$, $\mu + \nu \geq 1$. For $f \in \mathcal{S}(\mathbb{R}^d)$ the following conditions are equivalent:*

(a) $f \in S^\mu_\nu(\mathbb{R}^d)$.

(b) *There exist* $\epsilon > 0$, $\delta > 0$, $s \geq 0$ *such that* $\|f\|_{s,\nu;\epsilon} < \infty$ *and* $\|f\|_{\{s,\mu;\delta\}} < \infty$.

(c) *There exist* $\epsilon > 0$, $\delta > 0$, $s \geq 0$ *such that* $\|f\|_{s,\mu,\nu;\delta,\epsilon} < \infty$.

Proof. Note that (c) implies (b), since both $\|f\|_{s,\nu;\epsilon}$ and $\|f\|_{\{s,\mu;\delta\}}$ are estimated by $\|f\|_{s,\mu,\nu;\delta,\epsilon}$. In turn, (b) implies (a). In fact, if $\|f\|_{s,\nu;\epsilon} < \infty$, then each term of the sum in (6.1.37) is uniformly bounded, hence for $s \geq 0$,

$$\|x^\alpha f(x)\|_{L^2} \leq \|x^\alpha f(x)\|_{H^s} \lesssim \epsilon^{-|\alpha|}(\alpha!)^\nu, \quad \alpha \in \mathbb{N}^d,$$

and similarly from (6.1.38)

$$\|\partial^\beta f(x)\|_{L^2} \leq \|\partial^\beta f(x)\|_{H^s} \lesssim \delta^{-|\beta|}(\beta!)^\mu, \quad \beta \in \mathbb{N}^d.$$

Therefore we have proved (iv) in Theorem 6.1.6, which implies (a). Finally, we prove that (a) implies (c). To this end, we may start from (v) in Theorem 6.1.6. Assuming without loss of generality that $s \geq 0$ is an integer, we may estimate the Sobolev norm of $x^\alpha \partial^\beta f(x)$ as in (6.1.14), and then apply (6.1.15), (6.1.16). We obtain

$$\|x^\alpha \partial^\beta f(x)\|_{H^s} \lesssim C^{|\alpha|+|\beta|}(\alpha!)^\nu(\beta!)^\mu, \quad \alpha \in \mathbb{N}^d, \ \beta \in \mathbb{N}^d.$$

At this moment in (6.1.39) we choose $\epsilon = \delta = (2C)^{-1}$ and then

$$\|f\|_{s,\mu,\nu;\delta,\epsilon} = \sum_{\alpha,\beta} \left(\frac{1}{2}\right)^{|\alpha|+|\beta|} \frac{\|x^\alpha \partial^\beta f\|_{H^s}}{C^{|\alpha|+|\beta|}(\alpha!)^\nu(\beta!)^\mu} < \infty.$$

The proof of Proposition 6.1.11 is concluded. □

The case $\mu = \nu$ plays an important role in the applications, because of the invariance under Fourier transform. It is convenient for $S^\mu_\mu(\mathbb{R}^d)$ to reformulate (v), Theorem 6.1.6, in the following form.

Proposition 6.1.12. *A function* $f \in \mathcal{S}(\mathbb{R}^d)$ *belongs to* $S^\mu_\mu(\mathbb{R}^d)$, $\mu \geq \frac{1}{2}$, *if and only if there exists a constant* $C > 0$ *such that*

$$\|x^\alpha \partial^\beta f(x)\|_{L^2} \lesssim C^N N^{N\mu} \quad \text{for } |\alpha| + |\beta| \leq N, \ N = 0, 1, 2, \ldots . \tag{6.1.40}$$

Proof. From (6.1.40) we have

$$\|x^\alpha \partial^\beta f(x)\|_{L^2} \lesssim C^{|\alpha|+|\beta|}(|\alpha| + |\beta|)^{(|\alpha|+|\beta|)\mu}, \quad \alpha \in \mathbb{N}^d, \ \beta \in \mathbb{N}^d. \tag{6.1.41}$$

On the other hand, from (0.3.12), (0.3.5), (0.3.3) we obtain

$$(|\alpha| + |\beta|)^{|\alpha|+|\beta|} \leq (2ed)^{|\alpha|+|\beta|}\alpha!\beta!.$$

Applying this to the right-hand side of (6.1.41), we obtain (v) in Theorem 6.1.6, hence $f \in S^\mu_\mu(\mathbb{R}^d)$. In the opposite direction, starting from (v) in Theorem 6.1.6, for $|\alpha| + |\beta| \leq N$ we have

$$\left\| x^\alpha \partial^\beta f(x) \right\|_{L^2} \lesssim C^{|\alpha|+|\beta|} (\alpha! \beta!)^\mu \lesssim C^N N^{N\mu}, \quad N = 0, 1, 2, \ldots,$$

where we assume $C \geq 1$ and we apply (0.3.7). Proposition 6.1.12 is proved.   □

Starting from $S^\mu_\nu(\mathbb{R}^d)$ one can define spaces of temperate ultradistributions containing the Schwartz temperate distributions as a subclass. They will have a minor role in the following, and we treat them briefly. To be precise, limiting attention to the case $\mu = \nu > 1$, we first define a topology in $S^\mu_\mu(\mathbb{R}^d)$. Denote by $S^\mu_{\mu,C}(\mathbb{R}^d), C > 0$, the space of all functions $f \in \mathcal{S}(\mathbb{R}^d)$ such that

$$\sup_{\alpha,\beta} \sup_{x \in \mathbb{R}^d} C^{-|\alpha|-|\beta|} (\alpha! \beta!)^{-\mu} \left| x^\alpha \partial^\beta f(x) \right| < \infty. \tag{6.1.42}$$

From (vi) in Theorem 6.1.6 we have

$$S^\mu_\mu(\mathbb{R}^d) = \bigcup_{C>0} S^\mu_{\mu,C}(\mathbb{R}^d).$$

For any $C > 0$, the space $S^\mu_{\mu,C}(\mathbb{R}^d)$ is a Banach space endowed with the norm given by the left-hand side of (6.1.42). Therefore we can consider the space $S^\mu_\mu(\mathbb{R}^d)$ as an inductive limit of an increasing sequence of Banach spaces. Equivalent topologies come from equivalent definitions of $S^\mu_\mu(\mathbb{R}^d)$, so for example we may refer to (6.1.20) and define as norms, depending on the two parameters $C > 0, \epsilon > 0$:

$$\sup_{\beta} \sup_{x \in \mathbb{R}^d} C^{-|\beta|} (\beta!)^{-\mu} e^{\epsilon |x|^{\frac{1}{\mu}}} \left| \partial^\beta f(x) \right|. \tag{6.1.43}$$

We shall denote by $S^{\mu'}_\mu(\mathbb{R}^d)$ the dual space, i.e., the space of all linear continuous forms on $S^\mu_\mu(\mathbb{R}^d)$. To be definite: a linear form u on $S^\mu_\mu(\mathbb{R}^d)$ belongs to $S^{\mu'}_\mu(\mathbb{R}^d)$ if and only if for every $C > 0$ we have

$$|u(f)| \lesssim \sup_{\alpha,\beta} \sup_{x \in \mathbb{R}^d} C^{-|\alpha|-|\beta|} (\alpha! \beta!)^{-\mu} \left| x^\alpha \partial^\beta f(x) \right|$$

for all $f \in S^\mu_\mu(\mathbb{R}^d)$.

Given $u \in \mathcal{S}'(\mathbb{R}^d)$, the restriction of u on $S^\mu_\mu(\mathbb{R}^d) \subset \mathcal{S}(\mathbb{R}^d)$ is an element of $S^{\mu'}_\mu(\mathbb{R}^d)$. In this sense, we have $\mathcal{S}'(\mathbb{R}^d) \subset S^{\mu'}_\mu(\mathbb{R}^d)$. Under the preceding assumption $\mu > 1$, the theory of the temperate ultradistributions follows closely the theory of Schwartz. We do not give details, but limit ourselves to the following kernel theorem, which is a consequence of the nuclearity of the topology in $S^\mu_\mu(\mathbb{R}^d)$, and to the definition of the Fourier transform.

Theorem 6.1.13. *There exists an isomorphism between the space $\mathcal{L}(S_\mu^\mu(\mathbb{R}^d), S_\mu^{\mu'}(\mathbb{R}^d))$ of all linear continuous maps from $S_\mu^\mu(\mathbb{R}^d)$ to $S_\mu^{\mu'}(\mathbb{R}^d)$, and $S_\mu^{\mu'}(\mathbb{R}^{2d})$, which associates to every $T \in \mathcal{L}(S_\mu^\mu(\mathbb{R}^d), S_\mu^{\mu'}(\mathbb{R}^d))$ a distribution $K_T \in S_\mu^{\mu'}(\mathbb{R}^{2d})$ such that*

$$\langle Tf, g \rangle = \langle K_T, g \otimes f \rangle$$

for every $f, g \in S_\mu^\mu(\mathbb{R}^d)$. The temperate ultradistribution K_T is called the kernel of T.

The Fourier transform, defined as standard by duality $\widehat{u}(f) = u(\widehat{f})$, $u \in S_\mu^{\mu'}(\mathbb{R}^d)$, $f \in S_\mu^\mu(\mathbb{R}^d)$, extends to an isomorphism of $S_\mu^{\mu'}(\mathbb{R}^d)$. We finally observe that the definition of $S_\nu^\mu(\mathbb{R}^d)$ extends in a natural way to the cases when μ and ν take the values 0 or ∞. Namely $S_\infty^\mu(\mathbb{R}^d)$, denoted also by $S^\mu(\mathbb{R}^d)$, is the space of all $f \in S(\mathbb{R}^d)$ such that, for a suitable $C > 0$ and all α,

$$\left| x^\alpha \partial^\beta f(x) \right| \lesssim C^{|\beta|} (\beta!)^\mu, \quad x \in \mathbb{R}^d, \ \beta \in \mathbb{N}^d, \tag{6.1.44}$$

whereas $S_\nu^\infty(\mathbb{R}^d)$, denoted also by $S_\nu(\mathbb{R}^d)$, consists of all $f \in S(\mathbb{R}^d)$ such that, for all β,

$$\left| x^\alpha \partial^\beta f(x) \right| \lesssim C^{|\alpha|} (\alpha!)^\nu, \quad x \in \mathbb{R}^d, \ \alpha \in \mathbb{N}^d. \tag{6.1.45}$$

The class $S_0^\mu(\mathbb{R}^d)$, $\mu > 1$, is defined by replacing (6.1.1) with the assumption that f has compact support. It coincides with $G_0^\mu(\mathbb{R}^d)$, the space of the so-called compactly supported Gevrey functions. This is equivalent to saying that f belongs to $C_0^\infty(\mathbb{R}^d)$ and satisfies for a suitable $C > 0$ the estimates

$$\left| \partial^\beta f(x) \right| \lesssim C^{|\beta|} (\beta!)^\mu, \quad x \in \mathbb{R}^d, \ \beta \in \mathbb{N}^d. \tag{6.1.46}$$

Evidently, this implies $f \equiv 0$ if $\mu \leq 1$. The Gevrey spaces $G_0^\mu(\mathbb{R}^d)$, $\mu > 1$, are non-empty and one can construct in them cut-off functions and partitions of unity. Similarly we define $S_\nu^0(\mathbb{R}^d)$, which we may obtain by Fourier transforming $G_0^\nu(\mathbb{R}^d)$.

6.2 Γ-Operators and Semilinear Harmonic Oscillators

Let us consider the linear partial differential operators with polynomial coefficients in $\mathbb{R}^d$,

$$P = \sum_{|\alpha| + |\beta| \leq m} c_{\alpha\beta} x^\beta D_x^\alpha \tag{6.2.1}$$

where m is a positive integer and $c_{\alpha\beta}$ are given constants in $\mathbb{C}$. Consider the Γ-principal symbol:

$$p_m(x, \xi) = \sum_{|\alpha| + |\beta| = m} c_{\alpha\beta} x^\beta \xi^\alpha \tag{6.2.2}$$

and assume Γ-ellipticity

$$p_m(x, \xi) \neq 0 \quad \text{for } (x, \xi) \neq (0, 0). \tag{6.2.3}$$

We know from the results of Chapter 2 that $u \in \mathcal{S}'(\mathbb{R}^d)$ and $Pu = f \in \mathcal{S}(\mathbb{R}^d)$ imply $u \in \mathcal{S}(\mathbb{R}^d)$, in particular all the solutions $u \in \mathcal{S}'(\mathbb{R}^d)$ of the homogeneous equation $Pu = 0$ belong to $\mathcal{S}(\mathbb{R}^d)$, see Theorem 2.1.6. It is also useful to recall, cf. (2.1.27), that there exists a positive constant C^* such that, for every $u \in \mathcal{S}(\mathbb{R}^d)$,

$$\sum_{|\alpha|+|\beta| \leq m} \left\| x^\beta D_x^\alpha u \right\|_{L^2} \leq C^* (\|Pu\|_{L^2} + \|u\|_{L^2}). \tag{6.2.4}$$

Theorem 6.2.1. *Let* P *in* (6.2.1) *satisfy the* Γ-*ellipticity condition* (6.2.3). *If* $u \in \mathcal{S}'(\mathbb{R}^d)$ *is a solution* $Pu = f$ *with* $f \in S_\mu^\mu(\mathbb{R}^d)$, $\mu \geq \frac{1}{2}$, *then also* $u \in S_\mu^\mu(\mathbb{R}^d)$. *In particular,* $Pu = 0$ *and* $u \in \mathcal{S}'(\mathbb{R}^d)$ *imply* $u \in S_{\frac{1}{2}}^{\frac{1}{2}}(\mathbb{R}^d)$.

Proof. In view of Proposition 6.1.12, it will be sufficient to prove the estimates

$$\left\| x^\alpha \partial^\beta u(x) \right\|_{L^2} \leq C^{N+1} N^{N\mu}, \quad |\alpha| + |\beta| \leq N, \; N = 0, 1, 2, \ldots, \tag{6.2.5}$$

for some $C > 0$. We know that $Pu = f \in S_\mu^\mu(\mathbb{R}^d) \subset \mathcal{S}(\mathbb{R}^d)$ implies $u \in \mathcal{S}(\mathbb{R}^d)$. Then, choosing $C > 1$ sufficiently large, we have (6.2.5) for $N \leq 2m$.

Arguing by induction, assume that (6.2.5) is valid for all $N < M$, $M > 2m$, and prove it for $N = M$. For α, β satisfying $|\alpha| + |\beta| = M$ we write

$$x^\beta D_x^\alpha u = x^{\beta-\delta} x^\delta D_x^{\alpha-\gamma} D_x^\gamma u,$$

where we choose $\gamma \leq \alpha$, $\delta \leq \beta$ so that $|\gamma| + |\delta| = M - m$ and $|\alpha - \gamma| + |\beta - \delta| = m$. To be definite, in the case $|\beta| \geq m$, we may take $\theta \leq \beta$ such that $|\theta| = m$, and define consequently $\gamma = \alpha$, $\delta = \beta - \theta$. Otherwise, we have $|\alpha| \geq m$, since we are assuming $|\alpha| + |\beta| = M > 2m$; we may then take $\rho \leq \alpha$ with $|\rho| = m$ and define $\gamma = \alpha - \rho$, $\delta = \beta$. We have

$$\left\| x^\beta D_x^\alpha u \right\|_{L^2} \leq \left\| x^{\beta-\delta} D_x^{\alpha-\gamma} (x^\delta D_x^\gamma u) \right\|_{L^2} + \left\| x^{\beta-\delta} \left[x^\delta, D_x^{\alpha-\gamma} \right] D_x^\gamma u \right\|_{L^2}$$

$$\leq C^* \left(\left\| P(x^\delta D_x^\gamma u) \right\|_{L^2} + \left\| x^\delta D_x^\gamma u \right\|_{L^2} \right) + \left\| x^{\beta-\delta} \left[x^\delta, D_x^{\alpha-\gamma} \right] D_x^\gamma u \right\|_{L^2}$$

$$\leq C^* \left(\left\| x^\delta D_x^\gamma (Pu) \right\|_{L^2} + \left\| \left[P, x^\delta D_x^\gamma \right] u \right\|_{L^2} + \left\| x^\delta D_x^\gamma u \right\|_{L^2} \right)$$

$$+ \left\| x^{\beta-\delta} \left[x^\delta, D_x^{\alpha-\gamma} \right] D_x^\gamma u \right\|_{L^2},$$

where we have used (6.2.4). Since $Pu = f \in S_\mu^\mu(\mathbb{R}^d)$ and $|\gamma| + |\delta| = M - m$, from Proposition 6.1.12, we have, for some constant $C_1 > 1$,

$$\left\| x^\delta D_x^\gamma (Pu) \right\|_{L^2} \leq C_1^{M-m+1} (M-m)^{(M-m)\mu} \leq C_1^{M+1} M^{M\mu}.$$

Write explicitly, by using (6.2.1)

$$\left[P, x^\delta D_x^\gamma \right] = \sum_{|\tilde{\alpha}|+|\tilde{\beta}| \leq m} c_{\tilde{\alpha}\tilde{\beta}} \left[x^{\tilde{\beta}} D_x^{\tilde{\alpha}}, x^\delta D_x^\gamma \right].$$

Therefore, given $C_2 > 0$ such that $|c_{\tilde{\alpha}\tilde{\beta}}| \leq C_2$ for $|\tilde{\alpha}| + |\tilde{\beta}| \leq m$, we have

$$\left\| x^\beta D_x^\alpha u \right\|_{L^2} \leq C^* C_1^{M+1} M^{M\mu} + C^* C_2 \sum_{|\tilde{\alpha}|+|\tilde{\beta}|\leq m} \left\| \left[x^{\tilde{\beta}} D_x^{\tilde{\alpha}}, x^\delta D_x^\gamma \right] u \right\|_{L^2}$$
$$+ C^* \left\| x^\delta D_x^\gamma u \right\|_{L^2} + \left\| x^{\beta-\delta} \left[x^\delta, D_x^{\alpha-\gamma} \right] D_x^\gamma u \right\|_{L^2}. \qquad (6.2.6)$$

In (6.2.6) let us develop

$$\left[x^{\tilde{\beta}} D_x^{\tilde{\alpha}}, x^\delta D_x^\gamma \right] = \sum_{\substack{0\neq\sigma\leq\tilde{\alpha} \\ \sigma\leq\delta}} c^1_{\tilde{\alpha}\delta\sigma} x^{\delta+\tilde{\beta}-\sigma} D_x^{\gamma+\tilde{\alpha}-\sigma} - \sum_{\substack{0\neq\sigma\leq\tilde{\beta} \\ \sigma\leq\gamma}} c^2_{\tilde{\beta}\gamma\sigma} x^{\delta+\tilde{\beta}-\sigma} D_x^{\gamma+\tilde{\alpha}-\sigma},$$

where, with the help of (1.2.18) and Leibniz' formula, we may compute

$$c^1_{\tilde{\alpha}\delta\sigma} = \frac{(-i)^{|\sigma|}}{\sigma!} \frac{\tilde{\alpha}!}{(\tilde{\alpha}-\sigma)!} \frac{\delta!}{(\delta-\sigma)!},$$

and

$$c^2_{\tilde{\beta}\gamma\sigma} = \frac{(-i)^{|\sigma|}}{\sigma!} \frac{\tilde{\beta}!}{(\tilde{\beta}-\sigma)!} \frac{\gamma!}{(\gamma-\sigma)!}.$$

The constants $c^1_{\tilde{\alpha}\delta\sigma}, c^2_{\tilde{\beta}\gamma\sigma}$ can be roughly bounded from above by $C_3 M^{|\sigma|}$ for some constant C_3 depending only on the order m and the dimension d. Therefore,

$$\left\| \left[x^{\tilde{\beta}} D_x^{\tilde{\alpha}}, x^\delta D_x^\gamma \right] u \right\|_{L^2} \leq C_3 \sum_{\substack{0\neq\sigma\leq\tilde{\alpha},\sigma\leq\delta}} + \sum_{\substack{0\neq\sigma\leq\tilde{\beta},\sigma\leq\gamma}} M^{|\sigma|} \left\| x^{\delta+\tilde{\beta}-\sigma} D_x^{\gamma+\tilde{\alpha}-\sigma} u \right\|_{L^2}.$$
$$(6.2.7)$$

Observe at this moment that $|\gamma| + |\delta| = M - m$ and $|\tilde{\alpha}| + |\tilde{\beta}| \leq m$ imply

$$|\gamma + \tilde{\alpha} - \sigma| + |\delta + \tilde{\beta} - \sigma| \leq M - 2|\sigma|.$$

Then, from the inductive hypothesis we have

$$\left\| x^{\delta+\tilde{\beta}-\sigma} D_x^{\gamma+\tilde{\alpha}-\sigma} u \right\|_{L^2} \leq C^{M-2|\sigma|+1} (M - 2|\sigma|)^{(M-2|\sigma|)\mu} \leq C^M M^{M\mu} M^{-2|\sigma|\mu}.$$

Combining this last estimate with (6.2.7) and estimating the number of the terms in the sums by a constant C_4, depending only on m and on the dimension d, by the condition $\mu \geq \frac{1}{2}$, we conclude that

$$\left\| \left[x^{\tilde{\beta}} D_x^{\tilde{\alpha}}, x^\delta D_x^\gamma \right] u \right\|_{L^2} \leq C_3 C_4 C^M M^{M\mu}. \qquad (6.2.8)$$

By similar arguments, observing that $|\alpha - \gamma| \leq m$ and $|\beta - \delta| \leq m$, we may estimate the last term in the right-hand side of (6.2.6) as follows:

$$\left\| x^{\beta-\delta} \left[x^\delta, D_x^{\alpha-\gamma} \right] D_x^\gamma u \right\|_{L^2} \leq C_5 C^M M^{M\mu}. \qquad (6.2.9)$$

In (6.2.6), we also have

$$\left\|x^\delta D_x^\gamma u\right\| \le C^{M-m+1}(M-m)^{(M-m)\mu} \le C^M M^{M\mu}. \tag{6.2.10}$$

Inserting (6.2.8), (6.2.9) and (6.2.10) in (6.2.6) and denoting by C_6 the number of terms in the sum with $|\tilde{\alpha}| + |\tilde{\beta}| \le m$, we finally get

$$\left\|x^\beta D_x^\alpha u\right\|_{L^2} \le \left(C^* C_1^{M+1} + C^* C_2 C_3 C_4 C_6 C^M + C^* C^M + C_5 C^M\right) M^{M\mu}.$$

Hence, assuming the inductive constant C sufficiently large, we obtain

$$\left\|x^\beta D_x^\alpha u\right\|_{L^2} \le C^{M+1} M^{M\mu} \quad \text{for } |\alpha| + |\beta| \le M,$$

which gives the conclusion, i.e., $u \in S_\mu^\mu(\mathbb{R}^d)$. □

We have in particular from Proposition 6.1.8 that all the solutions $u \in \mathcal{S}'(\mathbb{R}^d)$ of the Γ-elliptic equation $Pu = 0$ extend as entire functions $u(x + iy)$ in $\mathbb{C}^d$, satisfying for suitable $\epsilon > 0$, $\delta > 0$:

$$|u(x + iy)| \lesssim e^{-\epsilon|x|^2 + \delta|y|^2}, \quad x \in \mathbb{R}^d, \ y \in \mathbb{R}^d. \tag{6.2.11}$$

A natural question, in view of the applications to Mathematical Physics, is whether this is valid for the solutions of the corresponding semilinear equations. Let us limit attention to the case of the Schrödinger harmonic oscillator:

$$-\Delta u + |x|^2 u - \lambda u = G[u], \tag{6.2.12}$$

where $\lambda \in \mathbb{R}$ and the nonlinear term is $u^k, k \ge 2$, or more generally

$$G[u] = L(u^k), \quad k \ge 2, \tag{6.2.13}$$

with

$$L = \sum_{|\alpha|+|\beta|\le 1} c_{\alpha\beta} x^\alpha D^\beta, \quad c_{\alpha\beta} \in \mathbb{C}. \tag{6.2.14}$$

As usual in the nonlinear case, we shall argue on solutions which already possess a certain regularity, expressed here in terms of the standard Sobolev spaces $H^s(\mathbb{R}^d)$.

Theorem 6.2.2. Let $u \in H^s(\mathbb{R}^d)$, $s > \frac{d}{2}$, be a solution of (6.2.12), (6.2.13), (6.2.14). Then $u \in S_{\frac{1}{2}}^1(\mathbb{R}^d)$.

Therefore, from Proposition 6.1.7 and (6.1.8) we have that $u(x)$ keeps the super-exponential decay of order 2, i.e., for a constant $\epsilon > 0$,

$$|u(x)| \lesssim e^{-\epsilon|x|^2} \quad \text{for } x \in \mathbb{R}^d, \tag{6.2.15}$$

whereas the extension to the complex domain $u(x + iy)$ is analytic only in a strip $\{x + iy \in \mathbb{C}^d : |y| < T\}$, for some $T > 0$, not entire in general.

To prove Theorem 6.2.2 we need some preliminary considerations. In the following identities and estimates, assume $u \in \mathcal{S}(\mathbb{R}^d)$ and when possible extend by density to $u \in H^s(\mathbb{R}^d)$. We denote for short $\|\cdot\|_s$ the norm in $H^s(\mathbb{R}^d)$. Write

$$H = -\Delta + |x|^2 \qquad (6.2.16)$$

and consider the inverse H^{-1} on $\mathcal{S}(\mathbb{R}^d)$. Appealing to the results of Section 2.2 we have $H^{-1} \in \mathrm{OP}\Gamma^{-2}(\mathbb{R}^d)$, hence for $p \in \mathbb{N}^d$, $q \in \mathbb{N}^d$ with $|p| + |q| \leq 2$,

$$H^{-1} \circ x^p D^q \in \mathrm{OP}\Gamma^0(\mathbb{R}^d).$$

On the other hand, operators in $\mathrm{OP}\Gamma^0(\mathbb{R}^d)$ are bounded on the standard Sobolev spaces, cf. Theorem 2.1.10, so we have proved the following smoothing property.

Proposition 6.2.3. *For every* $p \in \mathbb{N}^d$, $q \in \mathbb{N}^d$, *with* $|p| + |q| \leq 2$, *we have*

$$\left\| H^{-1} \circ x^p D^q u \right\|_s \leq C \left\| u \right\|_s$$

with C *depending on* $s \geq 0$.

We then introduce the notation, for $\mu > 0$, $\nu > 0$, $\alpha \in \mathbb{N}^d$, $\beta \in \mathbb{N}^d$, $\epsilon > 0$, $\delta > 0$:

$$[u^{(\alpha,\beta)}]_{\epsilon,\delta}^{\mu,\nu}(x) = \frac{\epsilon^{|\alpha|}\delta^{|\beta|}}{\alpha!^\nu \beta!^\mu} x^\alpha \partial_x^\beta u(x). \qquad (6.2.17)$$

We shall refer in the sequel to the norms (6.1.39); with the notation (6.2.17) they can be re-written

$$\| u \|_{s,\mu,\nu;\epsilon,\delta} = \sum_{\alpha \in \mathbb{N}^d, \beta \in \mathbb{N}^d} \left\| [u^{(\alpha,\beta)}]_{\epsilon,\delta}^{\mu,\nu} \right\|_s. \qquad (6.2.18)$$

In view of Proposition 6.1.11, we have that u belongs to $S_\nu^\mu(\mathbb{R}^d)$ if and only if $\| u \|_{s,\mu,\nu;\epsilon,\delta} < \infty$ for some $\epsilon > 0$, $\delta > 0$. We also define the partial sums

$$S_N^{\epsilon,\delta}[u] = \sum_{|\alpha|+|\beta|\leq N} \left\| [u^{(\alpha,\beta)}]_{\epsilon,\delta}^{\mu,\nu} \right\|_s, \qquad N = 0, 1, \dots, \qquad (6.2.19)$$

where the dependence on μ, ν and s is omitted in the notation for shortness.

To prove that $u \in S_\nu^\mu(\mathbb{R}^d)$, it is sufficient to prove that the sequence $S_N^{\epsilon,\delta}[u]$ remains bounded for some $\epsilon > 0$, $\delta > 0$, $s \geq 0$.

Proof of Theorem 6.2.2. First part. Write

$$H[u^{(\alpha,\beta)}]_{\epsilon,\delta}^{\mu,\nu} = [(Hu)^{(\alpha,\beta)}]_{\epsilon,\delta}^{\mu,\nu} + M_{\mu,\nu,\epsilon,\delta}^{\alpha,\beta}[u] \qquad (6.2.20)$$

with

$$M_{\mu,\nu,\epsilon,\delta}^{\alpha,\beta}[u] = \frac{\epsilon^{|\alpha|}\delta^{|\beta|}}{\alpha!^\nu \beta!^\mu}[H, x^\alpha \partial^\beta]u. \qquad (6.2.21)$$

Let u be a solution of (6.2.12), that is

$$Hu = \lambda u + G[u]. \tag{6.2.22}$$

Inserting this into (6.2.20) we obtain the identity

$$H[u^{(\alpha,\beta)}]_{\epsilon,\delta}^{\mu,\nu} = M_{\mu,\nu,\epsilon,\delta}^{\alpha,\beta}[u] + \lambda[u^{(\alpha,\beta)}]_{\epsilon,\delta}^{\mu,\nu} + [G[u]^{(\alpha,\beta)}]_{\epsilon,\delta}^{\mu,\nu}. \tag{6.2.23}$$

Finally we apply H^{-1} to both sides, and we have

$$[u^{(\alpha,\beta)}]_{\epsilon,\delta}^{\mu,\nu} = H^{-1}M_{\mu,\nu,\epsilon,\delta}^{\alpha,\beta}[u] + \lambda H^{-1}[u^{(\alpha,\beta)}]_{\epsilon,\delta}^{\mu,\nu} + H^{-1}[G[u]^{(\alpha,\beta)}]_{\epsilon,\delta}^{\mu,\nu}. \tag{6.2.24}$$

Assuming now $u \in H^s(\mathbb{R}^d)$ for a fixed $s > \frac{d}{2}$, we estimate the H^s-norm of the left-hand side of (6.2.24) by the sum of the H^s-norms of the terms in the right-hand side. From (6.2.19), (6.2.21) we then obtain

$$S_N^{\epsilon,\delta}[u] \leq A_N^{\epsilon,\delta}[u] + B_N^{\epsilon,\delta}[u] + C_N^{\epsilon,\delta}[u], \quad N = 0, 1, \dots, \tag{6.2.25}$$

where

$$A_N^{\epsilon,\delta}[u] = \sum_{0 < |\alpha|+|\beta| \leq N} \frac{\epsilon^{|\alpha|}\delta^{|\beta|}}{\alpha!^\nu \beta!^\mu} \left\| H^{-1}[H, x^\alpha \partial^\beta] u \right\|_s, \tag{6.2.26}$$

$$B_N^{\epsilon,\delta}[u] = \sum_{|\alpha|+|\beta| \leq N} \left\| \lambda H^{-1}[u^{(\alpha,\beta)}]_{\epsilon,\delta}^{\mu,\nu} \right\|_s, \tag{6.2.27}$$

$$C_N^{\epsilon,\delta}[u] = \sum_{|\alpha|+|\beta| \leq N} \left\| H^{-1}[G[u]^{(\alpha,\beta)}]_{\epsilon,\delta}^{\mu,\nu} \right\|_s. \tag{6.2.28}$$

We want now to prove by induction on N that the sequence $S_N^{\epsilon,\delta}[u]$ is bounded if $\mu \geq 1, \nu \geq \frac{1}{2}$ and ϵ, δ are sufficiently small. To this end we first estimate $A_N^{\epsilon,\delta}[u], B_N^{\epsilon,\delta}[u], C_N^{\epsilon,\delta}[u]$ in (6.2.26), (6.2.27), (6.2.28) in terms of $S_{N-1}^{\epsilon,\delta}[u]$, see the following lemmata.

Lemma 6.2.4. *Assume $\mu \geq \frac{1}{2}$, $\nu \geq \frac{1}{2}$. We can find $C_0 > 0$ such that*

$$A_N^{\epsilon,\delta}[u] \leq C_0(\epsilon + \delta)S_{N-1}^{\epsilon,\delta}[u],$$

for all $N = 1, 2, \dots$, and all ϵ, δ with $0 < \epsilon < 1$, $0 < \delta < 1$.

Lemma 6.2.5. *Assume $\mu \geq \frac{1}{2}$, $\nu \geq \frac{1}{2}$. We can find $C_1 > 0$ such that*

$$B_N^{\epsilon,\delta}[u] \leq C_1 \left(\|u\|_s + (\epsilon + \delta)S_{N-1}^{\epsilon,\delta}[u] \right),$$

for all $N = 1, 2, \dots$, and all ϵ, δ with $0 < \epsilon < 1$, $0 < \delta < 1$.

Lemma 6.2.6. *Assume* $\mu \geq 1$, $\nu \geq \frac{1}{2}$. *Let* k *be the integer in the nonlinearity* (6.2.13). *We can find* $C_2 > 0$ *such that*

$$C_N^{\epsilon,\delta}[u] \leq C_2 \big(\|u\|_s^k + (\epsilon + \delta)(S_{N-1}^{\epsilon,\delta}[u])^k \big),$$

for all $N = 1, 2, \ldots,$ *and all* ϵ, δ *with* $0 < \epsilon < 1$, $0 < \delta < 1$.

For sake of simplicity, we limit the proof of the previous lemmata to the one-dimensional case, the generalization to the d-dimensional case involving only notational complications. So in the sequel assume $\alpha \in \mathbb{N}$, $\beta \in \mathbb{N}$, $x \in \mathbb{R}$, etc.; also write here $\mathcal{D}$ for $\frac{d}{dx}$.

Proof of Lemma 6.2.4. In view of Proposition 6.2.3, the estimates are easy for a fixed N, by taking C_0 sufficiently large, depending on N. Then, it will not be restrictive to assume in the following $N > 8$. We first compute

$$- [H, x^\alpha \partial^\beta] = [\mathcal{D}^2, x^\alpha \mathcal{D}^\beta] - [x^2, x^\alpha \mathcal{D}^\beta]$$
$$= \alpha(\alpha - 1)x^{\alpha-2}\mathcal{D}^\beta + 2\alpha x^{\alpha-1}\mathcal{D}^{\beta+1} + \beta(\beta - 1)x^\alpha \mathcal{D}^{\beta-2} + 2\beta x^{\alpha+1}\mathcal{D}^{\beta-1}. \quad (6.2.29)$$

The terms with negative exponents do not appear, if $\alpha \leq 1$ or $\beta \leq 1$. We insert (6.2.29) in (6.2.26) and we estimate separately the four resulting terms. Consider initially

$$A_N^1 = \sum_{\alpha+\beta\leq N} \frac{\epsilon^\alpha \delta^\beta}{\alpha!^\nu \beta!^\mu} \alpha(\alpha - 1) \left\| H^{-1} \circ x^{\alpha-2}\mathcal{D}^\beta u \right\|_s. \quad (6.2.30)$$

Fix first attention on the terms of the sum with $\alpha \geq 4$. In this case we may write

$$H^{-1} \circ x^{\alpha-2}\mathcal{D}^\beta = H^{-1} \circ x^2 x^{\alpha-4}\mathcal{D}^\beta$$

and applying Proposition 6.2.3 we have

$$\left\| H^{-1} \circ x^{\alpha-2}\mathcal{D}^\beta u \right\|_s \leq C \left\| x^{\alpha-4}\mathcal{D}^\beta u \right\|_s.$$

Hence for $\epsilon < 1$,

$$\frac{\epsilon^\alpha \delta^\beta}{\alpha!^\nu \beta!^\mu} \alpha(\alpha - 1) \left\| H^{-1} \circ x^{\alpha-2}\mathcal{D}^\beta u \right\|_s \leq \epsilon C_0 \frac{\epsilon^{\alpha-4}\delta^\beta}{(\alpha-4)!^\nu \beta!^\mu} \left\| x^{\alpha-4}\mathcal{D}^\beta u \right\|_s \quad (6.2.31)$$

where we choose C_0 so that

$$C \frac{\alpha(\alpha - 1)}{\alpha^\nu(\alpha - 1)^\nu(\alpha - 2)^\nu(\alpha - 3)^\nu} \leq C_0. \quad (6.2.32)$$

Note that the left-hand side of (6.2.32) is bounded with respect to α since $\nu \geq \frac{1}{2}$. The sum for $\alpha + \beta \leq N$ of the terms in the right-hand side of (6.2.31) is estimated by $\epsilon C_0 S_{N-4}^{\epsilon,\delta}[u] \leq \epsilon C_0 S_{N-1}^{\epsilon,\delta}[u]$.

It remains to consider the terms in (6.2.30) with $\alpha = 2$ and $\alpha = 3$.

If $\alpha = 2$, we may assume $\beta \geq 2$ and write

$$H^{-1} \circ \mathcal{D}^\beta = H^{-1} \circ \mathcal{D}^2 \mathcal{D}^{\beta - 2}.$$

Applying Proposition 6.2.3 we have, for $\epsilon < 1$, $\delta < 1$,

$$\frac{\epsilon^2 \delta^\beta}{2^\nu \beta!^\mu} \left\| H^{-1} \circ \mathcal{D}^\beta u \right\|_s \leq \epsilon C_0 \frac{\delta^{\beta - 2}}{(\beta - 2)!^\mu} \left\| \mathcal{D}^{\beta - 2} u \right\|_s \qquad (6.2.33)$$

for a suitably large C_0. Since $2 + \beta \leq N$, the right-hand side of (6.2.33) is again a term of $\epsilon C_0 S_{N-4}^{\epsilon,\delta}[u]$. Similarly we argue for $\alpha = 3$. Summing up, we have proved that A_N^1 in (6.2.30) can be estimated by $\epsilon C_0 S_{N-1}^{\epsilon,\delta}[u]$, if C_0 is large enough.

Consider now

$$A_N^2 = \sum_{\alpha + \beta \leq N} \frac{\epsilon^\alpha \delta^\beta}{\alpha!^\nu \beta!^\mu} 2\alpha \left\| H^{-1} \circ x^{\alpha - 1} \mathcal{D}^{\beta + 1} u \right\|_s. \qquad (6.2.34)$$

Fixing first attention on the terms of the sum with $\alpha \geq 4$, we write

$$x^{\alpha - 1} \mathcal{D}^{\beta + 1} = x \mathcal{D} \circ x^{\alpha - 2} \mathcal{D}^\beta - (\alpha - 2) x^{\alpha - 2} \mathcal{D}^\beta \qquad (6.2.35)$$

and estimate consequently the right-hand side (6.2.34). By Proposition 6.2.3 we have, for $\epsilon < 1$,

$$\frac{\epsilon^\alpha \delta^\beta}{\alpha!^\nu \beta!^\mu} 2\alpha \left\| H^{-1} \circ x \mathcal{D} \circ x^{\alpha - 2} \mathcal{D}^\beta u \right\|_s \leq \epsilon C_0 \frac{\epsilon^{\alpha - 2} \delta^\beta}{(\alpha - 2)!^\nu \beta!^\mu} \left\| x^{\alpha - 2} \mathcal{D}^\beta u \right\|_s, \qquad (6.2.36)$$

where now we choose C_0 such that

$$2C \frac{\alpha}{\alpha^\nu (\alpha - 1)^\nu} \leq C_0.$$

We recognize in the right-hand side of (6.2.36) one of the terms of the sum $\epsilon C_0 S_{N-2}^{\epsilon,\delta}[u]$. On the other hand we have from the second term in the right-hand side of (6.2.35),

$$\frac{\epsilon^\alpha \delta^\beta}{\alpha!^\nu \beta!^\mu} 2\alpha(\alpha - 2) \left\| H^{-1} \circ x^{\alpha - 2} \mathcal{D}^\beta u \right\|_s$$

which we can treat exactly as (6.2.30). Discussing as before the cases $\alpha = 1$, $\alpha = 2$, $\alpha = 3$, we conclude

$$A_N^2 \leq \epsilon C_0 S_{N-1}^{\epsilon,\delta}[u].$$

Consider then

$$A_N^3 = \sum_{\alpha + \beta \leq N} \frac{\epsilon^\alpha \delta^\beta}{\alpha!^\nu \beta!^\mu} \beta(\beta - 1) \left\| H^{-1} \circ x^\alpha \mathcal{D}^{\beta - 2} u \right\|_s. \qquad (6.2.37)$$

Assume $\alpha \geq 4$, $\beta \geq 4$. It is convenient to write

$$x^\alpha \mathcal{D}^{\beta-2} = \mathcal{D}^2 \circ x^\alpha \mathcal{D}^{\beta-4} - 2\alpha \mathcal{D} \circ x^{\alpha-1} \mathcal{D}^{\beta-4} + \alpha(\alpha-1)x^{\alpha-2}\mathcal{D}^{\beta-4}. \quad (6.2.38)$$

Inserting in (6.2.37) we get three terms which we estimate as follows:

$$\frac{\epsilon^\alpha \delta^\beta}{\alpha!^\nu \beta!^\mu} \beta(\beta-1) \left\| H^{-1} \circ \mathcal{D}^2 \circ x^\alpha \mathcal{D}^{\beta-4} u \right\|_s$$

$$\leq \delta C_0 \frac{\epsilon^\alpha \delta^{\beta-4}}{\alpha!^\nu (\beta-4)!^\mu} \left\| x^\alpha \mathcal{D}^{\beta-4} u \right\|_s \quad (6.2.39)$$

where we may choose C_0 so that

$$2C \frac{\beta(\beta-1)}{\beta^\mu(\beta-1)^\mu(\beta-2)^\mu(\beta-3)^\mu} \leq C_0.$$

We use here the assumption $\mu \geq \frac{1}{2}$. As before, C is the constant in Proposition 6.2.3 and $\delta < 1$. Moreover:

$$\frac{\epsilon^\alpha \delta^\beta}{\alpha!^\nu \beta!^\mu} 2\alpha\beta(\beta-1) \left\| H^{-1} \circ \mathcal{D} \circ x^{\alpha-1} \mathcal{D}^{\beta-4} u \right\|_s$$

$$\leq \delta C_0 \frac{\epsilon^{\alpha-2} \delta^{\beta-4}}{(\alpha-2)!^\nu (\beta-4)!^\mu} \left\| x^{\alpha-2} \mathcal{D}^{\beta-4} u \right\|_s \quad (6.2.40)$$

where now we require

$$2C \frac{\alpha\beta(\beta-1)}{\alpha^\nu(\alpha-1)^\nu \beta^\mu(\beta-1)^\mu(\beta-2)^\mu(\beta-3)^\mu} \leq C_0.$$

Similarly

$$\frac{\epsilon^\alpha \delta^\beta}{\alpha!^\nu \beta!^\mu} \alpha(\alpha-1)\beta(\beta-1) \left\| H^{-1} \circ x^{\alpha-2} \mathcal{D}^{\beta-4} u \right\|_s$$

$$\leq \delta C_0 \frac{\epsilon^{\alpha-4} \delta^{\beta-4}}{(\alpha-4)!^\nu (\beta-4)!^\mu} \left\| x^{\alpha-4} \mathcal{D}^{\beta-4} u \right\|_s. \quad (6.2.41)$$

In the right-hand side of (6.2.39), (6.2.40), (6.2.41) we recognize terms from $\delta C_0 S_{N-4}^{\epsilon,\delta}[u]$. We leave to the reader the discussion of the case when $\alpha < 4$ or $\beta < 4$ in (6.2.37), as well as the proof of the estimate

$$A_N^4 = \sum_{\alpha+\beta \leq N} \frac{\epsilon^\alpha \delta^\beta}{\alpha!^\nu \beta!^\mu} 2\beta \left\| H^{-1} \circ x^{\alpha+1} \mathcal{D}^{\beta-1} u \right\|_s$$

$$\leq \delta C_0 S_{N-1}^{\epsilon,\delta}[u].$$

We then conclude

$$A_N^{\epsilon,\delta}[u] \leq A_N^1 + A_N^2 + A_N^3 + A_N^4 \leq C_0(\epsilon+\delta)S_{N-1}^{\epsilon,\delta}[u],$$

for a new constant C_0. Lemma 6.2.4 is proved. $\qquad\square$

Proof of Lemma 6.2.5. Proposition 6.2.3 yields, for some $C_1 > 0$,

$$\left\|\lambda H^{-1}([u^{(\alpha,\beta)}]^{\mu,\nu}_{\epsilon,\delta})\right\|_s \leq C_1 \frac{\epsilon}{\alpha^\nu}\left\|[u^{(\alpha-1,\beta)}]^{\mu,\nu}_{\epsilon,\delta}\right\|_s \qquad (6.2.42)$$

if $\alpha \geq 1$ and otherwise

$$\left\|\lambda H^{-1}([u^{(0,\beta)}]^{\mu,\nu}_{\epsilon,\delta})\right\|_s \leq C_1 \frac{\delta}{\beta\mu}\left\|[u^{(0,\beta-1)}]^{\mu,\nu}_{\epsilon,\delta}\right\|_s \qquad (6.2.43)$$

if $\beta \geq 1$. The sum of the terms in the right-hand side of (6.2.42), (6.2.43) is estimated by

$$C_1(\epsilon + \delta)S^{\epsilon,\delta}_{N-1}[u].$$

Taking into account the case $\alpha = 0$, $\beta = 0$ we obtain Lemma 6.2.5.   □

Proof of Lemma 6.2.6. Here we use the assumption $u \in H^s(\mathbb{R}^d)$, $s > \frac{d}{2}$. In fact, we know from Schauder's estimates that for $u \in H^s(\mathbb{R}^d)$, $v \in H^s(\mathbb{R}^d)$, $s > \frac{d}{2}$, we have

$$\|uv\|_s \leq \sigma \|u\|_s \|v\|_s \qquad (6.2.44)$$

for a constant $\sigma \geq 1$ depending on s. Hence $u^k \in H^s(\mathbb{R}^d)$ in (6.2.13). Therefore, in view of (6.2.14) and Proposition 6.2.3, the definition of $C^{\epsilon,\delta}_N[u]$ in (6.2.28) makes sense for $N = 0$.

Let us assume $N \geq 1$ and prove Lemma 6.2.6. As before, we limit now attention to the one-dimensional case. We begin by some inequalities for the nonlinear terms. Define:

$$E^\delta_N[u] = \sum_{\gamma \leq N} \frac{\delta^\gamma}{\gamma!^\mu}\|\mathcal{D}^\gamma u\|_s. \qquad (6.2.45)$$

Obviously $E^\delta_N[u] \leq S^{\epsilon,\delta}_N[u]$. We have for $\mu \geq 1$,

$$E^\delta_N[u^k] \leq \sigma^{k-1}(E^\delta_N[u])^k, \qquad (6.2.46)$$

where σ is the constant in (6.2.44). In fact, by Leibniz' rule and Schauder estimates:

$$\left\|\mathcal{D}^\gamma u^k\right\|_s \leq \sigma^{k-1} \sum_{\gamma_1+\ldots+\gamma_k=\gamma} \frac{\gamma!}{\gamma_1!\cdots\gamma_k!}\|\mathcal{D}^{\gamma_1}u\|_s\cdots\|\mathcal{D}^{\gamma_k}u\|_s. \qquad (6.2.47)$$

Hence we may write

$$\frac{\delta^\gamma}{\gamma!^\mu}\left\|\mathcal{D}^\gamma u^k\right\|_s \leq \sigma^{k-1}\sum_{\gamma_1+\ldots+\gamma_k=\gamma}\lambda^{(\gamma)}_{\gamma_1,\ldots,\gamma_k}\left(\frac{\delta^{\gamma_1}}{\gamma_1!^\mu}\|\partial^{\gamma_1}u\|_s\right)\cdots\left(\frac{\delta^{\gamma_k}}{\gamma_k!^\mu}\|\partial^{\gamma_k}u\|_s\right),$$
$$(6.2.48)$$

where

$$\lambda^{(\gamma)}_{\gamma_1,\ldots,\gamma_k} = \left(\frac{\gamma!}{\gamma_1!\cdots\gamma_k!}\right)^{1-\mu} \leq 1 \tag{6.2.49}$$

provided $\mu \geq 1$. From (6.2.48), (6.2.49) we obtain immediately (6.2.46). The same arguments give

$$S^{\epsilon,\delta}_N[u^k] \leq \sigma^{k-1}(S^{\epsilon,\delta}_N[u])^k. \tag{6.2.50}$$

In fact, it will be sufficient to write

$$\left\|x^\alpha \mathcal{D}^\beta u^k\right\|_s \leq \sigma^{k-1} \sum_{\beta_1+\ldots+\beta_k=\beta} \frac{\beta!}{\beta_1!\cdots\beta_k!} \left\|x^\alpha \mathcal{D}^{\beta_1} u\right\|_s \left\|\mathcal{D}^{\beta_2} u\right\|_s \cdots \left\|\mathcal{D}^{\beta_k} u\right\|_s$$

and proceed as before, to obtain

$$S^{\epsilon,\delta}_N[u^k] \leq \sigma^{k-1} S^{\epsilon,\delta}_N[u](E^\delta_N[u])^{k-1},$$

which gives in particular (6.2.50). Note that we do not need to strengthen the assumption $\nu \geq \frac{1}{2}$, for the validity of (6.2.50).

We can now estimate $C^{\epsilon,\delta}_N[u]$ in (6.2.28). Let us first consider the case when $G[u] = u^k$, $k \geq 2$. We have

$$\sum_{\alpha+\beta\leq N} \left\|H^{-1}[u^k]^{(\alpha,\beta)}_{(\epsilon,\delta)}\right\|_s = \left\|H^{-1}u^k\right\|_s + \sum_{\substack{\beta\leq N \\ \beta\neq 0}} \frac{\delta^\beta}{\beta!^\mu} \left\|H^{-1}\mathcal{D}^\beta u^k\right\|_s$$

$$+ \sum_{\substack{\alpha+\beta\leq N \\ \alpha\neq 0}} \frac{\epsilon^\alpha \delta^\beta}{\alpha!^\nu \beta!^\mu} \left\|H^{-1}(x^\alpha \mathcal{D}^\beta u^k)\right\|_s. \tag{6.2.51}$$

We estimate separately the three terms in the right-hand side. From Proposition 6.2.3 and (6.2.44) we have

$$\left\|H^{-1}u^k\right\|_s \leq C\left\|u^k\right\|_s \leq C\sigma^{k-1}\left\|u\right\|_s^k. \tag{6.2.52}$$

On the other hand from Proposition 6.2.3 and (6.2.46)

$$\sum_{\substack{\beta\leq N \\ \beta\neq 0}} \frac{\delta^\beta}{\beta!^\mu} \left\|H^{-1}\mathcal{D}^\beta u^k\right\|_s \leq C\frac{\delta}{\beta^\mu} \sum_{\substack{\beta\leq N \\ \beta\neq 0}} \frac{\delta^{\beta-1}}{(\beta-1)!^\mu} \left\|\mathcal{D}^{\beta-1} u^k\right\|_s$$

$$\leq C\delta E^\delta_{N-1}[u^k] \leq C\sigma^{k-1}\delta(E^\delta_{N-1}[u])^k. \tag{6.2.53}$$

Similarly from (6.2.50)

$$\sum_{\substack{\alpha+\beta\leq N \\ \alpha\neq 0}} \frac{\epsilon^\alpha \delta^\beta}{\alpha!^\nu \beta!^\mu} \left\|H^{-1}(x^\alpha \mathcal{D}^\beta u^k)\right\|_s \leq \frac{\epsilon}{\alpha^\nu}C \sum_{\substack{\alpha+\beta\leq N \\ \alpha\neq 0}} \frac{\epsilon^{\alpha-1}\delta^\beta}{(\alpha-1)!^\nu \beta!^\mu} \left\|x^{\alpha-1}\mathcal{D}^\beta u^k\right\|_s$$

$$\leq C\epsilon S^{\epsilon,\delta}_{N-1}[u^k] \leq C\sigma^{k-1}\epsilon(S^{\epsilon,\delta}_{N-1}[u])^k. \tag{6.2.54}$$

Applying (6.2.52), (6.2.53), (6.2.54) into (6.2.51), we conclude, for a suitable constant $C_2 > 0$,

$$\sum_{\alpha+\beta\leq N} \left\| H^{-1}[u^k]_{(\epsilon,\delta)}^{(\alpha,\beta)} \right\|_s \leq C_2(\|u\|_s^k + (\epsilon+\delta)(S_{N-1}^{\epsilon,\delta}[u])^k). \tag{6.2.55}$$

Let us now consider the case $G[u] = xu^k$. It will be not restrictive to assume $N > 2$. Writing

$$\mathcal{D}^\beta(xu^k) = x\mathcal{D}^\beta u^k + \beta\mathcal{D}^{\beta-1}u^k,$$

we have

$$\sum_{\alpha+\beta\leq N} \left\| H^{-1}[xu^k]_{(\epsilon,\delta)}^{(\alpha,\beta)} \right\|_s \leq \left\| H^{-1}(xu^k) \right\|_s + \sum_{0<\beta\leq N} \frac{\delta^\beta}{\beta!^\mu} \left\| H^{-1}\circ x\mathcal{D}^\beta u^k \right\|_s$$

$$+ \sum_{0<\beta\leq N} \frac{\delta^\beta}{\beta!^\mu}\beta \left\| H^{-1}\circ \mathcal{D}^{\beta-1}u^k \right\|_s$$

$$+ \sum_{\substack{\alpha+\beta\leq N \\ \alpha\neq 0}} \frac{\epsilon^\alpha\delta^\beta}{\alpha!^\mu\beta!^\nu} \left\| H^{-1}\circ x^{\alpha+1}\mathcal{D}^\beta u^k \right\|_s$$

$$+ \sum_{\substack{\alpha+\beta\leq N \\ \alpha\neq 0}} \frac{\epsilon^\alpha\delta^\beta}{\alpha!^\mu\beta!^\nu}\beta \left\| H^{-1}\circ x^\alpha\mathcal{D}^{\beta-1}u^k \right\|_s. \tag{6.2.56}$$

Estimating the five terms in the right-hand side by the previous arguments, we obtain respectively

$$\left\| H^{-1}(xu^k) \right\|_s \leq C\sigma^{k-1}\|u\|_s^k; \tag{6.2.57}$$

$$\sum_{0<\beta\leq N} \frac{\delta^\beta}{\beta!^\mu} \left\| H^{-1}\circ x\mathcal{D}^\beta u^k \right\|_s \leq C\frac{\delta}{\beta^\mu} \sum_{0<\beta\leq N} \frac{\delta^{\beta-1}}{(\beta-1)!^\mu} \left\| \mathcal{D}^{\beta-1}u^k \right\|_s$$

$$\leq C\sigma^{k-1}\delta(E_{N-1}^\delta[u])^k; \tag{6.2.58}$$

$$\sum_{0<\beta\leq N} \frac{\delta^\beta}{\beta!^\mu}\beta \left\| H^{-1}\circ \mathcal{D}^{\beta-1}u^k \right\|_s \leq C\delta\|u^k\|_s + C_0\delta \sum_{0<\beta\leq N} \frac{\delta^{\beta-2}}{(\beta-2)!^\mu} \left\| \mathcal{D}^{\beta-2}u^k \right\|_s$$

$$\leq C\sigma^{k-1}\delta\|u\|_s^k + C_0\sigma^{k-1}\delta(E_{N-1}^\delta[u])^k, \tag{6.2.59}$$

where

$$C\frac{\beta}{\beta^\mu(\beta-1)^\mu} \leq C_0;$$

$$\sum_{\substack{\alpha+\beta\leq N \\ \alpha\neq 0}} \frac{\epsilon^\alpha\delta^\beta}{\alpha!^\nu\beta!^\mu} \left\| H^{-1}\circ x^{\alpha+1}\mathcal{D}^\beta u^k \right\|_s \leq C\frac{\epsilon}{\alpha^\nu} \sum_{\alpha+\beta\leq N,\alpha\neq 0} \frac{\epsilon^{\alpha-1}\delta^\beta}{(\alpha-1)!^\nu\beta!^\mu} \left\| x^{\alpha-1}\mathcal{D}^\beta u^k \right\|_s$$

$$\leq C\sigma^{k-1}\epsilon(S_{N-1}^{\epsilon,\delta}[u])^k. \tag{6.2.60}$$

As for the last term, we split further

$$x^\alpha \mathcal{D}^{\beta-1} u^k = x\mathcal{D} \circ x^{\alpha-1}\mathcal{D}^{\beta-2} - (\alpha-1)x^{\alpha-1}\mathcal{D}^{\beta-2};$$

arguing as in the proof of Lemma 6.2.4, we may estimate by $C_0\sigma^{k-1}\epsilon S_{N-1}^{\epsilon,\delta}[u]$, for C_0 sufficiently large. Applying this and (6.2.57), (6.2.58), (6.2.59), (6.2.60) in (6.2.56), we conclude for a new constant $C_2 > 0$:

$$\sum_{\alpha+\beta\le N} \left\| H^{-1}[xu^k]_{(\epsilon,\delta)}^{\alpha,\beta} \right\|_s \le C_2(\|u\|_s^k + (\epsilon+\delta)(S_{N-1}^{\epsilon,\delta}[u])^k). \qquad (6.2.61)$$

The case of the nonlinearity $G[u] = \mathcal{D}u^k = ku^{k-1}\mathcal{D}u$ can be easily treated by following the same lines of the proof of (6.2.55). Summing up, we obtain Lemma 6.2.6. □

End of the proof of Theorem 6.2.2. Applying Lemmata 6.2.4, 6.2.5 and 6.2.6 to the right-hand side of (6.2.25), we obtain for a suitable constant $C_3 > 1$:

$$S_N^{\epsilon,\delta}[u] \le C_3\big(\|u\|_s + \|u\|_s^k + (\epsilon+\delta)S_{N-1}^{\epsilon,\delta}[u] + (\epsilon+\delta)(S_{N-1}^{\epsilon,\delta}[u])^k\big) \qquad (6.2.62)$$

with $N = 1, 2, \ldots$, and $S_0^{\epsilon,\delta}[u] = \|u\|_s$. So, taking δ and ϵ sufficiently small, the sequence $S_N^{\epsilon,\delta}[u]$ is bounded by $C_3(\|u\|_s + \|u\|_s^k + 1)$, say, and Theorem 6.2.2 is proved. □

Consider now the following nonlinear perturbation of the harmonic oscillator, in dimension $d = 1$, at the first eigenvalue $\lambda = 1$,

$$u'' - x^2 u + u = \left(\frac{d}{dx} - x\right)u^k, \quad k \ge 2, \qquad (6.2.63)$$

which is of the form (6.2.12), (6.2.13), (6.2.14). Theorem 6.2.2 applies and every solution $u \in H^s(\mathbb{R})$, $s > \frac{1}{2}$, belongs to $S_{\frac{1}{2}}^1(\mathbb{R})$. We shall check that such non-trivial solutions exist, expressed in terms of special functions, and prove the following:

Proposition 6.2.7. *There exist solutions* $u \in S_{\frac{1}{2}}^1(\mathbb{R})$ *of* (6.2.63) *which are not entire functions.*

To give an intuitive explanation of the loss of analyticity in $\mathbb{C}$, and the maintenance of super-exponential decay, we may observe that the nonlinearity in (6.2.13), (6.2.14) violates the symmetry with respect to the whole phase space variables. Indeed, the harmonic oscillator, and the class of Γ-elliptic operators, are invariant under the action of the Fourier transformation $\mathcal{F}$; this is congruent with the statement of Theorem 6.2.1. In fact, the Fourier transformation acts as an isomorphism

$$\mathcal{F}: S_\nu^\mu(\mathbb{R}^d) \to S_\mu^\nu(\mathbb{R}^d),$$

so in particular $\mathcal{F}\left(S^{\frac{1}{2}}_{\frac{1}{2}}(\mathbb{R}^d)\right) = S^{\frac{1}{2}}_{\frac{1}{2}}(\mathbb{R}^d)$. Otherwise, if we transform (6.2.63), setting $v = \mathcal{F}(u)$ and writing again x for the variable, we get the new equation with convolution in the right-hand side

$$v'' - x^2 v + v = -i(2\pi)^{\frac{k-1}{2}}\left(\frac{d}{dx} - x\right)\underbrace{(v * \cdots * v)}_{k \text{ times}}. \tag{6.2.64}$$

This admits solutions $v \in S^{\frac{1}{2}}_{1}(\mathbb{R}) = \mathcal{F}\left(S^{1}_{\frac{1}{2}}(\mathbb{R})\right)$, therefore now v is entire but the super-exponential decay is lost.

Proof. We may reduce to the first-order model

$$u' + xu = u^k. \tag{6.2.65}$$

In fact, composing the operators $\frac{d}{dx} \pm x$, cf. (2.2.20), we have

$$\left(\frac{d}{dx} - x\right)\left(\frac{d}{dx} + x\right)u = u'' - x^2 u + u.$$

Therefore, every solution $u \in H^2(\mathbb{R})$ of (6.2.65) is also a solution of (6.2.63). On the other hand, (6.2.65) is a Bernoulli equation, which we can treat explicitly. It will be convenient to refer to the complementary error function defined by

$$\mathrm{Erfc}(t) = \int_t^{+\infty} c^{-v^2}\,dv$$

which is positive and decreasing for $t \in \mathbb{R}$, with

$$\lim_{t \to -\infty} \mathrm{Erfc}(t) = \sqrt{\pi}, \quad \mathrm{Erfc}(0) = \frac{\sqrt{\pi}}{2}, \quad \lim_{t \to +\infty} \mathrm{Erfc}(t) = 0 \tag{6.2.66}$$

and asymptotic expansion for $t \to +\infty$,

$$\mathrm{Erfc}(t) \sim \frac{e^{-t^2}}{2t}\left(1 - \frac{1}{2t^2} + \cdots\right). \tag{6.2.67}$$

Setting $u(0) = u_0 > 0$ to fix ideas, we get the solution of (6.2.65), (6.2.63)

$$u(x) = e^{-\frac{x^2}{2}}\left[\lambda + \sqrt{2k - 2}\,\mathrm{Erfc}\left(\sqrt{\frac{k-1}{2}}\,x\right)\right]^{\frac{1}{1-k}} \tag{6.2.68}$$

with $\lambda = u_0^{1-k} - \sqrt{\frac{\pi(k-1)}{2}}$. Here and in the following, roots are defined to be positive for positive numbers, with continuous extension to the complex domain, i.e., we take principal branches. Let us test Theorem 6.2.2 on (6.2.63), by distinguishing three cases (see Figure 6.1 for the case $k = 2$).

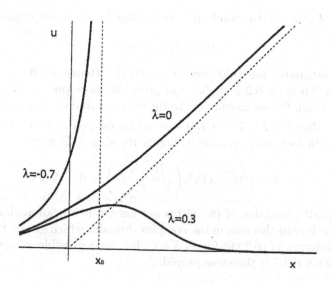

Figure 6.1: The solutions $u(x)$ in (6.2.68), with $k = 2$.

(i) $-\sqrt{\frac{\pi(k-1)}{2}} < \lambda < 0$, that is $\tilde{u}_0 < u_0 < +\infty$, with

$$\tilde{u}_0 = \left(\frac{\pi(k-1)}{2} \right)^{\frac{1}{2-2k}}.$$

Then, the solution blows up at the point $x_0 > 0$, defined uniquely by imposing

$$\sqrt{2k-2}\,\mathrm{Erfc}\left(\sqrt{\frac{k-1}{2}}x \right) = -\lambda,$$

cf. (6.2.66).

(ii) $\lambda = 0$, i.e., $u_0 = \tilde{u}_0$. Then the solution is well defined analytic in $\mathbb{R}$. The decay at $-\infty$ is superexponential, whereas from (6.2.67) we get

$$u(x) \sim x^{\frac{1}{k-1}} \quad x \to +\infty.$$

Note that $u \in \mathcal{S}'(\mathbb{R})$, but Theorem 6.2.2 cannot be applied since $u \notin H^s(\mathbb{R})$, $s > \frac{1}{2}$.

(iii) $\lambda > 0$, that is $0 < u_0 < \tilde{u}_0$. In this case, since

$$0 < \lambda < \lambda + \sqrt{2k-2}\,\mathrm{Erfc}\left(\sqrt{\frac{k-1}{2}}x \right)$$

in view of (6.2.66), the solution is well defined analytic in $\mathbb{R}$ and

$$0 < u(x) < \lambda^{\frac{1}{1-k}} e^{-\frac{x^2}{2}}.$$

Similar estimates are valid for $u'(x)$, $u''(x)$. Hence we have $u \in H^2(\mathbb{R})$, therefore Theorem 6.2.2 applies and gives the more precise information $u \in S^1_{\frac{1}{2}}(\mathbb{R})$. Then, the extension $u(z)$ to the complex domain is analytic in a strip $\{z \in \mathbb{C} : |\operatorname{Im} z| < T\}$, $T > 0$, but it is not an entire function, as evident from (6.2.68). In fact, $u(z)$ presents a singularity at $z_0 \in \mathbb{C}$ when

$$\lambda + \sqrt{2k - 2}\,\operatorname{Erfc}\left(\sqrt{\frac{k-1}{2}}\,z_0\right) = 0, \quad \lambda > 0. \tag{6.2.69}$$

The explicit discussion of (6.2.69) is not easy, but we can anyhow appeal to the great Picard theorem in the complex domain, which grants the existence of a solution z_0 of (6.2.69) for all $\lambda \in \mathbb{C}$, but for a possible exceptional value. Proposition 6.2.7 is therefore proved. $\qquad\square$

6.3 G-Pseudo-Differential Operators on $S^\mu_\nu(\mathbb{R}^d)$

In this section and in the sequel of the chapter we shall fix attention on G-elliptic equations. Basic examples are linear partial differential operators with polynomial coefficients in $\mathbb{R}^d$ of the form

$$P = \sum_{|\alpha| \le m, |\beta| \le n} c_{\alpha\beta} x^\beta D^\alpha_x \tag{6.3.1}$$

with $m > 0$, $n \ge 0$, satisfying the G-ellipticity condition

$$\left| \sum_{|\alpha| \le m, |\beta| \le n} c_{\alpha\beta} x^\beta \xi^\alpha \right| \ge C \langle \xi \rangle^m \langle x \rangle^n \quad \text{for } |x| + |\xi| \ge R \tag{6.3.2}$$

for constants $C > 0$, $R > 0$. We know from the results of Chapter 3 that all the solutions $u \in S'(\mathbb{R}^d)$ of the homogeneous equation $Pu = 0$ belong to $S(\mathbb{R}^d)$. To obtain a more precise result in terms of $S^\mu_\nu(\mathbb{R}^d)$, in this section we begin to present a class of analytic-type G-pseudo-differential operators. For P in (6.3.1), (6.3.2), the corresponding calculus will give that the solutions $u \in S'(\mathbb{R}^d)$ of $Pu = 0$ belong to $S^\theta_\theta(\mathbb{R}^d)$ for any $\theta > 1$.

Let μ, ν be real numbers such that $\mu \ge 1$, $\nu \ge 1$, and let $m \in \mathbb{R}$, $n \in \mathbb{R}$. Let $\theta \ge \max\{\mu, \nu\}$.

Definition 6.3.1. For every $C > 0$, we denote by $AG^{m,n}_{\mu\nu}(\mathbb{R}^d; C)$ the Banach space of all functions $p(x, \xi) \in C^\infty(\mathbb{R}^{2d})$ such that

$$\sup_{\alpha, \beta \in \mathbb{N}^d} \sup_{(x,\xi) \in \mathbb{R}^{2d}} C^{-|\alpha|-|\beta|} (\alpha!)^{-\mu} (\beta!)^{-\nu} \langle \xi \rangle^{-m+|\alpha|} \langle x \rangle^{-n+|\beta|} \left| D^\alpha_\xi D^\beta_x p(x, \xi) \right| < \infty,$$

$$\tag{6.3.3}$$

endowed with the norm $\| \cdot \|_C$ given by the left-hand side of (6.3.3).

We set

$$AG^{m,n}_{\mu\nu}(\mathbb{R}^d) = \varinjlim_{C \to +\infty} AG^{m,n}_{\mu\nu}(\mathbb{R}^d; C)$$

with the topology of inductive limit of an increasing sequence of Banach spaces.

Given a symbol $p \in AG^{m,n}_{\mu\nu}(\mathbb{R}^d)$, we can consider the associated pseudo-differential operator defined with the standard left quantization by

$$Pu(x) = p(x, D)u(x) = \int e^{ix\xi} p(x, \xi) \widehat{u}(\xi) \, d\xi, \quad u \in S^\theta_\theta(\mathbb{R}^d) \tag{6.3.4}$$

where $d\xi = (2\pi)^{-\frac{d}{2}} d\xi$. We denote by $OPAG^{m,n}_{\mu\nu}(\mathbb{R}^d)$ the space of all operators of the form (6.3.4) defined by a symbol $p \in AG^{m,n}_{\mu\nu}(\mathbb{R}^d)$. We set

$$OPAG_{\mu\nu}(\mathbb{R}^d) = \bigcup_{(m,n) \in \mathbb{R}^2} OPAG^{m,n}_{\mu\nu}(\mathbb{R}^d).$$

This is a subclass of the G-pseudo-differential operators in Chapter 3 on $S(\mathbb{R}^d)$, $S'(\mathbb{R}^d)$. Taking advantage of the estimates (6.3.3) we are able to prove continuity on $S^\theta_\theta(\mathbb{R}^d)$.

Theorem 6.3.2. *Let* $p \in AG^{m,n}_{\mu\nu}(\mathbb{R}^d)$ *and let* θ *be a real number such that* $\theta \geq \max\{\mu, \nu\}$. *Then, the operator* P *defined by* (6.3.4) *is a linear continuous operator from* $S^\theta_\theta(\mathbb{R}^d)$ *to* $S^\theta_\theta(\mathbb{R}^d)$ *and, when* $\theta > 1$, *it extends to a linear continuous map from* $S^{\theta\prime}_\theta(\mathbb{R}^d)$ *to* $S^{\theta\prime}_\theta(\mathbb{R}^d)$.

Proof. Let $u \in S^\theta_\theta(\mathbb{R}^d)$. Since $\mathcal{F}(S^\theta_\theta(\mathbb{R}^d)) = S^\theta_\theta(\mathbb{R}^d)$, we may start with u in a bounded subset F of the Banach space defined by the norm

$$\sup_{\beta} \sup_{\xi \in \mathbb{R}^d} A^{-|\beta|} (\beta!)^\theta e^{a|\xi|^{1/\theta}} |\partial^\beta \widehat{u}(\xi)| \tag{6.3.5}$$

for some $A > 0$, $a > 0$, cf. (6.1.43). It is sufficient to show that there exist positive constants A_1, B_1, C_1 such that, for every $\alpha, \beta \in \mathbb{N}^d$,

$$\sup_{x \in \mathbb{R}^d} |x^\alpha D^\beta_x Pu(x)| \leq C_1 A^{|\alpha|}_1 B^{|\beta|}_1 (\alpha! \beta!)^\theta \tag{6.3.6}$$

for all $u \in F$, with A_1, B_1, C_1 independent of $u \in F$. We have, for every $N \in \mathbb{N}$,

$$x^\alpha D^\beta_x Pu(x) = x^\alpha \sum_{\beta' \leq \beta} \binom{\beta}{\beta'} \int e^{ix\xi} \xi^{\beta'} D^{\beta-\beta'}_x p(x, \xi) \widehat{u}(\xi) \, d\xi$$

$$= x^\alpha \langle x \rangle^{-2N} \sum_{\beta' \leq \beta} \binom{\beta}{\beta'} \int e^{ix\xi} (1 - \Delta_\xi)^N \left[\xi^{\beta'} D^{\beta-\beta'}_x p(x, \xi) \widehat{u}(\xi) \right] d\xi.$$

By (6.3.3), (6.3.5), we easily obtain the estimate:

$$|x^\alpha D_x^\beta Pu(x)| \leq C_0 B_0^{|\beta|+2N} (2N!)^\theta \langle x \rangle^{|\alpha|+n-2N}$$

$$\times \sum_{\beta' \leq \beta} \binom{\beta}{\beta'} (\beta'!)^\theta (\beta - \beta'!)^\nu \int \langle \xi \rangle^m e^{-a|\xi|^{\frac{1}{\theta}}} d\xi$$

for some B_0, C_0 independent of $u \in F$. Choosing $N = \min\{r \in \mathbb{N} : 2r \geq |\alpha| + n\}$, we obtain that there exist $A_1, B_1, C_1 > 0$ such that (6.3.6) holds for all $u \in F$. This concludes the first part of the proof. To prove the second part, we observe that, for $u, v \in S_\theta^\theta(\mathbb{R}^d)$,

$$\int Pu(x)v(x)dx = \int \widehat{u}(\xi)p_v(\xi) d\xi$$

where

$$p_v(x, \xi) = \int e^{ix\xi} p(x, \xi) v(x) dx.$$

By the same argument of the first part of the proof, the map $v \mapsto p_v$ is linear and continuous from $S_\theta^\theta(\mathbb{R}^d)$ to itself. Then, we can define, for $u \in S_\theta^{\theta'}(\mathbb{R}^d)$,

$$Pu(v) = \widehat{u}(p_v), \quad v \in S_\theta^\theta(\mathbb{R}^d).$$

This map is linear and continuous from $S_\theta^{\theta'}(\mathbb{R}^d)$ to itself and it extends P.   □

From now on we shall assume $\mu > 1$, $\nu > 1$, hence $\theta > 1$.

By Theorems 6.1.13 and 6.3.2, we can associate to P a kernel $K_P \in S_\theta^{\theta'}(\mathbb{R}^{2d})$ given as standard by

$$K_P(x, y) = (2\pi)^{-d} \int e^{i(x-y)\xi} p(x, \xi) d\xi, \tag{6.3.7}$$

where (6.3.7) is understood in the sense of the Fourier transform of distributions, cf. (1.2.2). We can prove the following result of regularity for the kernel (6.3.7).

Theorem 6.3.3. Let $p \in AG_{\mu,\nu}^{m,n}(\mathbb{R}^d)$. For $k > 0$, define

$$\Omega_k = \{(x, y) \in \mathbb{R}^{2d}, \ |x - y| > k\langle x \rangle\}.$$

Then the kernel K_P defined by (6.3.7) is in $C^\infty(\Omega_k)$ and there exist positive constants C, a depending on k such that

$$|D_x^\beta D_y^\gamma K_P(x, y)| \leq C^{|\beta|+|\gamma|+1} (\beta! \gamma!)^\theta \exp\left[-a(|x|^{\frac{1}{\theta}} + |y|^{\frac{1}{\theta}})\right] \tag{6.3.8}$$

for every $(x, y) \in \overline{\Omega_k}$ and for every $\beta, \gamma \in \mathbb{N}^d$.

Lemma 6.3.4. *For any given* $R > 1$, *we may find a sequence* $\psi_N(\xi) \in C^\infty_0(\mathbb{R}^d)$, $N = 0, 1, 2, \ldots$, *such that* $\sum\limits_{N=0}^{\infty} \psi_N = 1$ *in* $\mathbb{R}^d$,

$$\text{supp } \psi_0 \subset \{\xi \in \mathbb{R}^d : \langle \xi \rangle \leq 3R\},$$

$$\text{supp } \psi_N \subset \{\xi \in \mathbb{R}^d : 2RN^\theta \leq \langle \xi \rangle \leq 3R(N+1)^\theta\}, \quad N = 1, 2, \ldots,$$

and

$$\left| D^\alpha_\xi \psi_N(\xi) \right| \leq C^{|\alpha|+1}(\alpha!)^\theta \left[R \sup\{N^\theta, 1\} \right]^{-|\alpha|}$$

for every $\alpha \in \mathbb{N}^d$ *and for every* $\xi \in \mathbb{R}^d$.

Proof. Let $\phi \in C^\infty_0(\mathbb{R}^d)$ such that $\phi(\xi) = 1$ if $\langle \xi \rangle \leq 2$, $\phi(\xi) = 0$ if $\langle \xi \rangle \geq 3$. Assume further $\phi \in G^\theta_0(\mathbb{R}^d)$, i.e., according to (6.1.46):

$$\left| D^\alpha_\xi \phi(\xi) \right| \leq C^{|\alpha|+1}(\alpha!)^\theta$$

for all $\alpha \in \mathbb{N}^d$ and for all $\xi \in \mathbb{R}^d$. We may then define

$$\psi_0(\xi) = \phi\left(\frac{\xi}{R}\right),$$

$$\psi_N(\xi) = \phi\left(\frac{\xi}{R(N+1)^\theta}\right) - \phi\left(\frac{\xi}{RN^\theta}\right), \quad N \geq 1. \qquad \square$$

Proof of Theorem 6.3.3. We can assume without loss of generality that p is in $AG^{0,0}_{\mu\nu}(\mathbb{R}^d)$. Let us consider a sequence $\{\psi_N\}_{N \geq 0}$ as in Lemma 6.3.4. We have, for $u, v \in S^\theta_\theta(\mathbb{R}^d)$,

$$\langle K_P, v \otimes u \rangle = \sum_{N=0}^{\infty} \langle K_N, v \otimes u \rangle$$

with

$$K_N(x, y) = (2\pi)^{-d} \int e^{i(x-y)\xi} p(x, \xi) \psi_N(\xi)\, d\xi$$

so we may decompose

$$K_P = \sum_{N=0}^{\infty} K_N.$$

Let $k > 0$ and $(x, y) \in \overline{\Omega}_k$. Let $h \in \{1, \ldots, d\}$ such that $|x_h - y_h| \geq \frac{k}{d}\langle x \rangle$. Then, for every $\alpha, \gamma \in \mathbb{N}^d$,

$$D^\alpha_x D^\gamma_y K_N(x, y) = (-1)^{|\gamma|}(2\pi)^{-d} \sum_{\beta \leq \alpha} \binom{\alpha}{\beta} \int e^{i(x-y)\xi}\xi^{\beta+\gamma}\psi_N(\xi) D^{\alpha-\beta}_x p(x, \xi)\, d\xi$$

$$= (-1)^{|\gamma|+N}(2\pi)^{-d}\sum_{\beta\leq\alpha}\binom{\alpha}{\beta}(x_h - y_h)^{-N}$$

$$\times \int e^{i(x-y)\xi}D_{\xi_h}^N\left[\xi^{\beta+\gamma}\psi_N(\xi)D_x^{\alpha-\beta}p(x,\xi)\right]d\xi.$$

Now, given $\zeta > 0$, we consider the operator

$$L = \frac{1}{m_{2\theta,\zeta}(x-y)}\sum_{j=0}^{\infty}\frac{\zeta^j}{(j!)^{2\theta}}(1-\Delta_\xi)^j,$$

where

$$m_{2\theta,\zeta}(x-y) = \sum_{j=0}^{\infty}\frac{\zeta^j}{(j!)^{2\theta}}\langle x-y\rangle^{2j}.$$

In view of the fact that $Le^{i(x-y)\xi} = e^{i(x-y)\xi}$, we can integrate by parts obtaining that

$$D_x^\alpha D_y^\gamma K_N(x,y) = (-1)^{|\gamma|+N}(2\pi)^{-d}\frac{(x_h-y_h)^{-N}}{m_{2\theta,\zeta}(x-y)}$$

$$\times\sum_{\beta\leq\alpha}\binom{\alpha}{\beta}\sum_{j=0}^{\infty}\frac{\zeta^j}{(j!)^{2\theta}}\int e^{i(x-y)\xi}\lambda_{hjN\alpha\beta\gamma}(x,\xi)d\xi$$

with

$$\lambda_{hjN\alpha\beta\gamma}(x,\xi) = (1-\Delta_\xi)^j D_{\xi_h}^N\left[\xi^{\beta+\gamma}\psi_N(\xi)D_x^{\alpha-\beta}p(x,\xi)\right]. \tag{6.3.9}$$

Let e_h be the h-th vector of the canonical basis of $\mathbb{R}^d$ and $\beta_h = \langle\beta,e_h\rangle$, $\gamma_h = \langle\gamma,e_h\rangle$. Developing in the right-hand side of (6.3.9) we obtain that

$$\lambda_{hjN\alpha\beta\gamma}(x,\xi) = \sum_{\substack{N_1+N_2+N_3=N \\ N_1\leq\beta_h+\gamma_h}}(-i)^{N_1}\frac{N!}{N_1!N_2!N_3!}\cdot\frac{(\beta_h+\gamma_h)!}{(\beta_h+\gamma_h-N_1)!}$$

$$\times(1-\Delta_\xi)^j\left[\xi^{\beta+\gamma-N_1 e_h}D_{\xi_h}^{N_2}\psi_N(\xi)D_{\xi_h}^{N_3}D_x^{\alpha-\beta}p(x,\xi)\right].$$

Hence

$$|\lambda_{hjN\alpha\beta\gamma}(x,\xi)| \leq \sum_{\substack{N_1+N_2+N_3=N \\ N_1\leq\beta_h+\gamma_h}}\frac{N!}{N_1!N_2!N_3!}\cdot\frac{(\beta_h+\gamma_h)!}{(\beta_h+\gamma_h-N_1)!}C_1^{|\alpha-\beta|+N_2+N_3+1}$$

$$\times(N_2!)^\theta(N_3!)^\mu[(\alpha-\beta)!]^\nu C_2^j(j!)^{2\theta}\left(\frac{1}{RN^\theta}\right)^{N_2}\langle\xi\rangle^{|\beta|+|\gamma|-N_1-N_3}.$$

We observe that, on the support of ψ_N, $2RN^\theta \leq \langle\xi\rangle \leq 3R(N+1)^\theta$. Thus, from standard factorial inequalities, since $\theta \geq \max\{\mu,\nu\}$, it follows that

$$|\lambda_{hjN\alpha\beta\gamma}(x,\xi)| \leq C_1^{|\alpha|+|\gamma|+1}(\alpha!\gamma!)^\theta C_2^j(j!)^{2\theta}\left(\frac{C_3}{R}\right)^N$$

with C_3 independent of R. Observe now that for every $c > 1$ there exist positive constants ϵ, c' such that, for $t > 0$,

$$\epsilon \exp\left[c't\right] \leq \sum_{j=0}^{\infty} \left(\frac{t^j}{j!}\right)^c. \tag{6.3.10}$$

Admitting (6.3.10) for a moment, and setting there $c = 2\theta$, $t = \zeta^{\frac{1}{2\theta}}\langle x - y \rangle^{\frac{1}{\theta}}$, we have that

$$|m_{2\theta,\zeta}(x - y)| \geq \epsilon \exp[c'\zeta^{\frac{1}{2\theta}}|x - y|^{\frac{1}{\theta}}].$$

From these estimates, choosing $\zeta < C_2^{-1}$, we deduce that

$$|D_x^\alpha D_y^\gamma K_N(x,y)| \leq C_1^{|\alpha|+|\gamma|+1}(\alpha!\gamma!)^\theta \left(\frac{C_4}{R}\right)^N \exp[-c'\zeta^{\frac{1}{2\theta}}|x-y|^{\frac{1}{\theta}}]$$

with $C_4 = C_4(k)$ independent of R. Choosing R sufficiently large and observing that $|x - y| \geq c''(\langle x \rangle + \langle y \rangle)$ on $\overline{\Omega}_k$, we obtain the estimates (6.3.8).

Let us now return to the proof of (6.3.10). It will be sufficient to argue for large t, say $t > 1$. Writing $s = 2^{1-c}t$, and denoting by $k = k(t)$ the smallest integer $k > 0$ such that $(2t)^k \leq k!$, we split

$$e^s = \sum_{0 \leq j < k} \frac{s^j}{j!} + \sum_{j \geq k} \frac{s^j}{j!}. \tag{6.3.11}$$

As for the second term in the right-hand side, it can be estimated by $e^s/4$, since, for $j \geq k$,

$$s^j/j! \leq \left(s^k/k!\right)\left(s^{j-k}/(j-k)!\right)$$
$$\leq 1/4\left(s^{j-k}/(j-k)!\right).$$

It will be then sufficient to estimate the first term in the right-hand side of (6.3.11). Actually, for $0 \leq j < k$ we have

$$\left(t^j/j!\right)^{c-1} \geq \left(1/2^j\right)^{c-1}$$

and multiplying both sides by $t^j/j!$ we get

$$\left(t^j/j!\right)^c \geq s^j/j!.$$

Applying this in (6.3.11) we obtain (6.3.10). This concludes the proof of Theorem 6.3.3 $\qquad\square$

Definition 6.3.5. A linear continuous operator from $S^\theta_\theta(\mathbb{R}^d)$ to $S^\theta_\theta(\mathbb{R}^d)$ is said to be $\theta-$regularizing if it extends to a linear continuous map from $S^{\theta\prime}_\theta(\mathbb{R}^d)$ to $S^\theta_\theta(\mathbb{R}^d)$.

We now consider formal sums of symbols. We set, for $t > 1$,

$$Q_t = \{(x, \xi) \in \mathbb{R}^{2d} : \langle x \rangle < t, \ \langle \xi \rangle < t\},$$

$$Q_t^e = \mathbb{R}^{2d} \setminus Q_t.$$

Definition 6.3.6. We denote by $FAG_{\mu\nu}^{m,n}(\mathbb{R}^d)$ the space of all the formal sums $\sum_{j \geq 0} p_j(x, \xi)$ such that $p_j(x, \xi) \in C^\infty(\mathbb{R}^{2d})$ for all $j \geq 0$ and there exist $B, C > 0$ such that

$$\sup_{j \geq 0} \sup_{\alpha, \beta \in \mathbb{N}^d} \sup_{(x,\xi) \in Q_{Bj\mu+\nu-1}^e} C^{-|\alpha|-|\beta|-2j} (\alpha!)^{-\mu} (\beta!)^{-\nu} (j!)^{-\mu-\nu+1}$$

$$\times \langle \xi \rangle^{-m+|\alpha|+j} \langle x \rangle^{-n+|\beta|+j} \left| D_\xi^\alpha D_x^\beta p_j(x, \xi) \right| < \infty. \tag{6.3.12}$$

We observe that every symbol $p \in AG_{\mu\nu}^{m,n}(\mathbb{R}^d)$ can be identified with an element of $FAG_{\mu\nu}^{m,n}(\mathbb{R}^d)$, by setting $p_0 = p$ and $p_j = 0$ for all $j \geq 1$.

Definition 6.3.7. We say that two sums $\sum_{j \geq 0} p_j$, $\sum_{j \geq 0} p_j'$ in $FAG_{\mu\nu}^{m,n}(\mathbb{R}^d)$ are equivalent, and we write

$$\sum_{j \geq 0} p_j \sim \sum_{j \geq 0} p_j',$$

if there exist constants $B, C > 0$ such that

$$\sup_{N \in \mathbb{N}} \sup_{\alpha, \beta \in \mathbb{N}^d} \sup_{(x,\xi) \in Q_{BN\mu+\nu-1}^e} C^{-|\alpha|-|\beta|-2N} (\alpha!)^{-\mu} (\beta!)^{-\nu} (N!)^{-\mu-\nu+1}$$

$$\times \langle \xi \rangle^{-m+|\alpha|+N} \langle x \rangle^{-n+|\beta|+N} \left| D_\xi^\alpha D_x^\beta \sum_{j < N} (p_j - p_j') \right| < \infty.$$

Theorem 6.3.8. *Given $\sum_{j \geq 0} p_j \in FAG_{\mu\nu}^{m,n}(\mathbb{R}^d)$, there exists a symbol $p \in AG_{\mu\nu}^{m,n}(\mathbb{R}^d)$ such that*

$$p \sim \sum_{j \geq 0} p_j \quad in \quad FAG_{\mu\nu}^{m,n}(\mathbb{R}^d).$$

Proof. Let $\varphi \in C^\infty(\mathbb{R}^{2d})$, $0 \leq \varphi \leq 1$, such that $\varphi(x, \xi) = 0$ if $(x, \xi) \in Q_2$, $\varphi(x, \xi) = 1$ if $(x, \xi) \in Q_3^e$ and

$$\sup_{(x,\xi) \in \mathbb{R}^{2d}} \left| D_\xi^\gamma D_x^\delta \varphi(x, \xi) \right| \leq C^{|\gamma|+|\delta|+1} (\gamma!)^\mu (\delta!)^\nu. \tag{6.3.13}$$

We define, for $R > 0$:

$$\varphi_0(x, \xi) \equiv 1 \quad \text{on } \mathbb{R}^{2d},$$

$$\varphi_j(x, \xi) = \varphi\left(\frac{x}{Rj^{\mu+\nu-1}}, \frac{\xi}{Rj^{\mu+\nu-1}} \right), \quad j \geq 1.$$

We want to prove that if R is sufficiently large, then

$$p(x,\xi) = \sum_{j\geq 0} \varphi_j(x,\xi) p_j(x,\xi)$$

is in $AG_{\mu\nu}^{m,n}(\mathbb{R}^d)$ and $p \sim \sum_{j\geq 0} p_j$ in $FAG_{\mu\nu}^{m,n}(\mathbb{R}^d)$.

Consider

$$D_\xi^\alpha D_x^\beta p(x,\xi) = \sum_{j\geq 0} \sum_{\substack{\gamma\leq\alpha \\ \delta\leq\beta}} \binom{\alpha}{\gamma}\binom{\beta}{\delta} D_\xi^{\alpha-\gamma} D_x^{\beta-\delta} p_j(x,\xi) D_\xi^\gamma D_x^\delta \varphi_j(x,\xi).$$

If $R \geq B$, where B is the same constant of Definition 6.3.6, we can apply the estimates (6.3.12) and obtain

$$\left| D_\xi^\alpha D_x^\beta p(x,\xi) \right| \leq C^{|\alpha|+|\beta|+1} \alpha! \beta! \langle\xi\rangle^{m-|\alpha|} \langle x\rangle^{n-|\beta|} \sum_{j\geq 0} H_{j\alpha\beta}(x,\xi),$$

where

$$H_{j\alpha\beta}(x,\xi) = \sum_{\substack{\gamma\leq\alpha \\ \delta\leq\beta}} \frac{(\alpha-\gamma)!^{\mu-1}(\beta-\delta)!^{\nu-1}}{\gamma!\delta!} C^{2j-|\gamma|-|\delta|}(j!)^{\mu+\nu-1}$$

$$\times \langle\xi\rangle^{|\gamma|-j}\langle x\rangle^{|\delta|-j}\left| D_\xi^\gamma D_x^\delta \varphi_j(x,\xi)\right|.$$

The condition (6.3.13) implies that

$$H_{j\alpha\beta}(x,\xi) \leq C^{|\alpha|+|\beta|+1}(\alpha!)^{\mu-1}(\beta!)^{\nu-1}\left(\frac{C_1}{R}\right)^j,$$

with C_1 independent of R. Choosing R sufficiently large, we obtain that p is in $AG_{\mu\nu}^{m,n}(\mathbb{R}^d)$. It remains to prove that $p \sim \sum_{j\geq 0} p_j$ in $FAG_{\mu\nu}^{m,n}(\mathbb{R}^d)$. Let N be a positive integer. We observe that, for $(x,\xi) \in Q_{3RN^{\mu+\nu-1}}^e$,

$$p(x,\xi) - \sum_{j<N} p_j(x,\xi) = \sum_{j\geq N} p_j(x,\xi)\varphi_j(x,\xi)$$

which we can estimate by arguing as above. $\qquad\square$

Proposition 6.3.9. *Let* $p \in AG_{\mu\nu}^{0,0}(\mathbb{R}^d)$ *and let* $\theta \geq \mu+\nu-1$. *If* $p \sim 0$ *in* $FAG_{\mu\nu}^{0,0}(\mathbb{R}^d)$, *then the operator* P *is* $\theta-$*regularizing.*

To prove this proposition, we need the following preliminary result.

Lemma 6.3.10. *Let* $M, r, \varrho, \overline{B}$ *be positive numbers,* $\varrho \geq 1$. *We define*

$$h(\lambda) = \inf_{0\leq N\leq \overline{B}\lambda^{\frac{1}{\varrho}}} \frac{M^{rN}(N!)^r}{\lambda^{rN/\varrho}}, \qquad \lambda \in \mathbb{R}_+. \tag{6.3.14}$$

Then there exist positive constants C, τ such that

$$h(\lambda) \le C e^{-\tau \lambda^{\frac{1}{\varrho}}}, \quad \lambda \in \mathbb{R}_+. \tag{6.3.15}$$

Proof. Consider first

$$H(\lambda) = \inf_{N \in \mathbb{N}} \frac{M^{rN}(N!)^r}{\lambda^{rN/\rho}},$$

which can be easily estimated by arguing as in the proof of Proposition 6.1.5. In fact we have, for $N = 0, 1, \dots$ and $\lambda \in \mathbb{R}_+$,

$$H(\lambda)^{1/r} \le M^N (N!) \lambda^{-N/\rho}$$

and therefore for $\tau' < r$ the function of $\lambda \in \mathbb{R}_+$,

$$H(\lambda)^{1/r} \exp\left[\tau'(rM)^{-1}\lambda^{1/\rho}\right] = \sum_{N=0}^{\infty} (\tau'/r)^N M^{-N}(N!)^{-1}\lambda^{N/\rho} H(\lambda)^{1/r}$$

is bounded. Hence

$$H(\lambda) \le C \exp\left[-\tau'(rM)^{-1}\lambda^{1/\rho}\right], \quad \lambda \in \mathbb{R}_+,$$

for a positive constant C. To get the estimate (6.3.15) for $h(\lambda)$ observe that the sequence

$$M^{rN}(N!)^r \lambda^{-rN/\rho}$$

is increasing for $N \ge M^{-1}\lambda^{1/\rho}$. If we take in (6.3.14) a larger constant M, such that $M^{-1} \le \overline{B}$, then $h(\lambda) = H(\lambda)$. Lemma 6.3.10 is therefore proved. $\square$

Proof of Proposition 6.3.9. It is sufficient to prove that the kernel

$$K_P(x, y) = (2\pi)^{-d} \int e^{i(x-y)\xi} p(x, \xi)\, d\xi \tag{6.3.16}$$

is in $S_\theta^\theta(\mathbb{R}^{2d})$. This will easily imply that P is θ-regularizing.

If $p \sim 0$, by Definition 6.3.7, there exist positive constants B_1, C_1 such that, for every $(x, \xi) \in \mathbb{R}^{2d}$:

$$\left| D_\xi^\alpha D_x^\beta p(x, \xi) \right| \le C_1^{|\alpha|+|\beta|+1}(\alpha!)^\mu (\beta!)^\nu \langle \xi \rangle^{-|\alpha|} \langle x \rangle^{-|\beta|}$$

$$\times \quad \inf_{0 \le N \le B_1(\langle \xi \rangle + \langle x \rangle)^{\frac{1}{\mu+\nu-1}}} \frac{C^{2N}(N!)^{\mu+\nu-1}}{\langle \xi \rangle^N \langle x \rangle^N}.$$

Applying Lemma 6.3.10, we obtain

$$\left| D_\xi^\alpha D_x^\beta p(x, \xi) \right| \le C_2^{|\alpha|+|\beta|+1}(\alpha!\beta!)^\theta \exp\left[-\sigma(|x|^{\frac{1}{\theta}} + |\xi|^{\frac{1}{\theta}})\right] \tag{6.3.17}$$

for some positive constants C_2, σ. Therefore, $p \in S_\theta^\theta(\mathbb{R}^{2d})$. Applying (6.3.17) in (6.3.16), we easily obtain that also $K_P \in S_\theta^\theta(\mathbb{R}^{2d})$. $\square$

To reach the expected results of regularity in $S^\theta_\theta(\mathbb{R}^d)$ for the G-elliptic equations, we need now to improve the symbolic calculus of Chapter 3. Namely, we study the stability of the classes $OPAG^{m,n}_{\mu\nu}(\mathbb{R}^d)$ under transposition, composition and construction of parametrices.

Proposition 6.3.11. *Let* $P = p(x, D) \in OPAG^{m,n}_{\mu\nu}(\mathbb{R}^d)$ *and let* ${}^tP$ *be the transposed operator defined by*

$$\langle {}^t Pu, v \rangle = \langle u, Pv \rangle, \quad u \in S^{\theta\prime}_\theta(\mathbb{R}^d), \ v \in S^\theta_\theta(\mathbb{R}^d). \tag{6.3.18}$$

Then, ${}^tP = Q + R$, *where* R *is a* $\theta-$*regularizing operator for* $\theta \geq \mu + \nu - 1$ *and* $Q = q(x, D)$ *is in* $OPAG^{m,n}_{\mu\nu}(\mathbb{R}^d)$ *with*

$$q(x, \xi) \sim \sum_{j \geq 0} \sum_{|\alpha|=j} (\alpha!)^{-1} \partial^\alpha_\xi D^\alpha_x p(x, -\xi)$$

in $FAG^{m,n}_{\mu\nu}(\mathbb{R}^d)$.

Theorem 6.3.12. *Let* $P = p(x, D) \in OPAG^{m,n}_{\mu\nu}(\mathbb{R}^d)$, $Q = q(x, D) \in OPAG^{m',n'}_{\mu\nu}(\mathbb{R}^d)$. *Then* $PQ = T + R$ *where* R *is* $\theta-$*regularizing for* $\theta \geq \mu + \nu - 1$ *and* $T = t(x, D)$ *is in* $OPAG^{m+m',n+n'}_{\mu\nu}(\mathbb{R}^d)$ *with*

$$t(x, \xi) \sim \sum_{j \geq 0} \sum_{|\alpha|=j} (\alpha!)^{-1} \partial^\alpha_\xi p(x, \xi) D^\alpha_x q(x, \xi)$$

in $FAG^{m+m',n+n'}_{\mu\nu}(\mathbb{R}^d)$.

To prove Proposition 6.3.11 and Theorem 6.3.12, it is convenient to introduce more general classes of symbols, called *amplitudes* in the sequel.

Let μ, ν be real numbers such that $\mu > 1, \nu > 1$ and let $(m, n, l) \in \mathbb{R}^3$.

Definition 6.3.13. *For* $C > 0$, *we shall denote by* $\Pi^{m,n,l}_{\mu\nu}(\mathbb{R}^d; C)$ *the Banach space of all functions* $a(x, y, \xi) \in C^\infty(\mathbb{R}^{3d})$ *such that*

$$\sup_{\alpha,\beta,\gamma \in \mathbb{N}^d} \sup_{(x,y,\xi) \in \mathbb{R}^{3d}} C^{-|\alpha|-|\beta|-|\gamma|} (\alpha!)^{-\mu} (\beta!\gamma!)^{-\nu}$$

$$\times \langle \xi \rangle^{-m+|\alpha|} \langle x \rangle^{-n+|\beta|} \langle y \rangle^{-l+|\gamma|} |D^\alpha_\xi D^\beta_x D^\gamma_y a(x, y, \xi)| < \infty.$$

We set

$$\Pi^{m,n,l}_{\mu\nu}(\mathbb{R}^d) = \varinjlim_{C \to +\infty} \Pi^{m,n,l}_{\mu\nu}(\mathbb{R}^d; C).$$

It is immediate to verify the following relations:

i) if $a(x, y, \xi) \in \Pi^{m,n,l}_{\mu\nu}(\mathbb{R}^d)$ for some $(m, n, l) \in \mathbb{R}^{3d}$, then the function $(x, \xi) \mapsto a(x, x, \xi)$ belongs to $AG^{m,n+l}_{\mu\nu}(\mathbb{R}^d)$;

ii) if $p \in AG^{m,n}_{\mu\nu}(\mathbb{R}^d)$ for some $(m, n) \in \mathbb{R}^2$, then $p(x, \xi) \in \Pi^{m,n,0}_{\mu\nu}(\mathbb{R}^d)$ and $p(y, \xi) \in \Pi^{m,0,n}_{\mu\nu}(\mathbb{R}^d)$.

Given $a \in \Pi_{\mu\nu}^{m,n,l}(\mathbb{R}^d)$, we can associate to a the pseudo-differential operator defined by

$$Au(x) = \int e^{i(x-y)\xi} a(x,y,\xi) u(y) \, dy \, d\xi, \quad u \in S_\theta^\theta(\mathbb{R}^d), \qquad (6.3.19)$$

with the standard meaning of iterated integrals.

Theorem 6.3.2 and Theorem 6.3.3 can be easily rephrased for operators (6.3.19), the kernel of A being now $K_A(x,y) = (2\pi)^{-d} \int e^{i(x-y)\xi} a(x,y,\xi) \, d\xi$.

In order to prove Proposition 6.3.11 and Theorem 6.3.12, we give first the following result.

Theorem 6.3.14. *Let A be an operator defined by an amplitude $a \in \Pi_{\mu\nu}^{m,n,l}(\mathbb{R}^d)$, $(m,n,l) \in \mathbb{R}^3$. Then we may write $A = P + R$, where R is a θ-regularizing operator for $\theta \geq \mu + \nu - 1$ and $P = p(x,D) \in OPAG_{\mu\nu}^{m,n+l}(\mathbb{R}^d)$ with $p \sim \sum\limits_{j\geq 0} p_j$, where*

$$p_j(x,\xi) = \sum_{|\alpha|=j} (\alpha!)^{-1} \partial_\xi^\alpha D_y^\alpha a(x,y,\xi)|_{y=x}. \qquad (6.3.20)$$

Proof. Let $\chi \in C^\infty(\mathbb{R}^{2d})$ such that

$$\chi(x,y) = \begin{cases} 1 & \text{if } |x-y| \leq \frac{1}{4}\langle x \rangle, \\ 0 & \text{if } |x-y| \geq \frac{1}{2}\langle x \rangle \end{cases}$$

and

$$|D_x^\beta D_y^\gamma \chi(x,y)| \leq C^{|\beta|+|\gamma|+1} (\beta! \gamma!)^\nu$$

for all $\beta, \gamma \in \mathbb{N}^d$ and $(x,y) \in \mathbb{R}^{2d}$. We may decompose a as the sum of two elements of $\Pi_{\mu\nu}^{m,n,l}(\mathbb{R}^d)$ writing

$$a(x,y,\xi) = \chi(x,y)a(x,y,\xi) + (1-\chi(x,y))a(x,y,\xi).$$

Furthermore, it follows from Theorem 6.3.3 that $(1-\chi(x,y))a(x,y,\xi)$ defines a θ-regularizing operator. Hence, possibly perturbing A with a θ-regularizing operator, we can assume that $a(x,y,\xi)$ is supported on $\left(\mathbb{R}^{2d} \setminus \Omega_{\frac{1}{2}}\right) \times \mathbb{R}^d$, where $\Omega_{\frac{1}{2}}$ is defined as in Theorem 6.3.3.

It is trivial to verify that $\sum\limits_{j\geq 0} p_j$ defined by (6.3.20) belongs to $FAG_{\mu\nu}^{m,n+l}(\mathbb{R}^{2d})$.

As in the proof of Theorem 6.3.8 we can find a sequence $\varphi_j \in C^\infty(\mathbb{R}^{2d})$ depending on a parameter R such that

$$p(x,\xi) = \sum_{j\geq 0} \varphi_j(x,\xi) p_j(x,\xi)$$

defines an element of $AG_{\mu\nu}^{m,n+l}(\mathbb{R}^d)$ for R large and $p \sim \sum\limits_{j\geq 0} p_j$ in $FAG_{\mu\nu}^{m,n+l}(\mathbb{R}^d)$, cf. (6.3.13) and subsequent formulas. Let $P = p(x,D)$. To prove the theorem it is sufficient to show that the kernel $K(x,y)$ of $A - P$ is in $S_\theta^\theta(\mathbb{R}^{2d})$.

We can write

$$a(x, y, \xi) - p(x, \xi) = (1 - \varphi_0(x, \xi))a(x, y, \xi)$$

$$+ \sum_{N=0}^{\infty} (\varphi_N - \varphi_{N+1})(x, \xi) \Big(a(x, y, \xi) - \sum_{j \leq N} p_j(x, \xi) \Big).$$

Consequently, we have

$$K(x, y) = \overline{K}(x, y) + \sum_{N=0}^{\infty} K_N(x, y),$$

where

$$\overline{K}(x, y) = (2\pi)^{-d} \int e^{i(x-y)\xi} (1 - \varphi_0(x, \xi))a(x, y, \xi) \, d\xi,$$

$$K_N(x, y) = (2\pi)^{-d} \int e^{i(x-y)\xi} (\varphi_N - \varphi_{N+1})(x, \xi) \Big(a(x, y, \xi) - \sum_{j \leq N} p_j(x, \xi) \Big) d\xi.$$

A power expansion in the second argument gives, for $N = 1, 2, \ldots,$

$$a(x, y, \xi) = \sum_{|\alpha| \leq N} (\alpha!)^{-1} (y - x)^\alpha \partial_y^\alpha a(x, x, \xi) + \sum_{|\alpha| = N+1} (\alpha!)^{-1} (y - x)^\alpha w_\alpha(x, y, \xi)$$

with

$$w_\alpha(x, y, \xi) = (N + 1) \int_0^1 \partial_y^\alpha a(x, x + t(y - x), \xi)(1 - t)^N dt.$$

In view of our definition of the $p_j(x, \xi)$'s, integrating by parts we obtain that

$$K_N(x, y) = W_N(x, y) + (2\pi)^{-d} \sum_{1 \leq |\alpha| \leq N} \sum_{0 \neq \beta \leq \alpha} \frac{1}{\beta!(\alpha - \beta)!}$$

$$\times \int e^{i(x-y)\xi} D_\xi^\beta (\varphi_N - \varphi_{N+1})(x, \xi)(D_\xi^{\alpha-\beta} \partial_y^\alpha a)(x, x, \xi) d\xi,$$

where

$$W_N(x, y) = (2\pi)^{-d} \sum_{|\alpha| = N+1} \sum_{\beta \leq \alpha} \frac{1}{\beta!(\alpha - \beta)!}$$

$$\times \int e^{i(x-y)\xi} D_\xi^\beta (\varphi_N - \varphi_{N+1})(x, \xi) D_\xi^{\alpha-\beta} w_\alpha(x, y, \xi) d\xi$$

for all $N = 1, 2, \ldots.$

Using an absolute convergence argument, we may re-arrange the sums under the integral sign. We also observe that

$$\sum_{N \geq |\alpha|} D_\xi^\beta (\varphi_N - \varphi_{N+1})(x, \xi) = D_\xi^\beta \varphi_{|\alpha|}(x, \xi).$$

Then we have

$$H = \overline{K} + \sum_{\alpha \neq 0} I_\alpha + \sum_{N=0}^{\infty} W_N,$$

where

$$I_\alpha(x,y) = (2\pi)^{-d} \sum_{0 \neq \beta \leq \alpha} \frac{1}{\beta!(\alpha-\beta)!} \int e^{i(x-y)\xi} D_\xi^\beta \varphi_{|\alpha|}(x,\xi) D_\xi^{\alpha-\beta} \partial_y^\alpha a(x,x,\xi) d\xi$$

and we may write $W_0(x,y)$ for $K_0(x,y)$. To conclude the proof, we want to show that $\overline{K}, \sum_{\alpha \neq 0} I_\alpha, \sum_{N=0}^{\infty} W_N \in S_\theta^\theta(\mathbb{R}^{2d})$. First of all, we have to estimate the derivatives of $\overline{K}$ for $(x,\xi) \in \operatorname{supp}(1-\varphi_0(x,\xi))$, i.e., for $\langle x \rangle \leq R, \langle \xi \rangle \leq R$. We have

$$\left| x^k y^h D_x^\delta D_y^\gamma \overline{K}(x,y) \right| = (2\pi)^{-d} \left| x^k y^h \sum_{\substack{\gamma_1+\gamma_2=\gamma \\ \delta_1+\delta_2+\delta_3=\delta}} \frac{\gamma! \delta!}{\gamma_1! \gamma_2! \delta_1! \delta_2! \delta_3!} \right.$$

$$\left. \times (-1)^{|\gamma_1|} \int e^{i(x-y)\xi} \xi^{\gamma_1+\delta_1} D_x^{\delta_2} D_y^{\gamma_2} a(x,y,\xi) D_x^{\delta_3} (1-\varphi_0(x,\xi)) d\xi \right|$$

$$\leq |x|^{|k|} |y|^{|h|} \sum_{\substack{\gamma_1+\gamma_2=\gamma \\ \delta_1+\delta_2+\delta_3=\delta}} \frac{\gamma! \delta!}{\gamma_1! \gamma_2! \delta_1! \delta_2! \delta_3!} C^{|\gamma_2|+|\delta_2|+|\delta_3|} (\gamma_2! \delta_2! \delta_3!)^\nu \langle x-y \rangle^{|\gamma_2+\delta_2|}$$

$$\times \langle x \rangle^n \langle y \rangle^l \int_{\langle \xi \rangle \leq R} \langle \xi \rangle^{|\gamma_1+\delta_1|} \langle \xi \rangle^m d\xi.$$

Now, $a(x,y,\xi)$ is supported on $\left(\mathbb{R}^{2d} \setminus \Omega_{\frac{1}{2}} \right) \times \mathbb{R}^d$ and in this region $|y| \leq \frac{3}{2}\langle x \rangle$, so there exist constants $C_1, C_2 > 0$ depending on R such that

$$\sup_{(x,y) \in \mathbb{R}^{2d}} \left| x^k y^h D_x^\delta D_y^\gamma \overline{K}(x,y) \right| \leq C_1 R^{|k|+|h|} C_2^{|\gamma|+|\delta|} (\gamma! \delta!)^\theta,$$

so $\overline{K} \in S_\theta^\theta(\mathbb{R}^{2d})$. Consider now

$$x^k y^h D_x^\delta D_y^\gamma I_\alpha(x,y) = (2\pi)^{-d} \sum_{0 \neq \beta \leq \alpha} \frac{1}{\beta!(\alpha-\beta)!} \sum_{\delta_1+\delta_2+\delta_3=\delta} \frac{\delta!}{\delta_1! \delta_2! \delta_3!} (-1)^{|\gamma|} x^k y^h$$

$$\times \int e^{i(x-y)\xi} \xi^{\gamma+\delta_1} D_x^{\delta_2} D_\xi^\beta \varphi_{|\alpha|}(x,\xi) D_\xi^{\alpha-\beta} D_x^{\delta_3} \partial_y^\alpha a(x,x,\xi) d\xi$$

$$= (2\pi)^{-d} \sum_{0 \neq \beta \leq \alpha} \frac{1}{\beta!(\alpha-\beta)!} \sum_{\delta_1+\delta_2+\delta_3=\delta} \frac{\delta!}{\delta_1! \delta_2! \delta_3!} (-1)^{|\gamma|} (-i)^h x^k$$

$$\times \int e^{-iy\xi} \partial_\xi^h \left[e^{ix\xi} \xi^{\gamma+\delta_1} D_x^{\delta_2} D_\xi^\beta \varphi_{|\alpha|}(x,\xi) D_\xi^{\alpha-\beta} D_x^{\delta_3} \partial_y^\alpha a(x,x,\xi) \right] d\xi.$$

We need the estimates for $(x,\xi) \in \operatorname{supp} D_\xi^\beta \varphi_{|\alpha|}(x,\xi) \subset \overline{Q}_{2R|\alpha|^{\mu+\nu-1}} \setminus Q_{R|\alpha|^{\mu+\nu-1}}$. Then, there exist $C_1, C_2, C_3 > 0$ such that

$$|x^k y^h D_x^\delta D_y^\gamma I_\alpha(x,y)| \leq C_1^{|h|+|k|+1} C_2^{|\alpha|} C_3^{|\gamma|+|\delta|} (k!h!\gamma!\delta!)^\theta (\alpha!)^\nu \langle x \rangle^{-|\alpha|}$$

$$\times \sum_{0 \neq \beta \leq \alpha} (\beta!)^{\mu-1} [(\alpha-\beta)!]^{\mu-1} \left(\frac{1}{R|\alpha|^{\mu+\nu-1}} \right)^{|\beta|} \int_{\langle \xi \rangle \leq 2R|\alpha|^{\mu+\nu-1}} \langle \xi \rangle^{-|\alpha-\beta|} d\xi$$

with C_2 independent of R. Now, if $(x,\xi) \in \overline{Q}_{2R|\alpha|^{\mu+\nu-1}} \setminus Q_{R|\alpha|^{\mu+\nu-1}}$, we have that

$$C_2^{|\alpha|} (\alpha!)^\nu \langle x \rangle^{-|\alpha|} \sum_{0 \neq \beta \leq \alpha} (\beta!)^{\mu-1} [(\alpha-\beta)!]^{\mu-1} \left(\frac{1}{R|\alpha|^{\mu+\nu-1}} \right)^{|\beta|}$$

$$\times \int_{\langle \xi \rangle \leq 2R|\alpha|^{\mu+\nu-1}} \langle \xi \rangle^{-|\alpha-\beta|} d\xi \leq \left(\frac{C_4}{R} \right)^{|\alpha|}$$

with C_4 independent of R. Finally, we conclude that

$$\sup_{(x,y) \in \mathbb{R}^{2d}} |x^k y^h D_x^\delta D_y^\gamma I_\alpha(x,y)| \leq C^{|h|+|k|+1} C_2^{|\gamma|+|\delta|} (k!h!\gamma!\delta!)^\theta \left(\frac{C_4}{R} \right)^{|\alpha|}.$$

Choosing $R > C_4$, we obtain that $\sum_{\alpha \neq 0} I_\alpha \in S_\theta^\theta(\mathbb{R}^{2d})$.

Arguing as for I_α, we can prove that also

$$\sup_{(x,y) \in \mathbb{R}^{2d}} |x^k y^h D_x^\delta D_y^\gamma W_N(x,y)| \leq C_1^{|h|+|k|+1} C_2^{|\gamma|+|\delta|} (h!k!\gamma!\delta!)^\theta \left(\frac{C}{R} \right)^N$$

with C independent of R, which gives, for R sufficiently large, that $\sum_{N=0}^\infty W_N$ is in $S_\theta^\theta(\mathbb{R}^{2d})$. This concludes the proof. $\qquad\square$

Proof of Proposition 6.3.11. By (6.3.18), ${}^t P$ is defined by

$${}^t Pu(x) = \int e^{i(x-y)\xi} p(y,-\xi)u(y)\, dy\, d\xi, \quad u \in S_\theta^\theta(\mathbb{R}^d).$$

Thus, ${}^t P$ is an operator of the form (6.3.19) with amplitude $p(y,-\xi)$. By Theorem 6.3.14, ${}^t P = Q + R$ where R is θ-regularizing and $Q = q(x,D) \in \mathrm{OPAG}_{\mu\nu}^{m,n}(\mathbb{R}^d)$, with

$$q(x,\xi) \sim \sum_{j \geq 0} \sum_{|\alpha|=j} (\alpha!)^{-1} \partial_\xi^\alpha D_x^\alpha p(x,-\xi). \qquad\square$$

Proof of Theorem 6.3.12. We can write $Q = {}^t({}^t Q)$. Then, by Proposition 6.3.11, $Q = Q_1 + R_1$, where R_1 is θ-regularizing and

$$Q_1 u(x) = \int e^{i(x-y)\xi} q_1(y,\xi)u(y)\, dy\, d\xi, \qquad (6.3.21)$$

with $q_1(y, \xi) \in AG_{\mu\nu}^{m',n'}(\mathbb{R}^{2d})$, $q_1(y, \xi) \sim \sum_\alpha (\alpha!)^{-1} \partial_\xi^\alpha D_y^\alpha q(y, -\xi)$. From (6.3.21) it follows that

$$\widehat{Q_1 u}(\xi) = \int e^{-iy\xi} q_1(y, \xi) u(y) \, dy, \quad u \in S_\theta^\theta(\mathbb{R}^d)$$

from which we deduce that

$$PQu(x) = \int e^{i(x-y)\xi} p(x, \xi) q_1(y, \xi) u(y) \, dy \, d\xi + PR_1 u(x).$$

We observe that $p(x, \xi) q_1(y, \xi) \in \Pi_{\mu\nu}^{m+m',n,n'}(\mathbb{R}^d)$; then we may apply Theorem 6.3.14 and obtain that

$$PQu(x) = Tu(x) + Ru(x)$$

where R is θ−regularizing and $T = t(x, D) \in OPAG_{\mu\nu}^{m+m',n+n'}(\mathbb{R}^d)$ with

$$t(x, \xi) \sim \sum_{j \geq 0} \sum_{|\alpha|=j} (\alpha!)^{-1} \partial_\xi^\alpha p(x, \xi) D_x^\alpha q(x, \xi)$$

in $FS_{\mu\nu}^{m+m',n+n'}(\mathbb{R}^d)$.   □

In Theorem 6.3.12, if $p \sim \sum_{j \geq 0} p_j$ in $FAG_{\mu\nu}^{m,n}(\mathbb{R}^d)$ and $q \sim \sum_{j \geq 0} q_j$ in $FAG_{\mu\nu}^{m',n'}(\mathbb{R}^d)$, then

$$t(x, \xi) \sim \sum_{j \geq 0} \sum_{|\alpha|+h+k=j} (\alpha!)^{-1} \partial_\xi^\alpha p_h(x, \xi) D_x^\alpha q_k(x, \xi) \quad \text{in } FS_{\mu\nu}^{m+m',n+n'}(\mathbb{R}^d).$$

To be definite, we restate the notion of ellipticity for elements of $OPAG_{\mu\nu}^{m,n}(\mathbb{R}^d)$. It coincides with the definition of G-ellipticity in Chapter 3.

Definition 6.3.15. A symbol $p \in AG_{\mu\nu}^{m,n}(\mathbb{R}^d)$ is said to be G-elliptic if there exist $B, C > 0$ such that

$$|p(x, \xi)| \geq C\langle\xi\rangle^m \langle x\rangle^n \quad \text{for all } (x, \xi) \in Q_B^e.$$

Theorem 6.3.16. *If $p \in AG_{\mu\nu}^{m,n}(\mathbb{R}^d)$ is G-elliptic and $P = p(x, D)$, then there exists $E \in OPAG_{\mu\nu}^{-m,-n}(\mathbb{R}^d)$ such that $EP = I + R_1$, $PE = I + R_2$, where R_1, R_2 are θ−regularizing operators, for $\theta \geq \mu + \nu - 1$.*

Proof. Let $e_0(x, \xi)$ be fixed such that

$$e_0(x, \xi) = p(x, \xi)^{-1} \quad \text{for } (x, \xi) \in Q_B^e,$$

and define by induction, for $j \geq 1$,

$$e_j(x, \xi) = -e_0(x, \xi) \sum_{0 < |\alpha| \leq j} (\alpha!)^{-1} \partial_\xi^\alpha e_{j-|\alpha|}(x, \xi) D_x^\alpha p(x, \xi).$$

It is easy to verify that $\sum_{j\geq 0} e_j(x,\xi) \in FAG_{\mu\nu}^{-m,-n}(\mathbb{R}^d)$. Applying Theorem 6.3.8, we can find $e \in AG_{\mu\nu}^{-m,-n}(\mathbb{R}^d)$ such that $e \sim \sum_{j\geq 0} e_j$. Denote by E the operator with symbol e. By construction, Theorem 6.3.12 implies that $EP - I$ and $PE - I$ are θ-regularizing operators, in view of Proposition 6.3.9. □

As an immediate consequence of Theorem 6.3.16, we obtain the following result of global regularity.

Corollary 6.3.17. *Let $p \in AG_{\mu\nu}^{m,n}(\mathbb{R}^d)$ be G-elliptic and let $f \in S_\theta^\theta(\mathbb{R}^d)$ for some $\theta \geq \mu + \nu - 1$. If $u \in S_\theta^{\theta'}(\mathbb{R}^d)$ is a solution of the equation*

$$Pu = f,$$

then $u \in S_\theta^\theta(\mathbb{R}^d)$.

Note that in Corollary 6.3.17, as well as in Theorem 6.3.16, we have $\theta > 1$.

To conclude this section, we apply Corollary 6.3.17 to the operators P with polynomial coefficients in (6.3.1). Their symbols can be regarded as elements of $AG_{1,1}^{m,n}(\mathbb{R}^d)$, with m, n as in (6.3.1). The condition (6.3.2) corresponds to G-ellipticity in Definition 6.3.15. Therefore, if $Pu = 0$ with $u \in \mathcal{S}'(\mathbb{R}^d)$, or even $u \in S_\theta^{\theta'}(\mathbb{R}^d)$, then we may conclude $u \in S_\theta^\theta(\mathbb{R}^d)$ for any $\theta > 1$, in particular

$$|u(x)| \leq Ce^{-\epsilon|x|^{1/\theta}}, \quad \theta > 1 \tag{6.3.22}$$

for suitable positive constants C and ϵ.

A simple example of an operator of this type is given by

$$P = (1 + |x|^{2k})(-\Delta + 1) + Q_1(x, D), \quad k \geq 1, \tag{6.3.23}$$

where $Q_1(x, D)$ is any first-order operator with polynomial coefficients of degree $2k - 1$. In turn, the operator L in (6.0.10) is a particular case of (6.3.23); this suggests that the estimates (6.3.22) should be valid for $\theta = 1$ as well. In fact, in Section 6.5 we shall improve Corollary 6.3.17, by obtaining $u \in S_1^1(\mathbb{R}^d)$ for G-elliptic symbols in $AG_{1,1}^{m,n}(\mathbb{R}^d)$.

6.4 A Short Survey on Travelling Waves

In the next Section 6.5 we shall discuss regularity in $S_1^1(\mathbb{R}^d)$ for semilinear equations, having as linear part G-elliptic pseudo-differential operators. The main motivation for such study comes from the theory of travelling (solitary) waves. In fact, a large part of these equations are of the above-mentioned type, with linear part given by G-elliptic partial differential equations with constant coefficients; non-local equations appear as well, whose linear parts are pseudo-differential operators with symbol in the classes $AG_{1,1}^{m,n}(\mathbb{R}^d)$ of the preceding Section 6.3. For travelling waves, both exponential decay and extension to the complex domain have relevance in Physics and they are exactly described by regularity in $S_1^1(\mathbb{R}^d)$, from the Mathematical point of view. Addressing non-experts, we give in the following a short discourse on travelling waves, adding some references.

The first documentation of the existence of shallow water waves appeared in 1834 when J. Scott Russell wrote one of the most cited papers about what later became known as soliton theory. Russell observed propagation of a solitary wave in the Glasgow-Edinburgh canal. In 1895 Korteweg and De Vries derived an equation describing shallow water waves, and gave the following interpretation of the solitary wave of Scott Russell. Ignoring some relevant physical aspects and simplifying parameters, we may write for short the KdV equation as

$$v_t + 2vv_x + v_{xxx} = 0, \tag{6.4.1}$$

where t is the time variable, x the point in the canal, $v(x,t)$ the height of the water (let us refer to Bona and Li [22], Porubov [166], Whitham [195] for a much more detailed presentation). Looking for a solitary wave solution, travelling forward with velocity $V > 0$, we impose $v(t,x) = u(x - Vt)$ in (6.4.1) and we obtain

$$\frac{d}{dx}(-Vu + u^2 + u'') = 0, \quad x \in \mathbb{R},$$

hence $u(x)$ satisfies $u'' - Vu + u^2 = \text{const}$. Assuming further const $= 0$, we are reduced to solving

$$Pu = u'' - Vu + u^2 = 0, \tag{6.4.2}$$

sometimes called Newton's equation. Equation (6.4.2) possesses explicit solutions in terms of special functions. If we impose $u(x) \to 0$ for $x \to \pm\infty$, we obtain simply translations of the function

$$u(x) = \frac{\frac{3}{2}V}{\text{Ch}^2\left(\frac{\sqrt{V}}{2}x\right)}, \tag{6.4.3}$$

where

$$\text{Ch}\,t = \frac{e^t + e^{-t}}{2}.$$

We emphasize two properties of $u(x)$ in (6.4.3): first, it can be extended as an analytic function in a strip of the form $\{z \in \mathbb{C} : |\text{Im}\,z| < a\}$ in the complex plane. The second property is the exponential decay for $x \to \pm\infty$. More precisely, these explicit solutions belong to the space $S_1^1(\mathbb{R})$.

After the discovery of the KdV equation, several related models were proposed. In particular recently, the theory of solitary waves had impressive developments, both concerning applicative aspects and mathematical analysis. Let us mention applications to internal water waves, nerve pulse dynamics, ion-acoustic waves in plasma, population dynamics, etc. In this order of ideas, we observe in particular that during the years 1990-2000, several papers were devoted to 5-th order and 7-th order generalizations of KdV, see for example Porubov [164], Chapter 1. The corresponding extension of equation (6.4.2) is of the type

$$\sum_{j=0}^{N} a_j u^{(j)} + Q(u) = 0, \tag{6.4.4}$$

where Q is a polynomial, $Q(u) = \sum_{j=2}^{M} b_j u^j$ and $a_0 = -V \neq 0$. Because of physical assumptions, the equation $\sum_{j=0}^{N} a_j \lambda^j = 0$ has no purely imaginary roots, and then all the solutions of the corresponding linear equation have exponential decay/growth. This condition can be read as G-ellipticity of the symbol of the linear part in (6.4.4): $p(\xi) = \sum_{j=0}^{N} a_j (i\xi)^j \neq 0$ for $\xi \in \mathbb{R}$, in particular $-\xi^2 - V \neq 0$ in (6.4.2), hence $|p(\xi)| \geq C \langle \xi \rangle^N$ for some $C > 0$. Non-trivial solutions u of (6.4.4) with $u(x) \to 0$ for $x \to \pm\infty$ may exist or not, according to the coefficients a_j, b_j, and when they exist, in general they do not have an explicit analytic expression. Exponential decay and holomorphic extensions are granted anyhow, see the next Section 6.5. Let us emphasize that, to reach the exponential decay, the boundedness of $u(x)$ is not sufficient as an initial assumption. We shall express in Section 6.5 a precise threshold in terms of Sobolev estimates; as counter-example, consider here the celebrated Burgers' equation (1948):

$$v_t + v_{xx} + 2vv_x = 0. \tag{6.4.5}$$

Imposing $v(t, x) = u(x - Vt)$ and arguing as before we obtain the Verhulst equation

$$u' - Vu + u^2 = 0 \tag{6.4.6}$$

which can be regarded as a particular case of (6.4.4). It admits the bounded solution

$$u(x) = \frac{V}{1 + e^{-Vx}}. \tag{6.4.7}$$

Assuming $V > 0$, we have exponential decay only for $x \to -\infty$, whereas $u(x) \to V \neq 0$ as $x \to +\infty$.

As d-dimensional generalization of (6.4.2), setting for simplicity $V = 1$, we may consider

$$-\Delta u + u = u^p \tag{6.4.8}$$

for an integer $p \geq 2$. Such equations in $\mathbb{R}^d$ have been widely studied. From the point of view of Mathematical Physics, they appear for example when considering nonlinear Schrödinger equations used in Plasma Physics and Nonlinear Optics. Travelling waves, in this case, have to be understood as stationary wave solutions, defined by time-modulation. From the point of view of Mathematical Analysis, we refer to the recent book of Ambrosetti and Malchiodi [4] for a collection of results of existence and uniqueness, or multiplicity, of the Sobolev solutions of (6.4.8) via

variational methods. Concerning exponential decay, we quote the following precise
result in Berestycki and Lions [12]. Assume $d \geq 3$. If $1 < p < \frac{d+2}{d-2}$, then (6.4.8)
has a positive radial solution $u(x) = U(|x|) \in H^1(\mathbb{R}^d)$. Such a solution satisfies,
for some $\epsilon > 0$,

$$U(r) \lesssim e^{-\epsilon r}, \quad r = |x|, \quad \text{as } r \to +\infty.$$

We refer to the next section for exponential decay and holomorphic extension
of general solutions of (6.4.8) and other semilinear G-elliptic partial differential
equations.

Non-local equations, i.e., nonlinear partial differential equations involving integral operators, have been proposed as models for different solitary wave phenom-
ena. Let us fix here attention on the so-called intermediate-long-wave equation,
see Joseph [124] and recent contributions by J. L. Bona, J. Albert and others:

$$v_t + 2vv_x - (Nv)_x + v_x = 0, \tag{6.4.9}$$

where N is the Fourier multiplier operator defined by

$$(Nv)\widehat{}(\xi) = \xi \, \text{Ctgh} \, \xi \, \widehat{v}(\xi). \tag{6.4.10}$$

We emphasize the analyticity of the function $\xi \text{Ctgh} \, \xi$. Looking for solutions
$v(t, x) = u(x - Vt)$ and arguing as before, we obtain the non-local equation

$$Nu + \gamma u = u^2 \tag{6.4.11}$$

where $\gamma = V - 1$. Under the assumption $V > 0$ a non-trivial solution $u \in S_1^1(\mathbb{R})$
can be computed explicitly. This also will be included in our results in the next
section, since the operator N in (6.4.10) can be seen as an example in the class of
Fourier multipliers

$$Pu(x) = \int e^{ix\xi} p(\xi) \widehat{u}(\xi) \, d\xi, \tag{6.4.12}$$

where $p(\xi)$ belongs to $AG_{1,1}^{m,0}(\mathbb{R}^d)$, $m > 0$:

$$|\partial_\xi^\alpha p(\xi)| \leq C^{|\alpha|+1} \alpha! \langle \xi \rangle^{m-|\alpha|}. \tag{6.4.13}$$

The assumption of G-ellipticity is formally as in Definition 6.3.15:

$$|p(\xi)| \geq c \langle \xi \rangle^m, \quad \xi \in \mathbb{R}^d, \tag{6.4.14}$$

for some $c > 0$. In particular the symbol of the linear part in (6.4.11)

$$p(\xi) = \xi \text{Ctgh} \, \xi + \gamma, \quad \gamma > -1, \tag{6.4.15}$$

satisfies (6.4.13), (6.4.14) with $m = 1$.

As a final example of a soliton equation, we mention the Benjamin-Ono
equation, see for example Amick and Toland [5]:

$$v_t + 2vv_x - (|D|v)_x = 0 \tag{6.4.16}$$

where $|D|$ is the Fourier multiplier operator defined by $(|D|u)\hat{}(\xi) = |\xi|\hat{u}(\xi)$. The corresponding equation for the wave profile $u(x)$, with $v(t, x) = u(x - Vt), V > 0$, is

$$|D|u + Vu = u^2, \tag{6.4.17}$$

with solution given by

$$u(x) = \frac{2V}{1 + V^2 x^2}. \tag{6.4.18}$$

This is the unique non-trivial solution in $L^2(\mathbb{R})$ of (6.4.17), modulo translations. The lack of exponential decay can be related to the singularity at the origin of the symbol $|\xi|$. Regretfully, the Benjamin-Ono equation will remain outside our arguments in the sequel. In fact, the organization of a pseudo-differential calculus predicting precisely the algebraic decay of (6.4.18) represents a challenging open problem.

6.5 Semilinear G-Equations

In this section we shall consider the semilinear equation

$$Pu = f + F[u], \tag{6.5.1}$$

where P is a pseudo-differential operator with symbol $p(x, \xi) \in AG_{\mu\nu}^{m,n}(\mathbb{R}^d)$, $\mu \geq 1$, $\nu \geq 1$, $m \geq 1$, $n \geq 0$, cf. Section 6.3. Let n be integer. We suppose that P is G-elliptic, according to Definition 6.3.15. We assume $f \in S_\nu^\mu(\mathbb{R}^d)$ and

$$F[u] = \sum_{h,l,\gamma,r} F_{hl\gamma r} x^h u^l (D^\gamma u)^r, \tag{6.5.2}$$

where $h \in \mathbb{N}^d$ with $0 \leq |h| \leq \max\{n - 1, 0\}$, $\gamma \in \mathbb{N}^d$ with $0 \leq |\gamma| \leq m - 1$ and $l, r \in \mathbb{N}$, $l + r \geq 2$.

We shall prove the following result.

Theorem 6.5.1. *Under the preceding hypotheses, assume that $u \in H^s(\mathbb{R}^d)$, $s > \frac{d}{2} + m - 1$, is a solution of (6.5.1). In the case $n = 0$ assume further $\langle x \rangle^{\epsilon_0} u \in H^s(\mathbb{R}^d)$, $s > \frac{d}{2} + m - 1$, for some $\epsilon_0 > 0$. Then $u \in S_\nu^\mu(\mathbb{R}^d)$.*

In particular, if $\mu = \nu = 1$, we conclude $u \in S_1^1(\mathbb{R}^d)$. For linear equations, i.e., $F[u] = 0$, this improves Corollary 6.3.17, where the threshold $\theta = \mu = \nu = 1$ was not reached. For the travelling waves in Section 6.4, see (6.4.4), (6.4.8), (6.4.11), we get the expected result of $S_1^1(\mathbb{R}^d)$ regularity.

To prove Theorem 6.5.1 we shall use Proposition 6.1.11, (b). Namely, to prove $u \in S_\nu^\mu(\mathbb{R}^d)$ it will be sufficient to obtain separately $\|u\|_{s,\nu;\epsilon} < \infty$ and $\|u\|_{\{s,\mu;T\}} < \infty$ for some $s > \frac{d}{2} + m - 1$, and sufficiently small $\epsilon > 0$, $T > 0$, cf. (6.1.37), (6.1.38). Equivalently, we shall prove the boundedness of the partial sums:

$$H_N^{s,\nu;\epsilon}[u] = \sum_{|k| \leq N} \frac{\epsilon^{|k|}}{k!^\nu} \|x^k u\|_s, \quad N = 0, 1, \ldots, \tag{6.5.3}$$

$$E_N^{s,\mu;T}[u] = \sum_{|j| \leq N} \frac{T^{|j|}}{j!^\mu} \|\partial_x^j u\|_s, \quad N = 0, 1, \ldots . \tag{6.5.4}$$

In the proof of Theorem 6.5.1 we shall not use the full pseudo-differential calculus of Section 6.3, but rely on the following technical proposition. Denote by E the pseudo-differential operator with symbol $e_0(x, \xi) = p(x, \xi)^{-1}$ for large (x, ξ), cf. the proof of Theorem 6.3.16. Since the cases $\mu = 1$, $\nu = 1$ were excluded in the symbolic calculus of Section 6.3, we have to refer in the following to μ', ν' satisfying $\mu' \geq \mu$, $\nu' \geq \nu$ and simultaneously $\mu' > 1$, $\nu' > 1$. We then have $e_0(x, \xi) \in AG_{\mu'\nu'}^{-m,-n}(\mathbb{R}^d)$, and from Theorem 6.3.12 we obtain $EP = I + R$ where $R \in OPAG_{\mu'\nu'}^{-1,-1}(\mathbb{R}^d)$. For such rough parametrix E of $P = p(x, D)$ we shall use the following proposition.

Proposition 6.5.2. Let $p \in AG_{\mu\nu}^{m,n}(\mathbb{R}^d)$, $\mu \geq 1$, $\nu \geq 1$, be G-elliptic and let $E \in OPAG_{\mu'\nu'}^{-m,-n}(\mathbb{R}^d)$ be the operator defined as before. For every $\alpha, \beta \in \mathbb{N}^d$ denote by $r_{\alpha\beta}$ the symbol of the operator $E(D_\xi^\alpha D_x^\beta p)(x, D)$. Then, for every $\sigma, \theta \in \mathbb{N}^d$, there exists a positive constant $C = C(\sigma, \theta)$ independent of α, β such that

$$|\partial_\xi^\theta \partial_x^\sigma r_{\alpha\beta}(x, \xi)| \leq C^{|\alpha|+|\beta|+1} \alpha!^\mu \beta!^\nu \langle\xi\rangle^{-|\alpha|-|\theta|} \langle x\rangle^{-|\beta|-|\sigma|} \tag{6.5.5}$$

for all $(x, \xi) \in \mathbb{R}^{2d}$.

Proof. By Theorem 6.3.12 we have that $r_{\alpha\beta} \sim \sum_{j \geq 0} r_{\alpha\beta j}$ in $FAG_{\mu'\nu'}^{-|\alpha|,-|\beta|}(\mathbb{R}^d)$ for some $\mu' > 1$, $\nu' > 1$, with

$$r_{\alpha\beta j}(x, \xi) = \sum_{|\gamma|=j} (\gamma!)^{-1} \partial_\xi^\gamma e(x, \xi) D_\xi^\alpha D_x^{\beta+\gamma} p(x, \xi).$$

Leibniz' rule gives

$$|\partial_\xi^\theta \partial_x^\sigma r_{\alpha\beta j}(x, \xi)| \leq C^{|\alpha|+|\beta|+|\theta|+|\sigma|+2j+1} \alpha!^\mu \beta!^\nu \theta!^{\nu'} \sigma!^{\mu'} j!^{\mu'+\nu'-1}$$
$$\times \langle\xi\rangle^{-|\alpha|-|\theta|-j} \langle x\rangle^{-|\beta|-|\sigma|-j}. \tag{6.5.6}$$

Thus we can apply Theorem 6.3.8, taking cut-off functions $\varphi_j(x, \xi)$ independent of α, β, and we obtain (6.5.5). $\qquad\square$

We start by deriving decay estimates. We first give a result for the linear case. With respect to Corollary 6.3.17 we now allow $\mu = 1$, $\nu = 1$.

Theorem 6.5.3. *Let $p \in AG_{\mu\nu}^{m,n}(\mathbb{R}^d)$ be G-elliptic and $f \in S_\nu^\mu(\mathbb{R}^d)$. Let $u \in \mathcal{S}'(\mathbb{R}^d)$ be a solution of the equation $Pu = f$. Then, for every $s \geq 0$ there exists $\epsilon > 0$ such that $\|u\|_{s,\nu;\epsilon} < \infty$.*

To prove the Theorem we need a preliminary result.

Lemma 6.5.4. *Let $p \in AG_{\mu\nu}^{m,n}(\mathbb{R}^d)$ be G-elliptic and let E be the operator in Proposition 6.5.2. Then, for every $s \geq 0$ there exists a positive constant A_s such that*

$$\|E[P, x^k]u\|_s \leq k!^\nu \sum_{0 \neq \beta \leq k} \frac{A_s^{|\beta|}}{(k-\beta)!^\nu} \|x^{k-\beta}u\|_s \qquad (6.5.7)$$

for every $k \in \mathbb{N}^d$, $k \neq 0$.

Proof. Fixing $k \in \mathbb{N}^d, k \neq 0$, and integrating by parts, we have:

$$x^k Pu(x) = \int e^{ix\xi} x^k p(x, \xi) \hat{u}(\xi) \, d\xi$$

$$= (-1)^{|k|} \int e^{ix\xi} D_\xi^k \left(p(x, \xi) \hat{u}(\xi) \right) d\xi$$

$$= \sum_{\beta \leq k} (-1)^{|\beta|} \binom{k}{\beta} (D_\xi^\beta p)(x, D)(x^{k-\beta}u),$$

from which it follows that

$$E[P, x^k]u = \sum_{0 \neq \beta \leq k} (-1)^{|\beta|+1} \binom{k}{\beta} E(D_\xi^\beta p)(x, D)(x^{k-\beta}u). \qquad (6.5.8)$$

We can now apply Proposition 6.5.2 and obtain that $E(D_\xi^\beta p)(x, D) = r_\beta(x, D)$ with $r_\beta(x, \xi)$ satisfying the estimate

$$|\partial_\xi^\theta \partial_x^\sigma r_\beta(x, \xi)| \leq C^{|\beta|} \beta!^\nu \langle\xi\rangle^{-|\beta|-|\theta|} \langle x\rangle^{-|\sigma|}$$

for some positive constant C independent of $\beta \neq 0$. Hence, we have, for every $s \in \mathbb{R}$:

$$\|r_\beta(x, D)\|_{\mathcal{B}(H^s)} \leq A_s^{|\beta|} \beta!^\nu, \qquad (6.5.9)$$

cf. Remark 1.5.6. Then, (6.5.8) and (6.5.9) give (6.5.7). $\qquad\square$

Proof of Theorem 6.5.3. In the linear case we know already that $u \in \mathcal{S}(\mathbb{R}^d)$. Moreover, with fixed $k \in \mathbb{N}^d$, $\epsilon > 0$, we can write

$$\frac{\epsilon^{|k|}}{k!^\nu} x^k Pu = \frac{\epsilon^{|k|}}{k!^\nu} x^k f,$$

from which we get

$$\frac{\epsilon^{|k|}}{k!^\nu} P(x^k u) = \frac{\epsilon^{|k|}}{k!^\nu} [P, x^k]u + \frac{\epsilon^{|k|}}{k!^\nu} x^k f. \qquad (6.5.10)$$

Applying to both members of (6.5.10) the operator E in Proposition 6.5.2 we obtain

$$\frac{\epsilon^{|k|}}{k!^\nu} x^k u = \frac{\epsilon^{|k|}}{k!^\nu} R(x^k u) + \frac{\epsilon^{|k|}}{k!^\nu} E[P, x^k]u + \frac{\epsilon^{|k|}}{k!^\nu} E(x^k f),$$

where $R \in OPAG_{\mu'\nu'}^{-1,-1}(\mathbb{R}^d)$, for some $\mu' > 1$, $\nu' > 1$. Taking now Sobolev norms and summing up for $|k| \leq N$, we have

$$H_N^{s,\nu;\epsilon}[u] \leq \|u\|_s + \sum_{0<|k|\leq N} \frac{\epsilon^{|k|}}{k!^\nu} \|R(x^k u)\|_s + \sum_{0<|k|\leq N} \frac{\epsilon^{|k|}}{k!^\nu} \|E[P, x^k]u\|_s$$

$$+ \sum_{0<|k|\leq N} \frac{\epsilon^{|k|}}{k!^\nu} \|E(x^k f)\|_s. \tag{6.5.11}$$

Let us estimate the terms in the right-hand side of (6.5.11). First of all we observe that if $k \in \mathbb{N}^d$, $k \neq 0$, then there exists $j_k \in \{1, \ldots, d\}$ with $k_{j_k} > 0$. Hence we have

$$\|R(x^k u)\|_s = \|R \circ x_{j_k}(x^{k-e_{j_k}} u)\|_s \leq C_s' \|x^{k-e_{j_k}} u\|_s$$

since $R \circ x_{j_k} \in OPAG_{\mu'\nu'}^{-1,0}(\mathbb{R}^d)$. Then, we have the estimate

$$\sum_{0<|k|\leq N} \frac{\epsilon^{|k|}}{k!^\nu} \|R(x^k u)\|_s \leq C_s' \epsilon \sum_{0<|k|\leq N} \frac{\epsilon^{|k|-1}}{k!^\nu} \|x^{k-e_{j_k}} u\|_s \leq C_s'' \epsilon H_{N-1}^{s,\nu;\epsilon}[u]. \tag{6.5.12}$$

By Lemma 6.5.4 there exists a positive constant A_s such that

$$\sum_{0<|k|\leq N} \frac{\epsilon^{|k|}}{k!^\nu} \|E[P, x^k]u\|_s \leq \sum_{0<|k|\leq N} \sum_{0\neq\beta\leq k} (\epsilon A_s)^{|\beta|} \frac{\epsilon^{|k-\beta|}}{(k-\beta)!^\nu} \|x^{k-\beta} u\|_s$$

$$\leq A_s' \epsilon H_{N-1}^{s,\nu;\epsilon}[u] \tag{6.5.13}$$

for ϵ sufficiently small.

Moreover, since $E \in OPAG_{\mu'\nu'}^{0,0}(\mathbb{R}^d)$, then obviously $\|E(x^k f)\|_s \leq C_s''' \|x^k f\|_s$ from which we obtain

$$\sum_{0<|k|\leq N} \frac{\epsilon^{|k|}}{k!^\nu} \|E(x^k f)\|_s \leq C_s''' H_N^{s,\nu;\epsilon}[f] \leq C_s''' \|\!|f|\!\|_{s,\nu;\epsilon} < \infty. \tag{6.5.14}$$

Hence, by (6.5.12), (6.5.13), (6.5.14), we obtain

$$H_N^{s,\nu;\epsilon}[u] \leq C_s \left(\|u\|_s + \epsilon H_{N-1}^{s,\nu;\epsilon}[u] + \|\!|f|\!\|_{s,\nu;\epsilon} \right).$$

Iterating this estimate and choosing ϵ sufficiently small, we conclude that $H_N^{s,\nu;\epsilon}[u] \leq C$ for some constant $C > 0$ independent of N. The assertion is then proved. $\square$

Let us now consider the general nonlinear equation in (6.5.1), that is

$$Pu = f + F[u], \tag{6.5.15}$$

with $F[u]$ as in (6.5.2). We need to treat separately the cases $n > 0$ and $n = 0$, since they require different assumptions on the a priori regularity of the solution and on the form of $F[u]$. Let us start from the case $n > 0$.

Theorem 6.5.5. *Let* $p \in AG_{\mu\nu}^{m,n}(\mathbb{R}^d)$, $m \geq 1$, $n > 0$, *be G-elliptic,* $f \in S_{\nu}^{\mu}(\mathbb{R}^d)$ *and* $F[u]$ *as in (6.5.2). Assume that* $u \in H^s(\mathbb{R}^d)$, $s > d/2 + m - 1$ *is a solution of the equation (6.5.15). Then, there exists* $\epsilon > 0$ *such that* $\|u\|_{s,\nu;\epsilon} < \infty$.

To prove Theorem 6.5.5 we shall proceed as in the proof of Theorem 6.5.3 but first we need to estimate the nonlinear term. Since $n > 0$ we can assume without loss of generality that $F[u] = x^h u^\ell (\partial_x^\gamma u)^r$ for some $\ell, r \in \mathbb{N}$, $\gamma, h \in \mathbb{N}^d$ with $\ell + r \geq 2$, $0 \leq |h| \leq n - 1$, $0 \leq |\gamma| \leq m - 1$.

Lemma 6.5.6. *Let* $p \in AG_{\mu\nu}^{m,n}(\mathbb{R}^d)$ *satisfy the assumptions of Theorem 6.5.5 and let* $u \in H^s(\mathbb{R}^d)$, $s > d/2 + |\gamma|$ *for some* $\gamma \in \mathbb{N}^d$ *with* $0 \leq |\gamma| \leq m - 1$. *Then, for every pair of non-negative integers* ℓ, r *with* $\ell + r \geq 2$, $0 \leq |h| \leq n - 1$, *the following estimate holds:*

$$\sum_{0 < |k| \leq N} \frac{\epsilon^{|k|}}{k!^\nu} \|E(x^k (x^h u^\ell (\partial_x^\gamma u)^r))\|_s \leq C_s'' \epsilon \|u\|_s^{\ell+r-1} H_{N-1}^{s,\nu;\epsilon}[u] \tag{6.5.16}$$

for every $N \in \mathbb{N}$, $N \neq 0$, *and for some positive constant* C_s'' *independent of* N.

Proof. With fixed $k \in \mathbb{N}^d$, $k \neq 0$, there exists $j_k \in \{1, \ldots, d\}$ such that $k_{j_k} > 0$. Then, since $E \circ x^{h+e_{j_k}}$ is an operator of orders $-|\gamma|$, 0, we can write

$$\|E(x^{k+h} u^\ell (\partial_x^\gamma u)^r)\|_s \leq C_s \|x^{k-e_{j_k}} u^\ell (\partial_x^\gamma u)^r\|_{s-|\gamma|}.$$

Now, if $\ell \geq 1$, applying Schauder's estimates (0.2.7), we get

$$\|x^{k-e_{j_k}} u^\ell (\partial_x^\gamma u)^r\|_{s-|\gamma|} \leq C_s' \|x^{k-e_{j_k}} u\|_{s-|\gamma|} \cdot \|u\|_{s-|\gamma|}^{\ell-1} \cdot \|\partial_x^\gamma u\|_{s-|\gamma|}^r$$
$$\leq C_s'' \|x^{k-e_{j_k}} u\|_s \cdot \|u\|_s^{\ell+r-1}.$$

Hence we have

$$\sum_{0 < |k| \leq N} \frac{\epsilon^{|k|}}{k!^\nu} \|E(x^k (x^h u^\ell (\partial_x^\gamma u)^r))\|_s \leq C_s'' \epsilon \|u\|_s^{\ell+r-1} \sum_{0 < |k| \leq N} \frac{\epsilon^{|k|-1}}{(k - e_{j_k})!^\nu} \|x^{k-e_{j_k}} u\|_s$$
$$\leq C_s''' \epsilon \|u\|_s^{\ell+r-1} H_{N-1}^{s,\nu;\epsilon}[u].$$

If $\ell = 0$, then by the assumption $\ell + r \geq 2$ we have $r \geq 2$. Hence similarly:

$$\|x^{k-e_{j_k}} (\partial_x^\gamma u)^r\|_{s-|\gamma|} \leq C_s' \|x^{k-e_{j_k}} \partial_x^\gamma u\|_{s-|\gamma|} \cdot \|u\|_s^{r-1}.$$

Now we use the identity:

$$x^{k-e_{j_k}} \partial_x^\gamma u = \sum_{\substack{i \le k-e_{j_k} \\ i \le \gamma}} \frac{(k-e_{j_k})!}{(k-e_{j_k}-i)!} \binom{\gamma}{i} (-1)^{|i|} \partial_x^{\gamma-i}(x^{k-e_{j_k}-i}u),$$

generalizing (6.2.38). Then, since $|\gamma| \le m-1$, we obtain the estimate

$$\frac{\epsilon^{|k|}}{k!^\nu} \|E(x^k(x^h(\partial_x^\gamma u)^r))\|_s \le C''_s \epsilon \|u\|_s^{r-1} \sum_{\substack{i \le k-e_{j_k} \\ i \le \gamma}} \frac{\epsilon^{|k|-1}}{(k-e_{j_k}-i)!^\nu} \|x^{k-e_{j_k}-i}u\|_s,$$

which gives again (6.5.16). $\square$

Proof of Theorem 6.5.5. Repeating the argument in the proof of Theorem 6.5.3, from (6.5.15) we obtain that

$$H_N^{s,\nu;\epsilon}[u] \le \|u\|_s + \sum_{0<|k|\le N} \frac{\epsilon^{|k|}}{k!^\nu} \|R(x^k u)\|_s + \sum_{0<|k|\le N} \frac{\epsilon^{|k|}}{k!^\nu} \|E[P,x^k]u\|_s$$

$$+ \sum_{0<|k|\le N} \frac{\epsilon^{|k|}}{k!^\nu} \|E(x^k f)\|_s + \sum_{0<|k|\le N} \frac{\epsilon^{|k|}}{k!^\nu} \|E(x^{k+h}u^\ell(\partial_x^\gamma u)^r)\|_s.$$

Then by (6.5.12), (6.5.13), (6.5.14) and by Lemma 6.5.6, we obtain

$$H_N^{s,\nu;\epsilon}[u] \le C_s \left(\|u\|_s + \epsilon H_{N-1}^{s,\nu;\epsilon}[u] + \|f\|_{s,\nu;\epsilon} + \epsilon \|u\|_s^{\ell+r-1} H_{N-1}^{s,\nu;\epsilon}[u] \right)$$

and we can conclude as in Theorem 6.5.3. $\square$

Let us now consider the case $n = 0$. In this situation, the parametrix E of P is an operator of orders $-m, 0$, so there is no gain in the decay estimates when we apply E to both the sides of the equation. For this reason we need to strengthen the assumptions on the a priori decay of u.

Theorem 6.5.7. *Let $p \in AG_{\mu\nu}^{m,0}(\mathbb{R}^d)$ be G-elliptic, $f \in S_\nu^\mu(\mathbb{R}^d)$ and $F[u]$ as in (6.5.2). Assume that u is a solution of (6.5.15) and that there exists $\epsilon_o > 0$ such that $\langle x \rangle^{\epsilon_o} u \in H^s(\mathbb{R}^d)$, $s > d/2 + m - 1$. Then, there exists $\epsilon > 0$ such that $\|u\|_{s,\nu;\epsilon} < \infty$.*

Since $n = 0$ we can assume without loss of generality that $F[u] = u^\ell(\partial_x^\gamma u)^r$ for some $\ell, r \in \mathbb{N}$, with $\ell + r \ge 2$, $0 \le |\gamma| \le m - 1$.

Lemma 6.5.8. *Under the assumptions of Theorem 6.5.7, we have $\langle x \rangle^{\epsilon_o+\rho} u \in H^s(\mathbb{R}^d)$ for every $\rho \le \min\{1, (\ell+r-1)\epsilon_o\}$.*

Proof. By (6.5.15) we have

$$\langle x \rangle^{\epsilon_o + \rho} P u = \langle x \rangle^{\epsilon_o + \rho} f + \langle x \rangle^{\epsilon_o + \rho} u^\ell (\partial_x^\gamma u)^r,$$

from which

$$\langle x \rangle^{\epsilon_o + \rho} u = E(\langle x \rangle^{\epsilon_o + \rho} f) + E[P, \langle x \rangle^{\epsilon_o + \rho}] u$$
$$+ R(\langle x \rangle^{\epsilon_o + \rho} u) + E(\langle x \rangle^{\epsilon_o + \rho} u^\ell (\partial_x^\gamma u)^r) \qquad (6.5.17)$$

where E, R are the operators defined before Proposition 6.5.2. Since $f \in S_\nu^\mu(\mathbb{R}^d)$ and $E \in OPAG_{\mu'\nu'}^{-m,0}(\mathbb{R}^d)$, then the Sobolev norm of the first term in the right-hand side of (6.5.17) is finite. Furthermore, since $E[P, \langle x \rangle^{\epsilon_o + \rho}]$ has orders -1, $\epsilon_o + \rho - 1$, we have

$$\|E[P, \langle x \rangle^{\epsilon_o + \rho}] u\|_s \leq C_s \|\langle x \rangle^{\epsilon_o + \rho - 1} u\|_s < \infty$$

since $\rho \leq 1$. Concerning the third term we have that $\|R(\langle x \rangle^{\epsilon_o + \rho} u)\|_s \leq C_s \|\langle x \rangle^{\epsilon_o} u\|_s$ because R has orders $-1, -1$, and $\rho \leq 1$. Also, since $|\gamma| < m$ we have

$$\|E(\langle x \rangle^{\epsilon_o + \rho} u^\ell (\partial_x^\gamma u)^r)\|_s \leq C_s \|\langle x \rangle^{\epsilon_o + \rho} u^\ell (\partial_x^\gamma u)^r\|_{s - |\gamma|}.$$

Assume for instance $\ell > 0$. Then, applying Schauder's estimates we get

$$\|\langle x \rangle^{\epsilon_o + \rho} u^\ell (\partial_x^\gamma u)^r\|_{s - |\gamma|} \leq C_s' \|\langle x \rangle^{\epsilon_o} u\|_{s - |\gamma|} \cdot \|\langle x \rangle^{\frac{\rho}{\ell + r - 1}} u\|_{s - |\gamma|}^{\ell - 1}$$
$$\times \|\langle x \rangle^{\frac{\rho}{\ell + r - 1}} \partial_x^\gamma u\|_{s - |\gamma|}^r$$
$$\leq C_s'' \|\langle x \rangle^{\epsilon_o} u\|_s \cdot \|\langle x \rangle^{\frac{\rho}{\ell + r - 1}} u\|_s^{\ell + r - 1} < \infty$$

since $\rho \leq \epsilon_o(\ell + r - 1)$. Similarly, we can treat the case $\ell = 0$ taking account of the condition $\ell + r \geq 2$. $\qquad \square$

Iterating Lemma 6.5.8 we obtain that $\langle x \rangle^\tau u \in H^s(\mathbb{R}^d)$ for any $\tau > 0$.

Lemma 6.5.9. *Let $p \in AG_{\mu\nu}^{m,0}(\mathbb{R}^d)$ satisfy the assumptions of Theorem 6.5.7 and let $u \in H^s(\mathbb{R}^d)$, $s > d/2 + |\gamma|$ for some $\gamma \in \mathbb{N}^d$ with $0 \leq |\gamma| \leq m - 1$. Then, for every pair of non-negative integers ℓ, r with $\ell + r \geq 2$, the following estimate holds:*

$$\sum_{0 < |k| \leq N} \frac{\epsilon^{|k|}}{k!^\nu} \|E(x^k u^\ell (\partial_x^\gamma u)^r)\|_s \leq C_s'' \epsilon \left\| \langle x \rangle^{\frac{1}{\ell + r - 1}} u \right\|_s^{\ell + r - 1} H_{N-1}^{s,\nu;\epsilon}[u] \qquad (6.5.18)$$

for every $N \in \mathbb{N}$, $N \neq 0$.

Proof. With fixed $k \in \mathbb{N}^d$, $k \neq 0$, there exists $j_k \in \{1, \ldots, d\}$ such that $k_{j_k} > 0$. If $\ell \geq 1$, applying Schauder's estimates we have

$$\frac{\epsilon^{|k|}}{k!^\nu} \|E(x^k u^\ell (\partial_x^\gamma u)^r)\|_s \leq C_s \frac{\epsilon^{|k|}}{k!^\nu} \|x^k u^\ell (\partial_x^\gamma u)^r\|_{s - |\gamma|}$$
$$\leq C_s' \epsilon \frac{\epsilon^{|k| - 1}}{(k - e_{j_k})!^\nu} \|x^{k - e_{j_k}} u\|_{s - |\gamma|} \cdot \|x_{j_k} u^{\ell - 1} (\partial_x^\gamma u)^r\|_{s - |\gamma|}$$

$$\leq C_s'' \epsilon \frac{\epsilon^{|k|-1}}{(k-e_{j_k})!^{\nu}} \left\| x^{k-e_{j_k}} u \right\|_s \cdot \left\| \langle x \rangle^{\frac{1}{\ell+r-1}} u \right\|_s^{\ell+r-1}$$

from which (6.5.18) directly follows. The case $\ell = 0$, $r \geq 2$ can be treated similarly. □

Proof of Theorem 6.5.7. The assertion can be proved repeating readily the argument in the proof of Theorem 6.5.5 and using Lemma 6.5.9 instead of Lemma 6.5.6. □

We pass now to examine the regularity of the solutions of (6.5.15).

Lemma 6.5.10. *Let $p \in AG_{\mu\nu}^{m,n}(\mathbb{R}^d)$ be G-elliptic and let E be the operator defined in Proposition 6.5.2. Then, for any $s \geq 0$, there exists a positive constant B_s such that*

$$\| E[P, \partial^j] u \|_s \leq j!^{\mu} \sum_{0 \neq \gamma \leq j} \frac{B_s^{|\gamma|}}{(j-\gamma)!^{\mu}} \| \partial_x^{j-\gamma} u \|_s. \tag{6.5.19}$$

Proof. We first observe that

$$\partial_x^j Pu = \sum_{\gamma \leq j} \binom{j}{\gamma} (\partial_x^{\gamma} p)(x, D) \partial_x^{j-\gamma} u.$$

Then

$$E[P, \partial_x^j] u = - \sum_{0 \neq \gamma \leq j} \binom{j}{\gamma} E(\partial_x^{\gamma} p)(x, D) \partial_x^{j-\gamma} u.$$

By Proposition 6.5.2, we have that $E(\partial_x^{\gamma} p)(x, D) = r_{\gamma}(x, D) \in OPAG_{\mu'\nu'}^{0,-|\gamma|}(\mathbb{R}^d)$ for some $\mu' > 1$, $\nu' > 1$ and

$$\| r_{\gamma}(x, D) \|_{B(H^s)} \leq B_s^{|\gamma|} \gamma!^{\mu}.$$

Hence, since $\mu \geq 1$, we have

$$\| E[P, \partial^j] u \|_s \leq \sum_{0 \neq \gamma \leq j} \binom{j}{\gamma} \gamma!^{\mu} B_s^{|\gamma|} \| \partial_x^{j-\gamma} u \|_s$$

$$\leq j!^{\mu} \sum_{0 \neq \gamma \leq j} \frac{B_s^{|\gamma|}}{(j-\gamma)!^{\mu}} \| \partial_x^{j-\gamma} u \|_s.$$

The lemma is then proved. □

Theorem 6.5.11. *Let $p \in AG_{\mu\nu}^{m,n}(\mathbb{R}^d)$ be G-elliptic and let $f \in S_{\nu}^{\mu}(\mathbb{R}^d)$. Let $u \in S'(\mathbb{R}^d)$ be a solution of the equation $Pu = f$. Then there exists $T > 0$ such that $\| u \|_{\{s,\mu;T\}} < \infty$.*

Proof. As in Theorem 6.5.3, we have indeed $u \in \mathcal{S}(\mathbb{R}^d)$ from the results of Chapters 1 and 3. Moreover, for any $T > 0$ and $j \in \mathbb{N}^d$ we have

$$\frac{T^{|j|}}{j!^\mu} \partial_x^j Pu = \frac{T^{|j|}}{j!^\mu} \partial_x^j f,$$

from which we get

$$\frac{T^{|j|}}{j!^\mu} P(\partial_x^j u) = \frac{T^{|j|}}{j!^\mu} [P, \partial_x^j] u + \frac{T^{|j|}}{j!^\mu} \partial_x^j f. \qquad (6.5.20)$$

Hence, arguing as in the proof of Theorem 6.5.3, we obtain

$$\frac{T^{|j|}}{j!^\mu} \partial_x^j u = \frac{T^{|j|}}{j!^\mu} R(\partial_x^j u) + \frac{T^{|j|}}{j!^\mu} E[P, \partial_x^j] u + \frac{T^{|j|}}{j!^\mu} E(\partial_x^j f)$$

for some operator R of orders $-1, -1$. Taking now Sobolev norms and summing up for $|j| \leq N$, we have

$$E_N^{s,\mu;T}[u] \leq \|u\|_s + \sum_{0 < |j| \leq N} \frac{T^{|j|}}{j!^\mu} \|R(\partial_x^j u)\|_s + \sum_{0 < |j| \leq N} \frac{T^{|j|}}{j!^\mu} \|E[P, \partial_x^j] u\|_s$$

$$+ \sum_{0 < |j| \leq N} \frac{T^{|j|}}{j!^\mu} \|E(\partial_x^j f)\|_s. \quad (6.5.21)$$

Now, for every $j \in \mathbb{N}^d$, $j \neq 0$, there exists $q_j \in \{1, \ldots, d\}$ such that $j_{q_j} > 0$. Then, the second term in the right-hand side of (6.5.21) can be estimated as follows:

$$\sum_{0 < |j| \leq N} \frac{T^{|j|}}{j!^\mu} \|R(\partial_x^j u)\|_s \leq C_s' T \sum_{0 < |j| \leq N} \frac{T^{|j|-1}}{j!^\mu} \|\partial_x^{j - e_{q_j}} u\|_s \leq C_s'' T E_{N-1}^{s,\mu;T}[u].$$

$$(6.5.22)$$

By Lemma 6.5.10 there exists a positive constant B_s such that

$$\sum_{0 < |j| \leq N} \frac{T^{|j|}}{j!^\mu} \|E[P, \partial_x^j] u\|_s \leq \sum_{0 < |j| \leq N} \sum_{0 \neq \gamma \leq j} (B_s T)^{|\gamma|} \frac{T^{|j-\gamma|}}{(j-\gamma)!^\mu} \|\partial_x^{j-\gamma} u\|_s$$

$$\leq B_s' T E_{N-1}^{s,\mu;T}[u] \qquad (6.5.23)$$

for T sufficiently small.

Moreover, we have

$$\sum_{0 < |j| \leq N} \frac{T^{|j|}}{j!^\mu} \|E(\partial_x^j f)\|_s \leq C_s''' E_N^{s,\mu;T}[f] \leq C_s''' \|f\|_{\{s,\mu;T\}} < \infty. \qquad (6.5.24)$$

Hence, by (6.5.22), (6.5.23), (6.5.24), we obtain

$$E_N^{s,\mu;T}[u] \leq C_s \left(\|u\|_s + T E_{N-1}^{s,\mu;T}[u] + \|f\|_{\{s,\mu;T\}} \right),$$

from which it follows that $\|u\|_{\{s,\mu;T\}} < \infty$. $\qquad \square$

As an immediate consequence of Proposition 6.1.11 and Theorems 6.5.3 and 6.5.11, we obtain the following result.

Theorem 6.5.12. *Let* $p \in AG_{\mu\nu}^{m,n}(\mathbb{R}^d)$, $\mu \geq 1$, $\nu \geq 1$ *be G-elliptic,* $f \in S_\nu^\mu(\mathbb{R}^d)$ *and let* $u \in S'(\mathbb{R}^d)$ *be a solution of the equation* $Pu = f$. *Then* $u \in S_\nu^\mu(\mathbb{R}^d)$. *In particular, if* $p \in AG_{1,1}^{m,n}(\mathbb{R}^d)$ *and* $Pu = 0$, *then* $u \in S_1^1(\mathbb{R}^d)$.

To treat the nonlinear case, we do not have here to distinguish the cases $n > 0$ and $n = 0$ as for the decay estimates. We can assume in general that $n \geq 0$.

Lemma 6.5.13. *Let* $p \in AG_{\mu\nu}^{m,n}(\mathbb{R}^d)$ *be G-elliptic and let* $u \in H^s(\mathbb{R}^d)$, $s > d/2 + |\gamma|$ *for some* $\gamma \in \mathbb{N}^d$ *with* $0 \leq |\gamma| \leq m - 1$, $0 \leq h \leq \max\{n - 1, 0\}$. *Then, for every pair of positive integers* ℓ, r *with* $\ell + r \geq 2$, *the following estimate holds:*

$$\sum_{0 < |j| \leq N} \frac{T^{|j|}}{j!^\mu} \|E(\partial_x^j(x^h u^\ell (\partial_x^\gamma u)^r))\|_s \leq C_s \left(\|u\|_s + T(E_{N-1}^{s,\mu;T}[u])^{\ell+r} \right). \qquad (6.5.25)$$

Proof. Fix $j \in \mathbb{N}^d$, $j \neq 0$. Then $j_{q_j} \neq 0$ for some $q_j \in \{1, \ldots, d\}$. Hence we can write

$$E(\partial_x^j(x^h u^\ell (\partial_x^\gamma u)^r)) = E\left(x^h \partial_{x_{q_j}} \partial_x^{j - e_{q_j}} (u^\ell (\partial_x^\gamma u)^r) \right)$$

$$+ \sum_{\substack{0 \neq j' \leq j \\ j' \leq h}} \frac{j!}{(j - j')!} \binom{h}{j'} E\left(x^{h-j'} \partial_x^{j-j'} (u^\ell (\partial_x^\gamma u)^r) \right). \qquad (6.5.26)$$

Now observe that $E \circ x^h \partial_{x_{q_j}}$ and $E \circ x^{h-j'}$ map continuously $H^{s-|\gamma|}(\mathbb{R}^d)$ into $H^s(\mathbb{R}^d)$. Then we have

$$\left\| E\left(x^h \partial_{x_{q_j}} \partial_x^{j - e_{q_j}} (u^\ell (\partial_x^\gamma u)^r) \right) \right\|_s \leq C_s \left\| \partial_x^{j - e_{q_j}} (u^\ell (\partial_x^\gamma u)^r) \right\|_{s-|\gamma|}.$$

Leibniz' formula gives

$$\frac{T^{|j|}}{j!^\mu} \left\| \partial_x^{j - e_{q_j}} (u^\ell (\partial_x^\gamma u)^r)) \right\|_{s-|\gamma|} \leq C_s' \frac{T^{|j|}}{j!^\mu} \sum_{j_1 + \ldots + j_{\ell+r} = j - e_{q_j}} \frac{j!}{j_1! \ldots j_{\ell+r}!}$$

$$\times \|\partial_x^{j_1} u\|_{s-|\gamma|} \cdots \|\partial_x^{j_\ell} u\|_{s-|\gamma|} \cdot \|\partial_x^{j_{\ell+1}} u\|_s \cdots \|\partial_x^{j_{\ell+r}} u\|_s$$

$$\leq C_s'' T \sum_{j_1 + \ldots + j_{\ell+r} = j - e_{q_j}} \prod_{k=1}^{\ell+r} \frac{T^{|j_k|}}{j_k!^\mu} \|\partial_x^{j_k} u\|_s,$$

from which we obtain

$$\sum_{0 < |j| \leq N} \frac{T^{|j|}}{j!^\mu} \left\| E\left(x^h \partial_{x_{q_j}} \partial_x^{j - e_{q_j}} (u^\ell (\partial_x^\gamma u)^r) \right) \right\|_{s-|\gamma|} \leq C_s''' T (E_{N-1}^{s,\mu;T}[u])^{\ell+r}.$$

The second term in the right-hand side of (6.5.26) can be estimated similarly. $\quad\square$

Theorem 6.5.14. *Let $p \in AG_{\mu\nu}^{m,n}(\mathbb{R}^d)$ be G-elliptic, $f \in S_\nu^\mu(\mathbb{R}^d)$ and $F[u]$ of the form (6.5.2). Assume that $u \in H^s(\mathbb{R}^d)$, $s > d/2 + m - 1$, is a solution of (6.5.15). Then there exists $T > 0$ such that $\|u\|_{\{s,\mu;T\}} < \infty$.*

Proof. The proof can be performed arguing as for Theorem 6.5.11 and estimating the nonlinear term by Lemma 6.5.13. We leave the details to the reader. □

Finally, by Theorems 6.5.5, 6.5.7, 6.5.14 and Proposition 6.1.11 we directly obtain Theorem 6.5.1.

Notes

The classes $S_\nu^\mu(\mathbb{R}^d)$ were first studied by Gelfand and Shilov, see [90] and the references there to earlier literature. The motivation in [90] was to introduce general ultradistributions, and to treat in this frame existence and uniqueness for parabolic initial-value problems. Later, the classes $S_\nu^\mu(\mathbb{R}^d)$ were widely used, under different names and notations, mainly in connection with applications to partial differential equations, see for example Avantaggiati [8], Biagioni and Gramchev [16], Mitjagin [147], Pilipovic and Teofanov [162], Cordero, Pilipovic, Rodino and Teofanov [56]. In particular, our presentation in Section 6.1 is inspired by the recent results of Chung, Chung and Kim [48], Gröchenig and Zimmermann [100]. For more details on the temperate ultradistributions $S_\mu^{\mu'}(\mathbb{R}^d)$ and similar classes, see for example Pilipovic [161]. For other results about the uncertainty principle, we refer to Gröchenig and Zimmermann [99].

The Gevrey classes $G_0^\mu(\mathbb{R}^d)$, $\mu > 1$, and their duals, spaces of the Gevrey ultradistributions, play an important role in the study of the local properties of the solutions of partial differential equations, see Rodino [172], Mascarello and Rodino [142] and the references therein. A detailed study of these spaces has been left outside the analysis of Section 6.1 for the sake of brevity.

Concerning the results on exponential decay, cf. Section 6.2 and sequel, the main interest comes historically from Quantum Mechanics, where the decay of the eigenfunctions has been intensively studied, see for instance Agmon [3], Hislop and Sigal [112] and the references quoted therein for $-\Delta + V(x)$ with general potentials. Solutions of semilinear second-order equations in $\mathbb{R}^d$ which decay to zero at infinity are usually called homoclinics in the literature and there are several results about their existence and multiplicity via variational methods, see for example Ambrosetti and Malchiodi [4], Le Bris and Lions [132]. Note that in our results we take already as granted the existence of homoclinics, and prove regularity in $S_\nu^\mu(\mathbb{R}^d)$. Concerning in particular the results of Section 6.2, our main reference is Cappiello, Gramchev and Rodino [41], where Theorem 6.2.2 is proved in the more general case when the linear part is given by an operator of the form (6.2.1), (6.2.2), (6.2.3). Concerning the proof of Proposition 6.2.7, we refer to Abramowitz, Stegun [1] and Tricomi [192] for details on the error function. Our arguments in Section 6.2 are based on inductive a priori estimates and perturbative methods in $S^\mu_\nu(\mathbb{R}^d)$; for different approaches which give similar results, see Martinez [140] on semi-classical analysis by micro-local techniques, and Rabier [165], Rabinovich [166], Buzano [28] combining abstract functional analysis and pseudo-differential techniques. The main references for Section 6.3 are the papers on G-pseudo-differential calculus of Cappiello [39], [40], Cappiello and Rodino [45], Cappiello, Gramchev and Rodino [42].

For preceding works on pseudo-differential operators of analytic-type see the bibliography of [172]; we refer in particular to the local calculus of Boutet de Monvel and Krée [24], and Zanghirati [201]. Concerning Section 6.5, we refer to the papers of Bona and Li [22], Biagioni and Gramchev [16], Cappiello, Gramchev and Rodino [42], [43], [44]. Finally, concerning the classical theory of asymptotic integration, which we used somewhere in this chapter, we again refer the reader to Wasow [194].

Bibliography

[1] M. Abramowitz and I. A. Stegun, *Handbook of mathematical functions with formulas, graphs and mathematical tables*, Nat. Bureau of Standards Appl. Math. Series **55**, Washington, DC, 1964.

[2] M. Adler, *On a trace functional for formal pseudo-differential operators and the symptectic structure of Korteweg-de Vries type equations*, Invent. Math., **50** (1978/1979), 219–248.

[3] S. Agmon, *Lectures on exponential decay of solutions of second order elliptic equations: bounds on eigenfunctions of N-body Schrödinger operators*, Math. Notes **29**, Princeton University Press, Princeton, NJ, 1982.

[4] A. Ambrosetti and A. Malchiodi, *Perturbation methods and semilinear elliptic problems on $\mathbb{R}^n$*, Progress in Mathematics **240**, Birkhäuser, Basel, 2006.

[5] C. J. Amick and J. F. Toland, *Uniqueness and related analytic properties for the Benjamin-Ono equation: a nonlinear Neumann problem in the plane*, Acta Math., **167** (1991), 107–126.

[6] M. F. Atiyah and I. M. Singer, *The index of elliptic operators: I*, Ann. of Math., **87** (1968), 484–530.

[7] M. F. Atiyah, R. Bott and V. K. Patodi, *On the heat equation and the index theorem*, Invent. Math., **19** (1973), 279–330.

[8] A. Avantaggiatti, *S-spaces by means of the behaviour of Hermite-Fourier coefficients*, Boll. Un. Mat. Ital., **6** (1985), n. 4-A, 487–495.

[9] R. Beals, *Characterization of pseudodifferential operators and applications*, Duke Math. J., **44** (1977), no. 1, 45–57.

[10] R. Beals, *Correction to "Characterization of pseudodifferential operators and applications"*, Duke Math. J., **46** (1979), no. 1, 215.

[11] R. Beals and C. Fefferman, *On local solvability of linear partial differential equations*, Ann. of Math., **97** (1973), 482–498.

[12] H. Berestycki and P.-L. Lions, *Nonlinear scalar field equations I, II*, Arch. Rational Mech. Anal., **82** (1983), 313–375.

[13] F. A. Berezin, *Wick and anti-Wick symbols of operators*, Mat. Sb. (N.S.), **86** (1971), 578–610.

[14] F. A. Berezin and M. A. Shubin, *Symbols of operators and quantization*, Colloquia mathematica Societatis Janos Bolyai **5** (1972), Hilbert space operators, Tihany (Hungary), 21–52.

[15] F. A. Berezin and M. A. Shubin, *Lecture on quantum mechanics*. Moskow State University, 1972 (Russian).

[16] H. A. Biagioni and T. Gramchev, *Fractional derivative estimates in Gevrey classes, global regularity and decay for solutions to semilinear equations in* $\mathbb{R}^n$, J. Differential Equations, **194** (2003), 140–165.

[17] P. Boggiatto, *Spazi di Sobolev associati ad un poliedro ed operatori pseudodifferenziali multi-quasi-ellittici in* $\mathbb{R}^n$, Boll. Un. Mat. Ital., **7**-B (1993), 511–548.

[18] P. Boggiatto and E. Buzano, *Spectral asymptotics for multi-quasi-elliptic operators in* $\mathbb{R}^n$, Ann. Scuola Norm. Sup. Pisa, Cl. Sc., **24** (1997), 511–536.

[19] P. Boggiatto, E. Buzano and L. Rodino, *Global hypoellipticity and spectral theory*, Mathematical Research **92**, Akademie Verlag, Berlin, 1996.

[20] P. Boggiatto and F. Nicola, *Noncommutative residues for anisotropic pseudo-differential operators in* $\mathbb{R}^n$, J. Funct. Anal., **203** (2003), 305–320.

[21] P. Boggiatto and L. Rodino, *Quantization and pseudo-differential operators*, Cubo Mat. Edu., **5** (2003), 237–272.

[22] J. Bona and Y. Li, *Decay and analyticity of solitary waves*, J. Math. Pures Appl., **76** (1997), 377–430.

[23] J. M. Bony and J. Y. Chemin, *Espaces fonctionnels associés au calcul de Weyl-Hörmander*, Bull. Soc. Math. France, **122** (1994), 77–118.

[24] L. Boutet de Monvel and P. Krée, *Pseudodifferential operators and Gevrey classes*, Ann. Inst. Fourier, Grenoble, **17** (1967), 295–323.

[25] C. Bouzar and R. Chaili, *Gevrey vectors of multi-quasi-elliptic systems*, Proc. Amer. Math. Soc., **131** (2003), 1565–1572.

[26] E. Buzano, *Analytic semigroups generated by globally hypoelliptic operators*, Integral Transforms Spec. Funct., **17** (2006), 157–163.

[27] E. Buzano, *Weyl formula for globally hypoelliptic operators in* $\mathbb{R}^n$, in "Microlocal analysis and spectral theory", NATO Adv. Sci. Inst. Ser. C Math Phys. Sci., **490**, Kluwer Acad. Publ., Dordrecht, 1997, 263–306.

[28] E. Buzano, *Super-exponential decay of solutions to differential equations in* $\mathbb{R}^d$, Operator Theory: Advances and Applications **172**, Birkhäuser, Basel, 2006, 117–133.

[29] E. Buzano and F. Nicola, *Pseudo-differential Operators and Schatten-von Neumann Classes*, Advances in Pseudo-Differential Operators (Boggiatto P., Ashino R. and Wong M. W., eds.), Operator Theory: Advances and Applications **155**, Birkhäuser, 2004, 117–130.

[30] E. Buzano and F. Nicola, *Compex powers of hypoelliptic pseudodifferential operators*, J. Funct. Anal., **245** (2007), 353–378.

[31] E. Buzano and J. Toft, *Continuity and compactness properties of pseudo-differential operators*, Fields Institute Communications, **52** (2007), 239–253.

[32] E. Buzano and J. Toft, *Schatten-von Neumann properties in the Weyl calculus*, Preprint 2008, available at arXiv:0809.1207.

[33] E. Buzano and A. Ziggioto, *Weyl formula for hypoelliptic operators of Schrödinger type*, Proc. Amer. Math. Soc., **131** (2003), 265–274.

[34] A. P. Calderón and R. Vaillancourt, *A class of bounded pseudo-differential operators*, Proc. Nat. Acad. Sci. U.S.A., **69** (1972), 1185–1187.

[35] D. Calvo, *Cauchy problem in multi-anisotropic Gevrey classes for weakly hyperbolic operators*, Boll. Un. Mat. Ital., **B, 9** (2006), 21–50.

[36] D. Calvo and L. Rodino, *Gelfand-Shilov classes of multi-anisotropic type*, Funct. Approx. Comment. Math., **40** (2009), 297–307.

[37] D. Calvo and B.-W. Schulze, *Operators on corner manifolds with exit to infinity*, J. Partial Differential Equations, **19** (2006), 147–192.

[38] I. Camperi, *Global hypoellipticity and Sobolev estimates for generalized SG-pseudo-differential operators*, Rend. Sem. Mat. Univ. Pol. Torino, **66** (2008), 99–112.

[39] M. Cappiello, *Gelfand spaces and pseudodifferential operators of infinite order in $\mathbb{R}^n$*, Ann. Univ. Ferrara Sez. VII (N.S.), **48** (2002), 75–97.

[40] M. Cappiello, *Fourier integral operators of infinite order and applications to SG-hyperbolic equations*, Tsukuba J. Math., **28** (2004), 311–361.

[41] M. Cappiello, T. Gramchev and L. Rodino, *Super-exponential decay and holomorphic extensions for semilinear equations with polynomial coefficients*, J. Funct. Anal., **237** (2006), 634–654.

[42] M. Cappiello, T. Gramchev and L. Rodino, *Gelfand-Shilov spaces, pseudo-differential operators and localization operators*, in "Modern Trends in Pseudo-Differential Operators", J. Toft, M.-W. Wong, H. Zhu Editors, Birkhäuser, Basel, 2007, 297–312.

[43] M. Cappiello, T. Gramchev and L. Rodino, *Exponential decay and regularity for SG-elliptic operators with polynomial coefficients*, in "Hyperbolic problems and regularity questions", Trends Math., Birkhäuser, Basel, 2007, 49–58.

[44] M. Cappiello, T. Gramchev and L. Rodino, *Semilinear pseudo-differential equations and travelling waves*, Fields Institute Communications, **52** (2007), 213–238.

[45] M. Cappiello and L. Rodino, *SG-pseudodifferential operators and Gelfand-Shilov spaces*, Rocky Mountain J. Math., **36** (2006), 1117–1148.

[46] L. Cattabriga, *Su una classe di polinomi ipoellittici*, Rend. Sem. Mat. Univ. Padova, **36** (1966), 285–309.

[47] L. Cattabriga, *Moltiplicatori di Fourier e teoremi di immersione per certi spazi funzionali I, II*, Ann. Scuola Norm. Sup. Pisa, Cl. Sc., **24** (1970), 111–158 and **25** (1971), 325–346.

[48] J. Chung, S. Y. Chung and D. Kim, *Characterization of the Gelfand-Shilov spaces via Fourier transforms*, Proc. Amer. Math. Soc., **124** (1996), 2101–2108.

[49] A. Connes, *The action functional in non-commutative geometry*, Comm. Math. Phys., **177** (1988), 673–683.

[50] A. Connes, *Noncommutative geometry*, Academic Press, Inc., San Diego, CA, 1994.

[51] R. R. Coifman and Y. Meyer, *Au delà des opérateurs pseudo-différentiels*. Astérisque **57**, 1978.

[52] E. Cordero and K. Gröchenig, *Time-frequency analysis of localization operators*, J. Funct. Anal., **205** (2003), 107–131.

[53] E. Cordero and K. Gröchenig, *Necessary conditions for Schatten class localization operators*, Proc. Amer. Math. Soc., **133** (2005), 3573–3579.

[54] E. Cordero and F. Nicola, *Pseudodifferential operators on L^p, Wiener amalgam and modulation spaces*, Int. Math. Res. Notices, to appear (available at arXiv:0904.1691).

[55] E. Cordero, F. Nicola and L. Rodino, *Time-frequency analysis of Fourier integral operators*, Commun. Pure Appl. Anal., **9** (2010), 1–21.

[56] E. Cordero, S. Pilipovic, L. Rodino and N. Teofanov, *Localization operators and exponential weights for modulation spaces*, Mediterr. J. Math., **2** (2005), n. 4, 381–394.

[57] E. Cordero and L. Rodino, *Short-time Fourier transform analysis of localization operators*, Contemp. Math., **451** (2008), 47–68.

[58] H. O. Cordes, *A global parametrix for pseudo-differential operators over $\mathbb{R}^n$, with applications*, Reprint, SFB 72, Universität Bonn, 1976.

[59] H. O. Cordes, *The technique of pseudodifferential operators*, London Math. Soc. Lecture Notes Ser. **202**, Cambridge University Press, Cambridge, 1995.

[60] A. Córdoba and C. Fefferman, *Wave packets and Fourier integral operators*, Comm. Partial Differential Equations, **3** (1978), 979–1005.

[61] S. Coriasco, *Fourier integral operators in SG-classes, I. Composition theorems and action on SG Sobolev spaces*, Rend. Sem. Mat. Univ. Pol. Torino, **57** (1999), 249–302.

[62] S. Coriasco, *Fourier integral operators in SG-classes, II. Applications to SG hyperbolic Cauchy problems*, Ann. Univ. Ferrara Sez. VII (N.S.), **44** (1998), 81–122.

[63] S. Coriasco and L. Maniccia, *Wave front set at infinity and hyperbolic linear operators with multiple characteristics*, Ann. Global Anal. Geom., **24** (2003), 375–400.

[64] S. Coriasco and P. Panarese, *Fourier integral operators defined by classical symbols with exit behaviour*, Math. Nachr., **242** (2002), 61–78.

[65] S. Coriasco and L. Rodino, *Cauchy problem for SG-hyperbolic equations with constant multiplicities*, Ricerche Mat., **48** (Suppl.) (1999), 25–43.

[66] S. Coriasco, E. Schrohe and J. Seiler, *Bounded imaginary powers of differential operators on manifolds with conical singularities*, Math. Z., **244** (2003), 235–269.

[67] I. Daubechies, *Time-frequency localization operators: a geometric phase space approach*, IEEE Trans. Inform. Theory, **34** (1988), 605–612.

[68] A. Dasgupta and M. W. Wong, *Spectral theory of SG pseudodifferential operators on $L^p(\mathbb{R}^n)$*, Studia Math., **187** (2008), 185–197.

[69] A. Dasgupta and M. W. Wong, *Essential self-adjointess and global hypoellipticity of the twisted Laplacian*, Rend. Sem. Mat. Univ. Pol. Torino, **66** (2008), 75–85.

[70] G. De Donno, *Generalized Vandermonde determinants for reversing Taylor's formula and application to hypoellipticity*, Tamkang J. Math., **38** (2007), 183–189.

[71] G. De Donno and A. Oliaro, *Local solvability and hypoellipticity for semilinear anisotropic partial differential equations*, Trans. Amer. Math. Soc., **355** (2003), 3405–3432.

[72] N. Dencker, *The resolution of the Nirenberg-Treves conjecture*, Ann. of Math., **163** (2006), 405–444.

[73] J. Dixmier, *Existence de traces non normales*, C.R. Acad. Sc. Paris, Série A, **262** (1966), 1107–1108.

[74] G. Dore and A. Venni, *On the closedness of the sum of two closed operators*, Math. Z., **196** (1987), 189–201.

[75] J. J. Duistermaat and V. W. Guillemin, *The spectrum of positive elliptic operators and periodic bicharacteristics*, Invent. Math., **29** (1975), 39–79.

[76] B. V. Fedosov, F. Golse, E. Leichtnam and E. Schrohe, *The noncommutative residue for manifolds with boundary*, J. Funct. Anal., **142** (1996), 1–31.

[77] C. Fefferman, *The uncertainty principle*. Bull. Amer. Math. Soc., **9** (1983), 129–206.

[78] C. Fefferman and D. H. Phong, *On positivity of pseudo-differential operators*, Proc. Nat. Acad. Sci. USA, **75** (1978), 4673–4674.

[79] C. Fefferman and D. H. Phong, *The uncertainty principle and Sharp Gårding inequality*, Comm. Pure Appl. Math., **34** (1981), 285–331.

[80] H. G. Feichtinger, *Modulation spaces on locally compact abelian groups*, Technical report, University of Vienna, 1983.

[81] H. G. Feichtinger and K. Nowak, *A first survey of Gabor multipliers*, in H. G. Feichtinger and T. Strohmer, editors, "Advances in Gabor Analysis", Birkhäuser, Boston, 2003.

[82] G. B. Folland, *Harmonic analysis in phase space*. Princeton Univ. Press, Princeton, NJ, 1989.

[83] J. Friberg, *Multi-quasi-elliptic polynomials*, Ann. Scuola Norm. Sup. Pisa, Cl. Sc., **21** (1967), 239–260.

[84] K. O. Friedrichs, *Pseudo-differential operators: an introduction*, Lecture notes, Courant Institute of Mathematical Sciences, New-York, 1970.

[85] G. Garello and A. Morando, *L^p-bounded pseudo-differential operators and regularity for multi-quasi-elliptic equations*, Integral Equations Operator Theory, **51** (2005), 501–517.

[86] C. Garetto, *Pseudo-differential operators in algebras of generalized functions and global hypoellipticity*, Acta Appl. Math., **80** (2004), 123–174.

[87] C. Garetto, T. Gramchev and M. Oberguggenberger, *Pseudodifferential operators with generalized symbols and regularity theory*, Electron. J. Differential Equations, **116** (2005), 1–43.

[88] C. Garetto and G. Hörmann, *Microlocal analysis of generalized functions: pseudodifferential techniques and propagation of singularities*, Proc. Edinburgh Math. Soc., **48** (2005), 603–629.

[89] V. Gayral, J. M. Gracia-Bondía, B. Iochum, T. Schücker and J.C. Vàrilly, *Moyal planes are spectral triples*, Comm. Math. Phys., **246** (2004), 569–623.

[90] I. M. Gel'fand and G. E. Shilov, *Generalized Functions II*, Academic Press, New York, 1968.

[91] S. Gindikin and L. Volevich, *A certain class of hypoelliptic polynomials*, Math. USSR Sbornik, **75** (1968), 400–416.

[92] S. Gindikin and L. Volevich, *The method of the Newton's polyhedron in the theory of partial differential equations*, Kluwer Academic Publishers, Dordrecht, 1992.

[93] I. C. Gohberg, M. G. Krein, *Introduction to the theory of linear nonselfadjoint operators*, Translations of Mathematical Monographs **18**, American Mathematical Society, Providence, R.I., 1969.

[94] I. S. Gradshteyn and I. M. Ryzhik, *Table of integrals, series, and products*, Academic Press, New York, 1980.

[95] T. Gramchev, S. Pilipovic and L. Rodino, *Classes of degenerate elliptic operators in Gelfand-Shilov spaces*, in "New Developments in Pseudo-differential Operators", Operator Theory: Advances and Applications **189**, Birkhäuser 2009, 15–31.

[96] T. Gramchev and P. Popivanov, *Partial differential equations. Approximate solutions in scales of function spaces*, Mathematical Research **108**, Wiley, Berlin, 2000.

[97] K. Gröchenig, *Foundations of time-frequency analysis*, Birkhäuser, Boston, 2001.

[98] K. Gröchenig, *Time-frequency analysis of Sjöstrand's class*, Rev. Mat. Iberoamericana, **22** (2006), 703–724.

[99] K. Gröchenig and G. Zimmermann, *Hardy's theorem and the short-time Fourier transform of Schwartz functions*, J. London Math. Soc., **63** (2001), 205–214.

[100] K. Gröchenig and G. Zimmermann, *Spaces of test functions via the STFT*, J. Funct. Spaces Appl., **2** (2004), 25–53.

[101] G. Grubb, *On the logarithm component in trace defect formulas*, Comm. Partial Differential Equations, **30** (2005), 1671–1716.

[102] G. Grubb, *Distributions and operators*, Graduate Texts in Mathematics **252**, Springer, New York, 2009.

[103] V. V. Grushin, *Pseudodifferential operators in $\mathbb{R}^n$ with bounded symbols*, Funkcional. Anal. i Prilozen, **4** (1970), 37–50.

[104] V. V. Grushin, *A certain class of hypoelliptic operators*, Mat. Sb., **83** (1970), 456–473; Math. USSR Sb., **12** (1970), 458–476.

[105] V. V. Grushin, *A certain class of elliptic pseudodifferential operators that are degenerate on a submanifold*, Mat. Sb., **84** (1971), 163–195; Math. USSR Sb., **13** (1971), 155–185.

[106] V. Guillemin, *A new proof of Weyl's formula on the asymptotic distribution of eigenvalues*, Adv. in Math., **55** (1985), 131–160.

[107] V. Guillemin, *Residue traces for certain algebras of Fourier integral operators*, J. Funct. Anal., **115** (1993), 391–417.

[108] G. Harutyunyan and B.-W. Schulze, *Elliptic mixed, transmission, and singular crack problems*, EMS Tracts in Mathematics **4**, Zürich, 2008.

[109] B. Helffer, *Théorie spectrale pour des opérateurs globalement elliptiques*, Astérisque **112**, Soc. Math. de France, 1984.

[110] B. Helffer and L. Rodino, *Elliptic pseudo-differential operators degenerate on a symplectic submanifold*, Bull. Amer. Math. Soc., **82** (1976), 619–622.

[111] B. Helffer and L. Rodino, *Opérateurs differentiels ordinaires intervenant dans l'étude de l'hypoellipticité*, Boll. Un. Mat. Ital., **14-B** (1977), 491–522.

[112] P. D. Hislop and I. M. Segal, *Introduction to spectral theory*, Applied Mathematical Sciences **113**, Springer-Verlag, Berlin, 1996.

[113] L. Hörmander, *On the theory of general partial differential operators*, Acta Math., **94** (1955), 161–248.

[114] L. Hörmander, *On interior regularity of the solutions of partial differential equations*, Comm. Pure Appl. Math., **11** (1958), 197–218.

[115] L. Hörmander, *Pseudo-differential operators*, Comm. Pure Appl. Math., **18** (1965), 501–517.

[116] L. Hörmander, *The spectral function of an elliptic operator*, Acta Math., **121** (1968), 193–218.

[117] L. Hörmander, *The Weyl calculus of pseudodifferential operators*, Comm. Pure Appl. Math., **32** (1979), 360–444.

[118] L. Hörmander, *On the asymptotic distribution of the eigenvalues of pseudodifferential operators in $\mathbb{R}^n$*, Ark. Mat., **17** (1979), 297–313.

[119] L. Hörmander, *The analysis of linear partial differential operators, I, II, III*, Springer-Verlag, 1983, 1985.

[120] G. Hörmann, M. Oberguggenberger and S. Pilipovic, *Microlocal hypoellipticity of linear partial differential operators with generalized functions as coefficients*, Trans. Amer. Math. Soc., **358** (2006), 3363-3383.

[121] I. L. Hwang, *The L^2-boundedness of pseudo-differential operators*, Trans. Amer. Math. Soc., **302** (1987), 55–76.

[122] V. Ivrii, *Microlocal analysis and precise spectral asymptotics*, Springer Monographs in Mathematics, Springer-Verlag, Berlin, 1998.

[123] A. Jeffrey and M. N. B. Mohamad, *Exact solutions to the Kdv-Burgers' equation*, Wave Motion, **14** (1991), 369–375.

[124] R. I. Joseph, *Solitary waves in a finite depth fluid*, J. Phys., A **10** (1977), 225–227.

[125] D. Kapanadze and B.-W. Schulze, *Crack theory and edge singularities*, Kluwer Acad. Publ, Dordrecht, 2003.

[126] C. Kassel, *Le residue non commutatif [d'apres M. Wodzicki]*, Astérisque **177-178**, Soc. Math. de France, 1989, 199–229.

[127] J. J. Kohn and L. Nirenberg, *An algebra of pseudo-differential operators*, Comm. Pure Appl. Math., **18** (1965), 269–305.

[128] H. Kumano-go, *Algebras of pseudo-differential operators*. J. Fac. Sci. Univ. Tokyo, Sect. I **17** (1970), 31–50.

[129] H. Kumano-go, *Pseudodifferential operators*, MIT Press, Cambridge, MA, 1981.

[130] H. Kumano-go and C. Tsutsumi, *Complex powers of hypoelliptic pseudo-differential operators with applications*, Osaka J. Math., **10** (1973), 147–174.

[131] R. Lauter and S. Moroianu, *Fredholm theory for degenerate pseudodifferential operators on manifolds with fibered boundaries*, Comm. Partial Differential Equations, **26** (2001), 233–283.

[132] C. Le Bris and P.-L. Lions, *From atoms to crystals: a mathematical journey*, Bull. Amer. Math. Soc., **42** (2005), 291–363.

[133] N. Lerner, *Coherent states and evolution equations*, in "General theory of partial differential equations and microlocal analysis" (Trieste, 1995), Pitman Res. Notes Math. Ser. **349**, Longman, Harlow, 1996, 123–154.

[134] P. I. Lizorkin, (L_p, L_q)-*multipliers of Fourier integrals*, Dokl. Akad. Nauk SSSR, **152** (1963), 808–811.

[135] P. Loya, *The structure of the resolvent of elliptic pseudodifferential operators*, J. Funct. Anal., **184** (2001), 77–135.

[136] P. Loya, *Complex powers of differential operators on manifolds with conical singularities*, J. Anal. Math., **89** (2003), 31–56.

[137] L. Maniccia, *Asintotica spettrale per una classe di operatori ellittici in $\mathbb{R}^n$*, Ph.D. Thesis, Università di Bologna, 1999.

[138] L. Maniccia and P. Panarese, *Eigenvalue asymptotics for a class of md-elliptic ψdo's on manifolds with cylindrical exits*, Ann. Mat. Pura Appl., **181** (2002), 283–308.

[139] Ju. I. Manin, *Algebraic aspects of nonlinear differential equations*, J. Soviet Math., **11** (1979), 1–122; Russian original in Itogi Nauki i Tekhniki, ser. Sovremennye Problemy Matematiki, **11** (1978).

[140] A. Martinez, *Estimates on complex interactions in phase space*, Math. Nachr., **167** (1994), 203–254.

[141] C. Martínez Carracedo and M. Sanz Alix, *The theory of fractional powers of operators*, North-Holland Mathematics Studies **187**, Elsevier Science B.V., Amsterdam, 2001.

[142] M. Mascarello and L. Rodino, *Partial differential equations with multiple characteristics*, Akademie Verlag-Wiley, Berlin, 1997.

[143] R. B. Melrose, *The Atiyah-Patodi-Singer index theorem*, Research Notes in Mathematics **4**, A K Peters Ltd., Wellesley, MA, 1993.

[144] R. B. Melrose, *Spectral and scattering theory for the Laplacian on asymptotically Euclidian spaces*, in "Spectral and scattering theory", M. Ikawa ed., Dekker, 1994, 85–130.

[145] R. B. Melrose, *Geometric scattering theory*, Cambridge University Press, Cambridge, 1995.

[146] R. B. Melrose and V. Nistor, *Homology of pseudo-differential operators I. Manifolds with boundary*, available at arXiv: funct-an/9606005.

[147] B. S. Mitjagin, *Nuclearity and other properties of spaces of type S*, Amer. Math. Soc. Transl., Ser. 2, **93** (1970), 45–59.

[148] A. Morando, L^p-*regularity for a class of pseudo-differential operators in* $\mathbb{R}^n$, J. Partial Differential Equations, **18** (2005), 241–262.

[149] M. Mughetti and F. Nicola, *On the generalization of Hörmander's inequality*, Comm. Partial Differential Equations, **30** (2005), 509–537.

[150] M. Mughetti and F. Nicola, *Hypoellipticity for a class of operators with multiple characteristics*, J. Anal. Math., **103** (2007), 377–396.

[151] F. Nicola, *Residue traces for a pseudo-differential operator algebra on foliated manifolds*, Integral Equations Operator Theory, **46** (2003), 473–487.

[152] F. Nicola, *Trace functionals for a class of pseudo-differential operators in* $\mathbb{R}^n$, Math. Phys. Anal. Geom., **6** (2003), 89–105.

[153] F. Nicola, *Lower bounds for pseudodifferential operators*, Ph.D. Thesis, University of Torino, January 2005, available at http://calvino.polito.it/~nicola/

[154] F. Nicola and L. Rodino, *Dixmier traceability for general pseudo-differential operators*, in "C^*-algebras and Elliptic Theory II", Trends in Mathematics, Birkhäuser, Basel, 2008.

[155] S. Paycha and S. Scott, *A Laurent expansion for regularized integrals of holomorphic symbols*, Geom. Funct. Anal., **17** (2007), 491–536.

[156] C. Parenti, *Operatori pseudo-differentiali in $\mathbb{R}^n$ e applicazioni*, Annali Mat. Pura Appl., **93** (1972), 359–389.

[157] C. Parenti and A. Parmeggiani, *A generalization of Hörmander's inequality-I*, Comm. Partial Differerential Equations, **25** (2000), 457–506.

[158] C. Parenti and L. Rodino, *Parametrices for a class of pseudo-differential operators, I,II*, Annali Mat. Pura Appl., **125** (1980), 221–254 and 255–278.

[159] A. Parmeggiani, *Introduction to the spectral theory of non-commutative oscillators*, COE Lecture Note **8**, Kyushu University, Fukuoka, 2008.

[160] A. Parmeggiani and W. Wakayama, *Oscillator representations and systems of ordinary differential equations*, Proc. Nat. Acad. Sci. USA, **98** (2001), 26–30.

[161] S. Pilipovic, *Tempered ultradistributions*, Boll. Un. Mat. Ital., VII Ser., B **2** (1988), n. 2, 235–251.

[162] S. Pilipovic and N. Teofanov, *Pseudodifferential operators on ultramodulation spaces*, J. Funct. Anal., **208** (2004), 194–228.

[163] B. Pini, *Osservazioni sulla ipoellitticità*, Boll. Un. Mat. Ital., **18** (1963), 420–432.

[164] A. V. Porubov, *Amplification of nonlinear strain in solids*, World Scientific, Singapore, 2003.

[165] P. J. Rabier, *Asymptotic behaviour of the solutions of linear and quasilinear elliptic equations on $\mathbb{R}^N$*, Trans. Amer. Math. Soc., **356** (2004), 1889–1907.

[166] V. S. Rabinovich, *Exponential estimates for eigenfunctions of Schrödinger operators with rapidly increasing and discontinuous potentials*, Contemp. Math., **364** (2004), 225–236.

[167] J. Ramanathan and P. Topiwala, *Time-frequency localization via the Weyl correspondence*, SIAM J. Math. Anal., **24** (1993), 1378–1393.

[168] M. Reed and B. Simon, *Methods of modern mathematical physics, I. Functional Analysis*, Academic Press Inc., San Diego, 1980.

[169] D. Robert, *Propriétés spectrales d'opérateurs pseudo-différentiels*, Comm. Partial Differential Equations, **3** (1978), 755–826.

[170] D. Robert, *Autour de l'approximmation semi-classique*, Birkhäuser, Boston, MA, 1987.

[171] R. Rochberg and K. Tachizawa, *Pseudodifferential operators, Gabor frames, and local trigonometric bases*, in "Gabor analysis and algorithms", Appl. Numer. Harmon. Anal., Birkhäuser Boston, Boston, MA, 1998, 171–192.

[172] L. Rodino, *Linear partial differential equations in Gevrey spaces*, World Scientific Publishing Co., Singapore, 1993.

[173] R. Schatten, *Norm ideals of completely continuous operators*, Second printing. Ergebnisse der Mathematik und ihrer Grenzgebiete, Band 27 Springer-Verlag, Berlin-New York, 1970.

[174] E. Schrohe, *Complex powers of elliptic pseudodifferential operators*, Integral Equations Operator Theory, **9** (1986), 337–354.

[175] E. Schrohe, *Spaces of weighted symbols and weighted Sobolev spaces on manifolds*, in "Pseudo-differential operators", Lecture Notes in Math. **1256**, Springer-Verlag, Berlin, 1987, 360–377.

[176] E. Schrohe, *Complex powers on noncompact manifolds and manifolds with singularities*, Math. Ann., **281** (1988), no. 3, 393–409.

[177] E. Schrohe, *Noncommutative residues and manifolds with conical singularities*, J. Funct. Anal., **150** (1997), 146–174.

[178] B.-W. Schulze, *Boundary values problems and singular pseudo-differential operators*, J. Wiley, Chichester, 1998.

[179] R. T. Seeley, *Complex powers of an elliptic operator*, in "Singular Integrals", Proc. Sympos. Pure Math. **10**, Amer. Math. Soc., 1967, 288–307.

[180] R. T. Seeley, *Analytic extension of the trace associated with elliptic boundary problems*, Amer. J. Math., **91** (1969), 963–983.

[181] R. T. Seeley, *The resolvent of an elliptic boundary problem*, Amer. J. Math., **91** (1969), 889–920.

[182] M. A. Shubin, *Pseudodifferential operators in $\mathbb{R}^n$*. Dokl. Akad. Nauk SSSR, **196** (1971), 316–319.

[183] M. A. Shubin, *Pseudodifferential operators and spectral theory*, Springer-Verlag, Berlin, 1987.

[184] E. M. Stein, *Singular integrals and differentiability properties of functions*, Princeton University Press, Princeton Math. Series **30**, Princeton, 1970.

[185] E. M. Stein, *Harmonic analysis: real-variable methods, orthogonality and oscillatory integrals*, Princeton University Press, Princeton, 1993.

[186] D. Tataru, *Strichartz estimates for second order hyperbolic operators with nonsmooth coefficients III*, J. Amer. Math. Soc., **15** (2002), 419–442.

[187] M. E. Taylor, *Partial Differential Equations*, II, Springer-Verlag, New-York, 1996.

[188] E. C. Titchmarsh, *The theory of functions*, 2nd edition, Oxford University Press, Oxford, 1932.

[189] F. Treves, *Linear partial differential equations with constant coefficients*, Gordon and Breach, Paris, 1966.

[190] F. Treves, *Topological vector spaces, distributions and kernels*, Academic Press, New York, 1967.

[191] F. Treves, *Introduction to pseudodifferential operators and Fourier integral operators*, I, II, Plenum Publ. Corp., New York, 1980.

[192] F. G. Tricomi, *Funzioni speciali*, Ed. Tirrenia, Torino, 1965.

[193] V. N. Tulovskiĭ, M. A. Shubin, *On the asymptotic distribution of eigenvalues of pseudodifferential operators in* $\mathbb{R}^n$, Math. USSR Sbornik, **21** (1973), no. 4, 565–583.

[194] W. Wasow, *Asymptotic expansions for ordinary differential equations*, Wiley Interscience Publ., New York, 1965.

[195] G. B. Whitham, *Linear and nonlinear waves*, J. Wiley and Sons Inc., New York, 1974.

[196] M. Wodzicki, *Non-commutative residue, Ch. I: Fundamentals*, in "K-Theory, Arithmetic and Geometry", Yu. I. Manin Editor, Lecture Notes in Math. **1289**, Springer-Verlag, Berlin, 1987, 320–399.

[197] M. W. Wong, *An introduction to pseudo-differential operators*, 2nd ed., World Scientific, Singapore, 1999.

[198] M. W. Wong, *Wavelets transforms and localization operators*, Operator Theory Advances and Applications **136**, Birkhäuser, 2002.

[199] M. W. Wong, *Weyl transforms, the heat kernel and Green function of a degenerate elliptic operator*, Ann. Global Anal. Geom., **28** (2005), 271–283.

[200] L. Zanghirati, *Iterati di una classe di operatori ipoellittici e classi generalizzate di Gevrey*, An. Funz. e Appl., Boll. Un. Mat. Ital. Suppl., **1** (1980), 177–195.

[201] L. Zanghirati, *Pseudodifferential operators of infinite order and Gevrey classes*, Ann. Univ. Ferrara Sez. VII (N.S.), **31** (1985), 197–219.

Index

Index of Notation

---

[1]The corresponding spaces of operators are denoted by the same name preceded by "OP", e.g. OP$E\Gamma^m_{\rho,\mathcal{P}}$, etc.